EDMUND HUSSERL

PHILOSOPHY OF ARITHMETIC
Psychological and Logical Investigations
with Supplementary Texts from 1887–1901

TRANSLATED BY

DALLAS WILLARD
*School of Philosophy, University of Southern California,
Los Angeles*

KLUWER ACADEMIC PUBLISHERS
DORDRECHT / BOSTON / LONDON

A C.I.P. Catalogue record for this book is available from the Library of Congress

ISBN 1-4020-1546-1

Published by Kluwer Academic Publishers,
P.O. Box 17, 3300 AA Dordrecht, The Netherlands.

Sold and distributed in North, Central and South America
by Kluwer Academic Publishers,
101 Philip Drive, Norwell, MA 02061, U.S.A.

In all other countries, sold and distributed
by Kluwer Academic Publishers,
P.O. Box 322, 3300 AH Dordrecht, The Netherlands.

Printed on acid-free paper

All Rights Reserved
© 2003 Kluwer Academic Publishers
No part of this work may be reproduced, stored in a retrieval system, or transmitted
in any form or by any means, electronic, mechanical, photocopying, microfilming, recording
or otherwise, without written permission from the Publisher, with the exception
of any material supplied specifically for the purpose of being entered
and executed on a computer system, for exclusive use by the purchaser of the work.

Printed in the Netherlands

CONTENTS

TRANSLATOR'S INTRODUCTION xiii

PHILOSOPHY OF ARITHMETIC:
PSYCHOLOGICAL AND LOGICAL INVESTIGATIONS
Volume One

FOREWORD 5
FIRST PART: THE AUTHENTIC CONCEPTS OF
MULTIPLICITY, UNITY AND WHOLE NUMBER 9
    INTRODUCTION 11
    Chapter I: THE ORIGINATION OF THE CONCEPT OF
    MULTIPLICITY THROUGH THAT OF THE COLLECTIVE
    COMBINATION 15
        The Analysis of the Concept of the Whole Number
            Presupposes that of the Concept of Multiplicity 15
        The Concrete Bases of the Abstraction Involved 16
        Independence of the Abstraction from the Nature of the
            Contents Colligated 17
        The Origination of the Concept of the Multiplicity through
            Reflexion on the Collective Mode of Combination 18
    Chapter II: CRITICAL DEVELOPMENTS 23
        The Collective Unification and the Unification of Partial
            Phenomena in the Total Field of Consciousness at a
            Given Moment 23
        The Collective "Together" and the Temporal
            "Simultaneously" 25
        Collection and Temporal Succession 26
        The Collective Synthesis and the Spatial Synthesis 35
            A. F. A. Lange's Theory. 35
            B. Baumann's Theory. 45

Colligating, Enumerating and Distinguishing 49
Critical Supplement 61
Chapter III: THE PSYCHOLOGICAL NATURE OF THE
COLLECTIVE COMBINATION 67
Review 67
The Collection as a Special Type of Combination 68
On The Theory of Relations 69
Psychological Characterization of the Collective
Combination 74
Chapter IV: ANALYSIS OF THE CONCEPT OF NUMBER
IN TERMS OF ITS ORIGIN AND CONTENT 81
Completion of the Analysis of the Concept of Multiplicity 81
The Concept 'Something' 84
The Cardinal Numbers and the Generic Concept of Number .. 85
Relationship Between the Concepts 'Cardinal Number'
and 'Multiplicity' 87
One and Something 88
Critical Supplement 89
Chapter V: THE RELATIONS "MORE" AND "LESS" 95
The Psychological Origin of These Relations 95
Comparison of Arbitrary Multiplicities, as well as of
Numbers, in Terms of More and Less 98
The Segregation of the Number Species Conditioned
upon the Knowledge of More and Less 99
Chapter VI: THE DEFINITION OF NUMBER-EQUALITY
THROUGH THE CONCEPT OF RECIPROCAL ONE-TO-
ONE CORRELATION 101
Leibniz's Definition of the General Concept of Equality 101
The Definition of Number-Equality 103
Concerning Definitions of Equality for Special Cases 105
Application to the Equality of Arbitrary Multiplicities 106
Comparison of Multiplicities of One Genus 108
Comparison of Multiplicities with Respect to their Number . 108
The True Sense of the Equality Definition under
Discussion 110
Reciprocal Correlation and Collective Combination 111
The Independence of Number-Equality from the Type
of Linkage 114

Chapter VII: DEFINITIONS OF NUMBER IN TERMS OF
EQUIVALENCE 117
   Structure of the Equivalence Theory...................................... 117
   Illustrations 120
      Critique 121
         Frege's Attempt... 123
         Kerry's Attempt.. 129
         Concluding Remark... 131

Chapter VIII: DISCUSSIONS CONCERNING UNITY
AND MULTIPLICITY... 133
   The Definition of Number as a Multiplicity of Units............. 133
   One as an Abstract, Positive Partial Content......................... 133
   One as Mere Sign.. 133
   One and Zero as Numbers... 136
   The Concept of the Unit and the Concept of the
      Number One.. 141
   Further Distinctions Concerning One and Unit..................... 143
   Sameness and Distinctness of the Units................................ 146
   Further Misunderstandings 157
   Equivocations of the Name "Unit" 159
   The Arbitrary Character of the Distinction between Unit
      and Multiplicity. The Multiplicity Regarded as One
      Multiplicity, as One Enumerated Unit, as One Whole. 162
   Herbartian Arguments... 164

Chapter IX: THE SENSE OF THE STATEMENT
OF NUMBER 169
   Contradictory Views 169
   Refutation, and the Position Taken 170

APPENDIX TO THE FIRST PART: The Nominalist Attempts
of Helmholtz and Kronecker... 179

SECOND PART: THE SYMBOLIC NUMBER CONCEPTS AND
THE LOGICAL SOURCES OF CARDINAL ARITHMETIC......... 189
   Chapter X: OPERATIONS ON NUMBERS AND THE
   AUTHENTIC NUMBER CONCEPTS 191
      The Numbers in Arithmetic Are Not Abstracta 191
      The Fundamental Activities on Numbers.............................. 192
      Addition 193

Partition 198
Arithmetic Does Not Operate with "Authentic" Number
  Concepts 200
Chapter XI: SYMBOLIC REPRESENTATIONS OF
MULTIPLICITIES 205
Authentic and Symbolic Representations 205
Sense Perceptible Groups 207
Attempts at an Explanation of How We Grasp Groups
  in an Instant 208
Symbolizations Mediated by the Full Process of
  Apprehending the Individual Elements 210
New Attempts at an Explanation of Instantaneous
  Apprehensions of Groups 211
Hypotheses 213
The Figural Moments 215
The Position Taken 223
The Psychological Function of the Focus upon
  Individual Members of the Group 225
What Is It that Guarantees the Completeness of the
  Traversive Apprehension of the Individuals in a Group? ... 226
Apprehension of Authentically Representable Groups
  through Figural Moments 228
The Elemental Operations on and Relations between
  Multiplicities Extended to Symbolically Represented
  Multiplicities 229
Infinite Groups 230
Chapter XII: THE SYMBOLIC REPRESENTATIONS
OF NUMBERS 235
The Symbolic Number Concepts and their Infinite
  Multiplicity 235
The Non-Systematic Symbolizations of Numbers 236
The Sequence of Natural Numbers 238
The System of Numbers 241
Relationship of the Number System to the Sequence
  of Natural Numbers 247
The Choice of the "Base Number" for the System 249
The Systematic of the Number Concepts and the
  Systematic of the Number Signs 251

The Process of Enumeration via Sense Perceptible Symbols 253
Expansion of the Domain of Symbolic Numbers
   through Sense Perceptible Symbolization 254
Differences between Sense Perceptible Means of
   Designation 257
The Natural Origination of the Number System 258
Appraisal of Number through Figural Moments 267
Chapter XIII: THE LOGICAL SOURCES OF ARITHMETIC .. 271
Calculation, Calculational Technique and Arithmetic 271
The Calculational Methods of Arithmetic and the
   Number Concepts 274
The Systematic Numbers as Surrogates for the
   Numbers in Themselves 275
The Symbolic Number Formations that Fall Outside
   the System, Viewed as Arithmetical Problems 275
The First Basic Task of Arithmetic 277
The Elemental Arithmetical Operations 277
Addition 279
Multiplication 283
Subtraction and Division 284
Methods of Calculation with the Abacus and in Columns.
   The Natural Origination of the Indic Numeral Calculation. 288
Influence of the Means of Designation upon the
   Formation of the Methods of Calculation 290
The Higher Operations 292
Mixing of Operations 294
The Indirect Characterization of Numbers by Means
   of Equations 296
Result: The Logical Sources of General Arithmetic 298
SELBSTANZEIGE – PHILOSOPHIE DER ARITHMETIK 301

SUPPLEMENTARY TEXTS (1887 – 1901)

A. ORIGINAL VERSION OF THE TEXT THROUGH CHAPTER IV:
   ON THE CONCEPT OF NUMBER: PSYCHOLOGICAL
   ANALYSES 305
   Introduction 305
   Chapter One 312

THE ANALYSIS OF THE CONCEPT OF NUMBER
AS TO ITS ORIGIN AND CONTENT 312
   Section 1: The Formation of the Concept of
   Multiplicity [Vielheit] out of That of
   the Collective Combination... 312
   Section 2: Critical Exposition of Certain Theories 318
   Section 3: Establishment of the "Psychological"
   Nature of the Collective Combination 344
   Section 4: The Analysis of the Concept of Number
   as to its Origin and Content... 352
APPENDIX TO "ON THE CONCEPT OF NUMBER:
PSYCHOLOGICAL ANALYSES" – THESES........................ 357
B. ESSAYS 359
   ESSAY I: ⟨ ON THE THEORY OF THE TOTALITY ⟩ 359
      ⟨ I. The Definition of the Totality ⟩ 359
      ⟨ II. Comparison of Numbers ⟩... 364
      ⟨ III. Addenda ⟩... 368
      ⟨ 1. Addendum to p. 367: Identity and Equality ⟩................. 368
      ⟨ 2. On the Definition of Number ⟩....................................... 369
      ⟨ IV. The Classification of the Cardinal Numbers ⟩ 369
      ⟨ V. Remark ⟩... 374
      ⟨ VI. Corrections ⟩ 375
      ⟨ VII. Addenda ⟩ 379
         ⟨ 1. Addendum to p. 369 ⟩... 379
         ⟨ 2. Addendum to p. 377 ⟩... 380
   ESSAY II: ⟨ ON THE CONCEPT OF THE OPERATION ⟩.. 385
      ⟨ I. Arithmetical Determinations of Number ⟩....................... 385
      ⟨ II. Combinations (or Operations) ⟩...................................... 397
         ⟨ 1. Division ⟩... 397
         ⟨ 2. On the Concept of Combination [Verknüpfung] ⟩........ 400
      ⟨ III. Addendum ⟩ 405
      On the Concept of Basic Operation 405
   ESSAY III: ⟨ DOUBLE LECTURE: ON THE
   TRANSITION THROUGH THE IMPOSSIBLE
   ("IMAGINARY") AND THE COMPLETENESS
   OF AN AXIOM SYSTEM ⟩ 409
      ⟨ I. For a Lecture before the Mathematical Society
      of Göttingen 1901 ⟩.. 409

⟨ 1. Introduction ⟩..409
⟨ 2. Theories Concerning the Imaginary ⟩.........................413
⟨ 3. The Transition through the Imaginary ⟩.....................427
APPENDIX I 453
APPENDIX II 459
APPENDIX III 464
⟨ Notes on a Lecture by Hilbert. ⟩...............................464
Husserl's Excerpts from an Exchange of Letters
between Hilbert and Frege. 468
ESSAY IV: ⟨ THE DOMAIN OF AN AXIOM SYSTEM/
AXIOM SYSTEM – OPERATION SYSTEM ⟩......................475
⟨ System of Numbers ⟩..477
⟨ Arithmetizability of a Manifold ⟩..............................479
⟨ On the Concept of an Operation System ⟩ 482
ESSAY V: ⟨ THE QUESTION ABOUT THE CLARIFICATION
OF THE CONCEPT OF THE "NATURAL" NUMBERS
AS "GIVEN," AS "INDIVIDUALLY DETERMINANT" ⟩...............493
ESSAY VI: ⟨ ON THE FORMAL DETERMINATION
OF MANIFOLD ⟩..497
INDEX 505

TRANSLATOR'S INTRODUCTION

This volume contains the main body of writings that emerged from Edmund Husserl's first philosophical project. That project was the clarification of the nature of number and of our knowledge of number. In the course of his research the project grew to include the elucidation of arithmetic as a necessary and effectual instrument for extending our knowledge onward into the infinite domain of numbers and number relationships. Husserl wanted to provide what, following D'Alembert, had long been called a "metaphysics of the calculus."[1] He sought to know exactly what the subject matter of mathematics is, and how mathematics – in the first instance, just arithmetic – manages to relate itself and the human mind to that subject matter.

The most comprehensive result of his earliest efforts was his first book, *Philosophie der Arithmetik*, first published in 1891. It is here translated along with his "Habilitationsschrift" – printed, but not properly published, in 1887 – and several related research documents and lecture notes associated with his first inquiries. Thus the writings here translated consist of all the materials published in *"Husserliana XII,"* Edmund Husserl, *Philosophie der Arithmetik (1890-1901)*[2], except for Abhandlungen I, II and III. Those "Abhandlungen" were translated in *Early Writings in the Philosophy of Logic and Mathematics*, volume V in the English edition of Husserl's "Collected Works."

My aim here is to orient the reader to the main questions addressed by Husserl in his first project, and to outline and briefly explicate major points in his answers to those questions. I shall

---

[1] P. 6 of this volume, and p. 3 of *Early Writings in the Philosophy of Logic and Mathematics*, Dordrecht: Kluwer Academic Publishers, 1994.

[2] Nijhoff: Den Haag, 1970. *"Husserliana XII"* must be supplemented by *"Husserliana XXI," Studien zur Arithmetik und Geometrie*, edited by Ingeborg Strohmeyer, Nijhoff: Den Haag, 1983, for a more complete understanding of Husserl's early work.

also discuss a few issues concerning his answers which, for one reason or another, may pose special difficulties or be of special interest to the contemporary reader.[3]

There are three main questions which Husserl addresses in his earliest writings:

1. What is number itself?
2. In what kind of cognitive act is number itself actually present to our minds?
3. How do the symbols and symbolic systems used in arithmetical thought enable us to represent, and to arrive at knowledge of, numbers and number relations that are not, and in most cases never could be, intuitively given to minds such as ours – never could *themselves* be actually present to our minds – and even enable us to have the most secure knowledge possible concerning many of the properties of and relationships between the larger numbers? (p. 203 below)

The "Habilitationsschrift," titled "Concerning the Concept of Number: Psychological Analyses" (pp. 305-356 below), and the "First Part" of the *Philosophy of Arithmetic* (described as "psychological investigations"), are mainly devoted to questions 1 and 2. The "Second Part" of *Philosophy of Arithmetic* (understood to be "logical investigations") and the "Essays" (pp. 359-504) are mainly devoted to question 3.

In Husserl's own opinion, his findings for questions 1 and 2 were generally satisfactory, and they formed a crucial part of his philosophical views for the remainder of his life. As to question 3, he never arrived at a wholly satisfactory result. His work on this question eventually pushed him (during the 1890s) beyond the philosophy of arithmetic – and, indeed, beyond the philosophy of mathematics – into perfectly general inquiries concerning the nature of knowledge, and especially of scientific knowledge, which he soon came to recognize as almost totally symbolic or non-intuitional in character.

His first comprehensive treatment of general epistemological problems was his *Logical Investigations* of 1900 and 1901, which developed from the 1894 paper, "Psychological Studies in the Ele-

---

[3] Many issues will seem to the reader to cry out for a detailed examination which simply cannot be provided in this context. See references to further work in footnote #42, below.

ments of Logic."⁴ The sense in which these were logical investigations, however, or "studies in the elements of logic," will not be obvious to the reader of 100 years later, and can be understood only from within the intellectual context of the 1880s and 1890s. But, in any case, his efforts to understand how the symbolic systems of arithmetic – and indeed those of mathematics in the broader sense he came to appreciate – serve to procure knowledge of the infinite domain of numbers were only partially successful, for reasons we shall see.

*

What, then, is Husserl's account of number "itself," of what number is in its essence, and of the kinds of cognitive processes or acts in which the essence of number ("number itself") is clearly and fully ("authentically") present to the mind? And especially, what is the nature of those inner relationships that essentially determine the type of whole that is "a number of things" – or "a multiplicity" or "totality," as Husserl also calls it?

To answer these questions is to give *an analysis of the concept of number*, as he understood "analysis." At this point, Husserl's method of "concept analysis" was to carefully trace out the course of experiences through which we come to exercise the particular concept, and therefore to "have" that concept. Those experiences constitute in this case what he calls *the psychological origin of the concept of number*. By reflectively living through them, he holds, we are led to the properties essential to any object that falls under that concept. Those properties then make up the *content* of the concept – that is, what the concept "contains" as object, what it is primarily and in the first instance *of* or *about*.

For purposes of understanding these early writings, we can think of a concept as a repeatable and shareable 'thought' – a universal of a type belonging exclusively to mental states. To *analyze* a certain concept requires that we discern what is of necessity *thought of* or *meant* – through the intentionalities

---

[4] Translated in *Early Writings*, pp. 139-170, also translated by R. Hudson and P. McCormick in *Husserl: Shorter Works*, Notre Dame, IN.: University of Notre Dame Press, 1981, pp. 126-142.

essentially involved – whenever that concept is deployed. The "content" of any concept consists, we have noted, of the essential properties of the objects that fall under that concept. Carefully attending to the concept's "origin" enables us to bring this "content" of the concept clearly before our mind. We can also clarify the concept repeatedly, as may be needed, by bringing the concept – in cases where this is actually possible – over against intuitively presented conceptual objects that fall under it. We 'match up' the concept or 'thought' with the essence of the conceptual object. Thus we are enabled, on his view, to *see* precisely how the concept correlates to the essential properties of those conceptual objects as such.

Carl Stumpf was Husserl's colleague and mentor in Halle during this period. It was Stumpf's view that we best discern the *content* of a concept by examining the concept's origin: by tracing the "idea" to the partial "ideas" it contains, and, finally, to the "impressions" (to use Hume's language) from whence it developed or "originated." That is, one cannot (at least in the philosophically interesting cases) simply focus one's reflexive attention upon the concept or thought in question and thereby accurately discern the intentionalities which go into its makeup. You have to see it *in relation to* its corresponding conceptual object. (One recalls the method of 'analysis' or clarification of concepts in "Linguistic Analysis," which worked by returning to the context where the *use* of the term was learned.)

As Stumpf had remarked in his *Über den psychologischen Ursprung der Raumvorstellung*, a book that was studied carefully by the young Husserl:

"The question 'Whence arises a representation?' is of course (though this is not always done) to be clearly distinguished from the other question, 'What is its knowledge content, once we have it?' However, these two questions are methodologically related, insofar as the question about the origin of a representation leads us to the separate parts of which it is composed, and therefore yields a more precise grasp of its content."[5]

---

[5] Leipzig: S. Hirzel, 1873, p. v; see also pp. 3-4. Cp. also Hermann Lotze, *Logic*, ed Bernard Bosanquet, Oxford: The Clarendon Press, 1884, pp. 7-8. Note that "representation" in this literature is often used interchangeably with "concept." But a representation

Now it is because Husserl was guided by this methodological strategy that he described his 1887 essay as dealing with the "... specific ... question about the content and origin of the concept of number" (p. 311 below), and as aiming at "the exhibition of the origin and content of the concepts *multiplicity* and *number*." (p. 352; cp. 81) By far the greater part of that essay was, in fact, devoted to discussions of the *origin* of the concept. Only four pages at the very end (pp. 353-356; cp. 84ff.) are given to a statement of the concept's content, and much of that brief statement really turns out to be a discussion of the "origin" of another concept involved, that of *something*. Husserl clearly assumed that the content of a concept – which would be the corresponding "essence" – would be obvious, once the concept's origin was made clear.

This strategy for concept analysis also makes clear why Chapter I of *Philosophy of Arithmetic* is titled, "The Origination of the Concept of Multiplicity through That of the Collective Combination." (That chapter as it appeared in the 1887 piece was titled "The Analysis of the Concept of Number as to Its Origin and Content.") "Collective combination" is the name Husserl gave to one of the two characteristics (the other is 'something') that make up the essence of those peculiar wholes which are multiplicities, totalities, or "a number of things." The last stage of the "psychological origin" of *number* is the abstraction of the concept from that essence, standing fully in view.

*

What, then, is the sequence of experiences through which number *itself* (the essence of the multiplicity or totality) becomes directly or "authentically" present to us, permitting us to gain,

---

might on occasion fail to be a concept. A concept was thought of as a *general* representation, in the twofold sense that it is of a kind, not an individual, and that it is intersubjectively shareable. Obviously all of this was in desperate need of philosophical clarification. (See Husserl's Vth "Logical Investigation," chapter VI.) John Dewey's 1891 paper, "How Do Concepts Arise from Percepts," is helpful in understanding how 'concepts' were thought of in that day. (*The Early Works*, III, London: Feffer & Simons, 1969, pp. 142-146.) Also, see Chapter I of F. H. Bradley, *The Principles of Logic*, London: Oxford University Press, 1958.

to clarify and to re-clarify (as needed) the *authentic* concept of number? The entire "First Part" of the *Philosophy of Arithmetic* is devoted to the "authentic" concept, and accordingly that Part is given the title: "The Authentic Concepts of Multiplicity, Unity and Cardinal Number." (p. 9) An "authentic" concept is a concept acquired through the immediate presence to our mind, in an adequate or full intuition, of the essence corresponding to (but *not* identical with) the concept. Concepts can and do "originate" (come into our possession) by other (symbolic or "inauthentic") means than such an immediate presence, as we shall see; but then they are, of course, *inauthentic* concepts.

Now in the case of the general concept of number ("multiplicity"), as well as of the concepts of (the very small) determinate numbers, the most common objects to which application is made are sets or groups of determinately given things, such as pencils on a desk or strokes of a clock. But it is clear that merely to be intuitively aware of some of the various items on a desk, for example, is not necessarily to intuit them *as* a "number" of things, or as a "totality" in the sense here in question. I can look at them without grouping them. To intuit them as a "number" of things, one must perform a characteristic, complex type of act, which we might describe as the *intuitive enumeration* of the objects in the group.

In such an act I serially consider certain objects from among those present to me on the occasion, with that distinct type of emphatic, purposive, and ordered noticing essential to the explicit counting of objects. In such a case a new and distinctive type of whole is, Husserl finds, intuitively present to me within my field of consciousness: the totality or multiplicity – a concrete unity of x number of objects. Such a unified "totality" appears to me with clear, nonspatial boundaries within my total field of consciousness – boundaries coinciding, I may observe, with the range of the characteristic, noticing acts in question. The wholes (totalities) thus bounded are not intuitively given in any type of mental act other than this sort of articulated considering just described. They are, in the language of that day, "objects of a higher order," and hence are only graspable in acts of thought which essentially involve an awareness of subordinate acts, those making up the ordered noticing mentioned. (p. 77)

Now we must carefully consider this point about the intuitability of totalities. It is the clear assumption of the second of Husserl's three main questions listed above, and it will certainly be the point of strongest resistance to his view of number. In the givenness of "a number of things" we have one instance of what he was later to call "categorial intuition," in the Sixth Chapter of "Investigation VI" of the *Logical Investigations* and elsewhere. The wording "categorial intuition" is perhaps new with "Investigation VI," but the viewpoint is present from 1887 on. In the early essay Husserl says: "In the totality there is a lack of any intuitive unification, as that sort of unification so clearly manifests itself in the metaphysical or continuous whole <See page 71>. And that is so, although a certain unity is present in the totality, and is perceivable with Evidence." (pp. 350) This goes hand in hand with his treatment of number in terms of form-concepts, or *categories*, in *Philosophy of Arithmetic*. (p. 89) All but the wording, however, is already present in the early essay of 1887.

This point must be kept constantly in view, if we are to understand Husserl's view of number: Number (with its essential constituents) *presents itself to* the human mind in characteristic types of perceptual experiences. At the opening of Chapter Two of *Philosophy of Arithmetic*, he states that "the shortest answer to the question about what kind of unification is present in the totality lies in a direct reference to the phenomena. And here we are genuinely dealing with ultimate facts." (p. 23; cp. 318) These "facts" are ultimate in the sense that if we do not *see* them, we cannot come to understand them correctly, because the peculiar type of unification in a mere "number" of things is unanalyzable and indefinable. There is nothing else that we can start from to arrive at it. It can only be seen and described.

So let us try once more, and more concretely, to *direct attention to* the basic phenomenon of number. If, for example, I attempt to count the trees in a certain area of a park, I (and the reader with me) must do something more than just be conscious of them, or even clearly see them – whether as spatially clumped together or taken individually. I must rather, *as* I view them, think in a characteristic manner: There is *that one and that one and that one and* . . . . As I go through these acts in which the things enumerated are "separately and specifically noticed," as Husserl says, there arises for me a division of the trees into those "already" enumerated and those not or "not yet" enumerated. His view is that this division is an objective fact intuitively given to me. If it does not *present itself to me* with some force and clarity, I simply cannot number the trees. But in that it does come before me as I count, the trees already enumerated appear "together," and in their unification with each other they stand "apart from" the remaining trees and other objects – of which I nonetheless may be quite conscious all along. The "number of things," the "totality" or "multiplicity" – a different one at each step as I count – is intuitively "constituted" (made present) for me in this type of thoughtful enumeration.

It is *this* "together" which is the ultimate phenomenal fact of number, and to which Husserl gives the name "collective combination." Once we have *seen* it, we come into possession of the concept of it by abstraction. All that remains is to show (as Husserl does in Chapter Two, pp. 23ff.) why this "together" cannot be the same as certain other relations that have been offered as an analysis of it – temporal simultaneity, temporal succession, mere difference, etc. (pp. 67-68) – and to find (Chapter Three, pp. 67ff.) what can be given by way of a description (not a definition) of the essence of this peculiar "together" relation that is 'itself present' under the conditions described. This completes Husserl's positive account of the *origin* of the relational element involved in the concept of number.

We might try to represent Husserl's account schematically as follows:

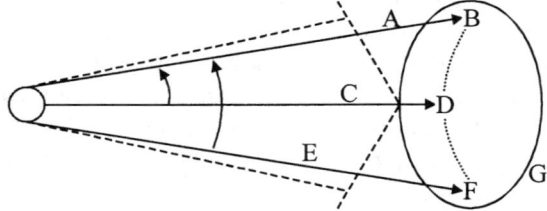

Here A through G are objects in a given field of consciousness. The unbroken lines to B, D, and F are those characteristic acts of noticing which are involved in enumerating. The small arrows crossing between the unbroken lines are awarenesses of earlier such acts built into subsequent such acts, ordering them into a *progression*. The diffuse arrow formed from broken lines is the founded consciousness of the higher order object, the totality BDF. The dotted lines connecting B to D to F are the "collective combinations" in virtue of which these three elements form a *number* of things, excluding the other objects in the field of consciousness. Please note that the totality as thus represented by the closed curved line, like the relations of "collective combination" which it contains, is not a *part* of the complex act in which it is grasped. Being "a number" of things is not restricted to mental acts or to the elements of mental acts, as the cases cited clearly indicate.

We can see, then, that Husserl made a disastrous choice of terminology in deciding to call the collective combination a "psychical" relation and describing it as having a "psychological nature." Having refuted various theories concerning the precise nature of the unifying relation in the totality, he attempts a positive account of it in Chapter Three (pp. 67ff.). Since, as we have noted, he assumes that the collective combination can have any type of object for its terms – for all objects are countable, and countable together with any other objects (p. 17; cp. 314-315, 350-351) – it cannot be distinguished in terms of a restricted class of terms. Thus, one has to find something in collective combinations themselves which can characterize them. They do in fact, Husserl believes, possess "noteworthy peculiarities which very essentially distinguish them in their phenomenal existence from all of the remaining kinds of relations" (p. 346). Thus we can "classify relations in terms of their particular phenomenal character" (p. 71).

As to what the precise character of the collective combination is, he takes his clue from Brentano's distinction between the psychical (mental) and the physical. The psychical 'relation' of intentionality comprises its object term in a peculiar, indefinable manner, marked by the term "intentional inexistence." This manner stands out clearly when compared with how, for example, the relation of similarity grasps its terms. This apparent difference is regarded by Husserl as the difference between a "psychical" relation and a physical relation. These latter are also called "content relations" by him, because something literally in the *sense* 'contents' which are *part* of the respective *act* directed upon "physical" relations corresponds to those relations.

But how is this difference to be spelled out further? Upon

closer examination (see pp. 72-77; 344-351), psychical relations all turn out to have three phenomenal features which can be stated. *First,* they admit of an *unlimited variety* in the types of terms which they take. Anything at all can be (i) an object of consciousness, (ii) collectively combined into a "number" of things, or (iii) stand in the "relation of 'distinctness' in the widest of senses . . . in the case of which two contents are brought into relation merely by means of an Evident, negative judgment" of non-identity. (p. 348; cp. 56-57 & 342) These are the three cases of "psychical" relations which Husserl brings up in his discussion here. They all obviously admit of unlimited variety in their terms. In that traditional sense they are "transcendentals."

Second, in order to be aware of things *as* "psychically" related, we must be aware – in some marginal and nonthetical manner, no doubt – of a prior act (or acts) of consciousness in relation to those things. To be aware of X *as* an object of consciousness, for example, I must be (marginally) aware of my simple consciousness of X; to be aware of it *as* one of a "number" of things, I must be aware of a prior emphatic and ordered noticing of it and of the others in the "number" concerned; and to be aware of two things *as* simply different or non-identical, I must, Husserl thinks, be aware of a prior act of judging them to be different. So in each of these cases we have an act of higher order and an object of higher order.

And *Third,* as a follower of Brentano, Husserl was something of a critical realist at this stage of his career. This meant, among other things, that every non-mental object was intended on the basis of an apperceptive grasp of something actually present in (a part or constituent of) consciousness. But then there is a problem with these "psychical" relations; for Husserl finds, as we have also noted, that there corresponds to them no sense content (sensum or image) in the mind itself. It is this third and final feature of psychical relations that leads Husserl to remark that we "could also quite appropriately call physical relations '*content relations*'." (p. 348; cp. 72.)[6] Now in the absence of such a sensuous *mental content,* the prior *mental acts* reflected upon in grasping a

---

[6] Note that "content" here has a very different meaning from when we speak of the "content" of a *concept.*

"psychical" relation function structurally as sense contents do in grasping physical relations: as, that is, a kind of apperceptive support of the intention or meaning directed upon the "psychical" relation. So acts directed upon the psychical relation are, once again, acts of higher order – "founded," in Husserl's later terminology (See the IIIrd "Logical Investigation"), on "lower" acts.

These three features give the "psychical" relation only a very weak connection with its terms. The relation does not immediately reside in the phenomena themselves – as with similarity, for example – but, so to speak, is external to them. Yet, Husserl says, "a certain unity is present in the totality, and is perceivable with Evidence." (p. 350) The "externality" of these "categorial" relations will turn out to be of vital importance for Husserl's mature thought.[7]

*

The above completes Husserl's account of the "origin" and abstractive basis ("content") of that *relational* concept which is an essential part of the concept of a "number" of things, of a mere totality or multiplicity. However, there is one more essential part of that concept, as already noted, and Husserl discusses it briefly – almost, it seems, as an afterthought. The question arises as to how the elements related by the collective combination in a "number" of things are to be referred to in the concept of number or totality? Since those elements must be utterly without restriction as to kind, there can "enter into the general concept of the multiplicity no peculiarities of content. However, ... parts must be somehow thought of in it" (p. 353), since it is a whole. (pp. 81-82)

Husserl's solution is to say that in the concept *number* things are referred to as mere "somethings." "There is," he says, "only one all-encompassing concept, that of the *something*." (p. 123) This accounts for "The complete diminishment of content which occurs with the abstraction of number." (p. 98) And: "The concept of *something* owes its origination to reflexion upon the

---

[7] On this point see my *Logic and the Objectivity of Knowledge*, Athens, OH.: Ohio University Press, 1984, pp. 242-243.

psychical act of representing, as the content [object] of which just any determinate object may be given" (p. 354; cp. 83-85). That is, it derives from the psychical 'relation' (i), mentioned above. The feature of being a "something" also has, as will have been suggested by the above discussion of "psychical" relations, no corresponding "partial content" (such as a sensum or image) present in the act directed upon it. And, like the collective combination, the character of "something" "belongs to the content of any concrete object only in . . . [an] external and non-literal fashion." (p. 84; cp. 150, 158, 164, 354)

In Sum, then: "We obtain the abstract multiplicity form belonging to a <small> group by diminishing each of its elements to a mere 'one' <or 'something'> and collectively grasping together the units thus originating. And we obtain the corresponding <specific> number by classifying the multiplicity form thus constructed as a two, a three, etc." (p. 109)

So the concept of number contains two subconcepts as parts. And Husserl thought, following Bolzano, that the collective combination was designated in common language by the conjunction "and." To grasp a "number" of things is, then, simply to grasp certain objects under the phenomenal character of mere "somethings" and as united by the phenomenal relation of a mere "and." Thus, the concept of a multiplicity or totality is just *something and something and something, etc.* A particular number, such as four, suppresses the "etc." at a specific point. A number is therefore not a class ("a class of similar classes," as various people have said), but rather a structural property which defines a class of similar classes. *A number is a specific, non-sense-perceptible, structural property of similar groups which are seen or conceived of as mere* "somethings" joined by mere "ands."

*

In fact, Husserl's account of the totality and of number is very close to, if not identical with, the one given earlier by Bolzano. We know from Husserl's autobiographical writings that between 1884 and 1886 he heard Brentano lecture on "the descriptive psychology of the continua, with detailed consideration of Bolzano's *Paradoxien des Unendlichen*."[8] And we find the essence and much of the form of Husserl's analysis of the *content* – but certainly not of the "origin" – of the concept of the totality in subsection 3 of Bolzano's book. It is clear from that book, as well as his *Wissenschaftslehre,* that the use to which Husserl puts the "and" and the "something" is substantially the same as Bolzano's – especially in relation to the analysis of number.

But some will say, with Werner Illemann – but most famously, Frege – that, unlike Bolzano, the word "totality" is "used by Husserl as a designation for a psychical <meaning mental> fact."[9] Now it must be admitted that there is much in Husserl's language to justify this interpretation – not the least of which is his decision to use the term "psychical" as a designation for the general *type* of relation of which collective combination then stands as one case. This was an extremely unfortunate selection of terminology, but if one looks closely at the texts, one sees that by "psychical" Husserl does not mean "part of a mind" or "intentional." Rather he means, as we have indicated: having a certain character (discussed above) which is most well-known as showing up in the intentional (Brentano's "psychical") nexus.

Clearly, the collective combination is not itself an *intentional* relation in the specific sense of Brentano. Its terms are not subject/object. One member of a totality is not *about* the other. It interrelates the members of a totality, but these members do not thereby cognize, do not refer to, each other. Nor are the only things collectively combinable (and therefore countable) mental acts or constituents of mental acts – as Husserl explicitly says, and shows by the examples he chooses. Even the sense contents caught up as constituents within the act of enumeration are not themselves united by collective combination. However, those who read Husserl seem to have found it practically impossible – no doubt given other confusions of thought and language present in the texts – to transfer the term "psychical" from its ordinary sense, designating the mental, to the more general sense proposed by

---

[8] Husserl's "Erinnerungen an Franz Brentano," in Oskar Kraus, ed., *Franz Brentano*, Munich: C. H. Beck, 1919, p. 157; see p. 344 of the English translation in *Husserl: Shorter Works*.

[9] Husserl's *Vor-phänomenologische Philosophie*, Leipzig: Hirzel, 1932, p. 13.

him, better indicated by the words "categorial" or "formal."

This was certainly made all the more difficult by Husserl's use of the term "content" to refer both to the *object* – either real or "merely intentional" – of a representation, concept or act, and to sense contents and other *constituents* or parts of a representation, concept or act. He seems not to have been aware in 1887-1891 of the problems embedded in his usage. In his (1903) review of Palagyi he admitted that for some time he did not know what to make of Bolzano's "objective" concepts and propositions.[10] By 1894, however, in his "Psychological Studies in the Elements of Logic," he had become aware of the difficulties and said, "I think that it is a good principle to avoid such equivocal names as 'representation,' so far as this is possible."[11] In chapter six of "Investigation V" of the *Logical Investigations* (1901) he singled out thirteen equivocations associated with the term *'Vorstellung'*. And in a note from September 25, 1906, he wrote concerning his *Philosophy of Arithmetic*: "How immature – how naïve and almost childlike – this work seems to me. Well, it was not without reason that my conscience punished me on its publication. It in fact came, in essentials, from 1886-1887. I was a mere beginner, without correct appreciation of philosophical problems, without proper training of my philosophical faculties."[12] He went on to say, in particular, that he had at that time no idea of how to bring together the world of the purely logical and the world of the act of consciousness. And it is precisely this inability which shows up in the obscure and equivocal talk of representations, concepts or acts and their 'contents' in the following texts, and thus creates the main problems of its interpretation.

*

It is nonetheless true to say, with Marvin Farber, "that Husserl means to name something objective when he speaks of totalities or pluralities."[13] Perhaps Farber was prompted in this statement by

---

[10] *Early Writings*, p. 201.

[11] *Early Writings*, p. 146.

[12] *Early Writings*, p. 490.

[13] Marvin Farber, *The Foundations of Phenomenology: Edmund Husserl and the Quest for a Rigorous Science of Philosophy*, New York: Paine-Whiteman Publishers, 1962, p. 26

later statements from Husserl, such as those from §22 and §23 of *Ideas I*: "To refer to <the number two> as a mental construct is thus an absurdity, an offence against the perfectly clear meaning of arithmetical speech." It is a "unique member of the number series, which, like all such members, is a non-temporal being." And: "What is engendered in the spontaneous act of abstracting is not the *essence*, but the consciousness of the essence." But that certainly was his *intent*, however badly understood and expressed, from the very beginning.

What then are we to make of the many statements such as this: "The unification <in the multiplicity> comes about ... only in the psychical act of interest and perception which picks out and combines the particular contents and can also only be perceived in reflexion upon that act." (p. 164)

The decision that must be taken in dealing with such passages is whether the reflexion in question takes the act of enumeration as its terminal object, and so must find the "somethings" and "ands" *in* that act, making it up; or the reflexion takes the act of enumeration as the apperceptive basis for an on-reaching, higher order intentionality carrying over to the totality of whatever things are being counted, that totality not being a part of the act or dependent for its nature or existence upon the act. In short, does the act of enumeration and the reflexion thereupon *make* the totality and its specific number, or only *make them present*?

Obviously I take it that Husserl's intent was always the latter, and that, whatever must be said about the many confusions of thought and language in his earliest works, he never held that number and the laws of number were in any usual sense of the word "psychical."

If Frege had studied the *Philosophy of Arithmetic* fairly and thoroughly, he could not have missed this point. He would never have said such things as that for Husserl "Everything becomes presentation. The references of words are presentations."[14] Husserl was, in fact, *never* guilty of "Psychologism" with respect to numbers and their laws, nor in any sense in which he later rejected and refuted it in the *"Prolegomena"* to the *Logical Investigations*

---

[14] P. 323 of Gottlob Frege, "Review of Dr. E. Husserl's *Philosophy of Arithmetic*," translated by E. W. Kluge, *Mind*, LXXXI, #323, (July, 1972), pp. 321-337.

or elsewhere. He certainly did, however, make the mistake of thinking that the symbolic techniques of formal systems, including mathematics, could be analyzed in terms of Brentano's psychology of "inauthentic representations" (p. 205n), discussed below. His change of views on this point is, I believe, that rejection of "a psychological foundation for logic" acknowledged in the "Forward" to the first edition of the *Logical Investigations*.[15]

By examining the act of enumeration through which a totality is rendered present, it is clear that no relation that is a constituent of that act is a mere "and." As we have said, collective combination is not what holds *acts* of any kind together, though of course they too can be counted. Nor does the obtaining of collective combination between two objects depend upon acts or on being thought of. The unity of the act of enumeration or of a thought of number simply is not, nor does it produce or sustain, the unity of the corresponding totality or number. The latter unity is one aspect of the categorial form of all that is, of the "world." And therefore no mere intuition of the mental act as such would present a totality or allow us to get the concept thereof.

It is the intended objective character of this "totality" that is well brought out in the *Philosophy of Arithmetic* where, as we have noted, Husserl places the concepts expressed by 'and' and 'something' (along with, of course, *multiplicity* and *number*) altogether beyond such qualitative distinctions as that between the mental and the physical. These concepts, he says, are the "most general and empty of content of all concepts" and "can with full right be designated as form concepts or categories." (p. 89) It seems to me that the texts here translated provide ample grounds for believing that even in his earliest publications a non-subjectivist analysis of number was intended and carried out.[16] Moreover, it is an analysis of number that is neither Logicist, Formalist, or

---

[15] On this specific point see pp. 111-118 of my *Logic and the Objectivity of Knowledge*, and my "Husserl on a Logic that Failed," *Philosophical Review*, LXXXIX, #1 (January 1980), pp. 46-64.

[16] For a fuller discussion of Husserl's theory of number, see Chapter Two of my *Logic and the Objectivity of Knowledge*, Athens, OH: Ohio University Press, 1984, as well as my "The Concept of Number," in *Husserl's Phenomenology: A Textbook*, edited by J. N. Mohanty and William R. McKenna, 1989, pp. 1-27.

Intuitionist, much less agnostic[17] or nihilist.[18]

*

So among the authentic concepts that are involved in our knowledge of number, and are foundational for the development of arithmetic, are *collective combination, "something"* (or *"a"*), *multiplicity, two, three,* etc. (perhaps up to the number twelve), and *number* as the genus of two, three, etc. But there are two more "authentic" concepts that play an essential role in our understanding of the domain of numbers; and specifically in the *order* of the number series, which of course is crucial for arithmetic. These are the concepts *more* and *less* ("larger" and "smaller") as applied to totalities and numbers. Analysis of these concepts incidentally requires further insight into the types of mental acts and processes required for our knowledge of numbers and number relations.

Knowledge of totalities and numbers beyond those that can be authentically given or fully presented rests entirely, according to Husserl, upon two techniques: enumeration and calculation. It is these techniques that allow us, with justification, to "consider number determinations and number distinctions <in general> to be the most rigorous in the domain of our knowledge." (p. 95) Enumeration and calculation are, in use, purely "mechanical" operations, but they work for knowledge because they are based upon "the elemental relations between numbers" (p. 95), which are *more, less,* and, of course, *equal.*

Inquiry into the "psychological origin" of the concepts *more* and *less* discovers our ability to partition one totality to find two, or to adjoin two totalities to find one. This is often done with concrete objects ("external contents," p. 96), such as pieces of furniture or books. We can select a totality of some of them, and then adjoin others to it, singly or in groups, to represent a larger

---

[17] Bertrand Russell says, only half humorously, "Mathematics may be defined as the subject in which we never know what we are talking about, nor whether what we are saying is true," in *Mysticism and Logic*, Garden City, NY.: Doubleday and Co., 1957, p. 71.

[18] Hartry Field, *Science without Numbers: a Defense of Nominalism*, Oxford: Basil Blackwell, 1980. By contrast, Marcus Giaquinto, "Knowing Numbers," *The Journal of Philosophy*, XCVIII, January 2002, pp. 5-18.

totality, or omit some members from it by partition, to represent a smaller one.

Here the "adjoining" and "omitting" may, though it need not, find support in physical or other relations between the objects. But essentially the same processes occur in cases where we collectively *think* sense-perceptible or non-sense-perceptible 'contents' together and no physical relations are possible. One of Husserl's more startling cases of 'a number of things' is "a feeling, an angel, the Moon, and Italy, etc." (p. 17)

As to what is *meant* by this adjoining and partitioning, that, according to Husserl, "certainly can only be shown and not defined. It is an elemental fact, to be described in no other way than by reference to the phenomena." (p. 96)

But then "facts" of the enlarging and narrowing of totalities do not by themselves "suffice to ground the relational concepts of the more and the less." To allow these relations *themselves* to stand authentically before us, a further fact of inner experience is required and is given. That is the *simultaneous* presence of distinct totalities in one act of consciousness.[19] It must be possible to see the one totality to be larger/smaller than the other. More and less are relations; and, as the cognitive grasp of "any relation requires that the terms be together in a single act of consciousness," so also for them. "The original and the augmented totality <must> be present to us simultaneously and in *one* act." (p. 96)

But even this is not enough to do justice to our consciousness of more and less between totalities. For the "enlarged" totality even appears as the *sum* of the two totalities, one of which is recognized as identical with the original totality, while the other represents the totality of the newly added contents. "If, for example, I augment the totality (A, B, C) to form (A, B, C, D, E), then the judgment that the second is more by D and E requires the simultaneous representation of: (A, B, C), (A, B, C, D, E) and (A, B, C; D, E), and indeed in *one* act." (pp. 96-97; cp. 193 & 208) Of course a similar point is to be made in the case of partitioning.

So we must and do represent totalities together as *one* totality "without thereby their separate unification being lost." And we do

---

[19] Here there emerges, driven by a specific inquiry, the major Husserlian theme of the polythetic act. See *Ideas I*, §§118-119 and elsewhere.

so in some cases *authentically*, "For otherwise the very idea of such composite structures would be absurd, and there would be no question of a comparison of multiplicities as to more and less." (p. 97)

In this passage there occurs one of Husserl's earliest discussions of – not just reference to – "psychical acts of higher order." (See *Ideas I*, §§118 and 133 for later discussions.) We return to the basic idea that some acts bear upon (are of or about) their object in virtue of a built in sub-awareness of other acts bearing upon other objects. This, of course, is a huge Husserlian theme. We have already seen it at work in his account of the "psychological origin" of the concepts of the totality and of number in general. One has to be 'marginally' aware of the acts of separate and specific noticing in the authentic enumeration of the members of a totality for there to be awareness of the totality as such.[20] Now in the "authentic representation" of one totality being more or less than another we have acts of a *fourth* order. They essentially occur in the "psychological origination" of the concepts *more* and *less*.

Here is how it goes: In being aware that (A, B, C, D, E) is more than (A, B, C) by (D, E), we represent it as a combination of the two latter totalities, not merely as *a* totality. This requires that I have separately and specifically noticed A, B, C, etc. (in first order acts), but have noticed them as "somethings" (second order acts); and then that I have noticed the totality they make up (third order), and finally have seen the totality of totalities (fourth order). So a fourth order act is required to *see* that one totality is more than another. (p. 97) This is his "psychological analysis" (p. 95) of the concepts *more* and *less* as to their origin. The "content" of the concepts is precisely the indefinable relation brought to full givenness by means of the experiences described in the psychological analysis. Once we have the concepts, they allow us to think of arbitrary multiplicities and numbers in terms of more or less. (pp. 98-99)

In every case, the recognition of such a relation between totali-

---

[20] See Sartre's use of this idea in *Being and Nothingness*, trans. Hazel Barnes, New York: Philosophical Library, 1965, p. liii; and see *Logic and the Objectivity of Knowledge*, pp. 56 and 83.

ties or numbers includes the knowledge of identity: the identity of one part of the more inclusive totality or number with the whole of the other. "If we judge that five is more by two than three, we then represent five divided up into two partial numbers, two and three" (p. 99), the three being identical on both sides of the "more"-by-two relation. "The multiplicity relations of equal, more and less essentially condition the origination of the number concepts." To know what the determinate numbers, two, three, etc. are "presupposes a comparison and differentiation of delimited multiplicities, thought *in abstracto*, in terms of more and less." (p. 100) One would not really have understanding of the number two unless he knew it was a member of a peculiar series organized in terms of more and less. And "two numbers are non-identical <they are more or less> if the one is identical with a part of the other."

With this Husserl completes his account of the *authentic* concepts involved in our knowledge of number and number relations as to their "origin and content." The *inauthentic* or "symbolic" concepts and propositions (judgments) make up all but a very small corner of our knowledge of numbers, and there would be no arithmetic at all if we were restricted to the authentic ones. But the inauthentic ones must all be constructed from authentic ones, for without the latter we would have no knowledge of what arithmetic is about and why it works. But before turning to his analysis of arithmetic we should notice a few of points he takes up in the remaining chapters of *Philosophy of Arithmetic* Part I.

*

Husserl's view is that the authentic concepts of the number domain are indefinable, and also that they are not in need of definition (p. 101), since they are abstracted from a clear view of the formal or categorial objectivities with which they deal. In Chapters VI and VII he criticizes various attempts to define number, and, especially, attempts to define *equality* of number as it applies to multiplicities. Sometimes identity is substituted for *equality*, he points out, and then the attempt is made to define identity (and thereby equality) by means of the *salva veritate* principle of Leibniz. (pp. 101-102) According to that principle, two things are identical provided that the one may be substituted for the other in any judgment without changing its truth value.

Husserl makes a number of objections to this procedure. First, identity is too strong to be substituted for equality of number. Two multiplicities could be equal in number but not identical. "So long as there is a trace of difference there will be judgments in which the things concerned cannot be substituted *salva veritate*." (p. 102) Secondly, identity (or equality) is what explains substitutability, not the other way around. If we were asked, "Why are the two contents equal?" we would never think of replying, "Because they admit of substitution in true judgments." (pp. 102-103) Finally, the substitution procedure presupposes sameness by speaking of substitution in "the same" judgment.

But others have tried to define equality and inequality (more/less) of number for multiplicities by means of a correlation between the members of the multiplicities concerned. Stolz says: "Two multiplicities are said to be *equal* to one another <in number>, provided that each single thing in the first can be correlated with one in the latter, and none of the latter remains uncorrelated." (p. 103) "Greater" and "smaller" are, then, a matter of some things remaining uncorrelated in one of the multiplicities.

Husserl's main response to this tactic is to point out that the equality of two multiplicities is in fact not conceptually equivalent with their mutually exhaustive one-one correlation. "To represent to oneself two equinumerous multiplicities, and to represent two multiplicities reciprocally correlated term-by-term, is not one and the same thing." (p. 104) What is before the mind in the two cases is simply different. It is true that multiplicities are equal in number if and only if the correlation in question is possible, and we may in certain cases carry out the correlation to verify equality. But that operation is not necessary in every case for us to recognize equality of number, "nor does there reside in it alone, where this happens, the essence of the comparative act." (p. 104) "The possibility of the reciprocal one-to-one correlation of two multiplicities is not the same thing as their equinumerosity, but rather only guarantees it." (p. 110) The "definition" of equality in question does not even provide a nominal definition of equality with respect to number, but only "formulates a *necessary and sufficient condition in the logical sense* . . . for the obtaining of equality."

(p. 110) That is its only possible use.

Husserl further holds that in a process of correlation between the members of two multiplicities, whatever specific type of relationship may be used to set the members in pairs is of no significance, even though such relationships often play an auxiliary role facilitating accurate comparison of groups of "external things," placing them over against one another in pairs. (pp. 113-114) One must not confuse such correlations, utilizing specific types of relationships, with the correlation itself. "The correlation as such is, in every case, a collective combination." (p. 112) Thus – a matter of considerable significance for outstanding mathematicians of his day – "the number equality of two multiplicities is independent of the type of linkage" (p. 115) that may obtain between the members on each side.

In Chapter VII Husserl proceeds on from his rejection of attempts to define the equinumerosity of multiplicities in terms of reciprocal one-to-one correlation to a consideration of attempts to define number itself in terms of equivalence. The former attempts have led, in his view, "to a total misconstrual of the concept of number itself." (p. 117)

If we take any arbitrary group M, we can think of all other conceivable groups that reciprocally correlate one-to-one with it. We can then "speak of the *class of groups*, K, belonging to the group M." (p. 118) We can add (or take away) an element of M and there will be a corresponding new class of equivalent groups. This process can go on without limit. The classification of all conceivable groups thus given is "the most rigorous one imaginable," and a principle of ordering them in terms of next lower or next higher is obvious.

From this it might be supposed that we reach the number concepts by a simple line of thought. Each class, as above described, takes in all conceivable groups of the same cardinal number. But this means that all those groups have a characteristic in common. What could that be? Surely nothing but the circumstance that "they stand in the relationship of mutual equivalence." (p. 119) Any such group can deputize for the whole class of groups. So let us use groups of "1" marks: 11, 111, 1111, etc. as representatives of the corresponding class of duos, triplets, etc. They can also stand in as representatives of the natural numbers. "The numbers form an ordered sequence corresponding to the series of classes." (p. 119)

Husserl describes this as a remarkable "attempt to derive the concept of number originally from that of number equality, and, by-passing all psychological analyses (which are always somewhat precarious), to gain insight into the foundational concepts of arithmetic." (p. 119) It would be a "purely logical construction of the number concept." (p. 112n) In short, it is a version of Logicism, which Husserl rejects.

In criticizing this view Husserl returns to points he made in the previous chapter, concerned with the equinumerosity of multiplicities. It is not true, he says, that "equivalence" (of groups) and "equal in number" are concepts with the same content. "Only this much is true, that their *extensions* are the same. If one identifies equivalence with equality of number, then it is of course natural to regard equivalence as also the source of the concept of number itself, and to conclude: the entirety of the groups equinumerous to one another (i.e., equivalent, belonging to one 'class') can surely have nothing else in common than equinumerosity defined in the manner indicated. Belonging to the class would therefore be what is essential for the number concept in question. To ascribe a number to a concrete group would consequently mean nothing other than to classify it in this sense." (pp. 121-122)

But what the equivalent groups have in common is not just equivalence, "but rather the same number in the true and authentic sense of the word." (p. 122) Equivalence holds precisely because of something in the nature of the totalities involved. Each group has within itself an internal structure of "somethings" unified by "collective combinations." Number and numbers do not consist of relationships between groups. If they did, "then surely every numerical assertion, instead of being directed upon the concretely present group as such, would always be directed upon its relationships to other groups.... But this is absolutely not the sense of a numerical assertion." (p. 122) We do not say there are four walnuts here because this group belongs to a huge or infinite class of groups that can be put into a mutual, one-to-one correspondence with each other. That simply never comes to mind in seeing the walnuts to be four, and it would be of no use if it did. What we *see*, and what is of interest, "is the fact that one nut and one nut and one nut and one nut is here." (p. 122) What we mean by four lies before us, present in the group. The (admitted) equivalence to so much as one other real or possible group does not, and is not so much as thought of.[21]

Although Husserl discussed the views of Stolz (pp. 120ff.) and Kerry (pp. 129ff.) in connection with such attempted definitions, they have no standing in contemporary discussions. It is the views of Frege that remain of some interest, and we should notice Husserl's response to Frege. Frege does not define number in terms of the equivalence of sets. Instead, as is well known, he takes number to be something that belongs to a concept, rather than to a set, and holds two concepts to be equinumerous or equal if and only if the objects falling under the one can be put into one-to-one correspondence to those falling under the other.

Thus, according to him, "The number which belongs to the concept F is the extension of the concept *equal to the concept F*."[22] In this definition we are therefore dealing with a second order concept, a concept of a concept. A range of concepts will have the predicate, or fall under the concept, *equal to the concept F*. For example, the concept *book on that shelf* may fall under the concept *equal to the concept: ball bearing in that wheel*. It will do so provided that the objects of which *ball bearing in that wheel* is true can be correlated one-to-one with the objects that fall under the concept *book on that shelf*. So the second order concept has as its extension, precisely, concepts – e.g., *book on that shelf, ball bearing in that wheel*, etc., etc. And what Frege means by his definition is that the extension of the second order concept corresponding to any concept F is the number of F. Two concepts G and F, have the *same number*, then, if the concepts which fall under the second order concept *equal to G* also fall under the second order concept *equal to F*, and conversely.

Frege provides similar definitions of *direction of a line* and *shape of a triangle*. "The direction of a line a is the extension of the concept *parallel to line a*." By now Husserl's response will be

---

[21] One recalls Husserl's identical argument against reducing the Ideal unity of the property to resemblance between members of a class, in the IInd "Logical Investigation," §5.

[22] Gottlob Frege, *The Foundations of Arithmetic*, trans. J. L. Austin, Oxford: Blackwell, 1953, p. 79.

predictable: "I am unable to find that this method represents an enrichment of logic. Its results are of a type that can only make us wonder how anyone could even provisionally take them to be correct. In fact, what this method allows us to define are not the contents of the concepts *direction, shape* and *number*, but rather their *extension*.... The direction of the straight line *a* would therefore be the totality of the straight lines parallel to *a*." (p. 128) And this last sentence quoted simply constitutes, to Husserl's mind, a *reductio ad absurdum* of Frege's general method.

If the number belonging to the concept *F* is defined as the extension of the concept *equinumerous with the concept F*, this means to Husserl that what we intend or mean (essentially have before the mind) in thinking of *the number* belonging to *F* is: *the totality or set of concepts which have extensions that can be correlated one-to-one with the extension of F*. Given this, he says that "Further commentary is surely pointless." Many persons have thought of – even intuitively grasped – the number four or five, but have had no idea of that which Frege mentions.

In his review of Husserl's book Frege replies, in effect, that the mathematician requires only a true material equivalence of the expressions in a definition, not a sameness of sense. In mathematical definitions, he remarks, "the expressions neither have the same sense nor evoke the same representations <images>.... Coincidence of extension is a necessary and sufficient condition for the fact that between the concepts there obtains that relation which corresponds to that of sameness in the case of objects."[23] But I think it unlikely that Husserl was ignorant of this fact, and am sure that he never intended to deny it. However he certainly did deny that such 'definitions' as may quite rightly satisfy the mathematician can also provide a philosophical clarification of the concept of number and of the corresponding essence.[24] On Husserl's view mathematics should in the end be intelligibly tied down to a definite subject matter, or at least to a relatively small range of distinct but analogous subject matters, by the explication of concepts and corresponding essences. In particular, number

---

[23] "Review of Dr. E. Husserl's *Philosophy of Arithmetic*," p. 327.

[24] See p. 193 in what follows, where the requirements of formal arithmetic are distinguished from those of philosophical understanding. Also pp. 138-139.

must be explained in such a way that we really do have insight into what it is. Frege's method, regarded as doing *this*, is what Husserl criticizes. For philosophical purposes he requires that a definition specify a clear sameness of concept and essence. From the coincidence of extensions, upon which Frege comes to rest, a sameness of concepts or essences simply does not follow, as is surely well known. Much less, then, does it provide a clarification of what the concept or essence in question may be.

Now it may be true, in Russell's words, that currently "we have <in arithmetic> the curious spectacle of a science which does not know what it is really dealing with." But on Husserl's view this situation is not unavoidable, and his intent is to change it. There is much more to number than arithmetic, and there would have been number and (admittedly, quite limited) knowledge of number had arithmetic never arisen.[25]

*

Chapter VIII of *Philosophy of Arithmetic* is a potpourri of philosophical topics mostly having to do with what "one" means, and how oneness and unity relate to and (in certain respects) seem to require reconciliation with multiplicity. The issues range from time-worn philosophical conundrums to problems raised by Husserl's novel understanding of what number is. This chapter contains some of the most interesting and profound philosophical discussions of the book, culminating in a listing of eight different senses of "unit" (or "unity"). (pp. 159-162)

First Husserl takes up the idea that being a unit means to be a certain *specific type* of concrete or abstract "content" – if it is only a written name or symbol such as "1." (pp. 133-135; cp. the "Nominalist Attempt" dealt with on 179-187) This type of view, he holds, can be rendered informative and correct only if the reference to a specific type of "content" is eliminated and "one" is taken as synonymous to "something" in his sense, explained above. He observes that "with definitions of this type very little is

---

[25] Professor Byeong-uk Yi has an excellent but, unfortunately, still unpublished paper, "Arithmetic Is Not All There Is to Number." But what more is there? See also his "Is Two a Property?" *Journal of Philosophy*, XCVI, April 1999, pp. 163-190.

accomplished. The difficulty lies in the phenomena, in their correct description, analysis and interpretation. It is only with reference to the phenomena that insight into the essence of the number concepts is to be won." (p. 136)

Then Husserl takes up the obvious point that one and zero are not numbers, on his understanding of "number." (pp. 136ff.) They incorporate no "collective combination," and in the case of zero not even any "something." "The designation of zero and one as 'numbers'," he says, "presents a *transposition* of that term to concepts of a different kind, even though they stand in close relationship with the literal numbers" and the transposition is driven by "scientific motives of the greatest importance." (p. 138)

"Zero and one are possible, and often enough actual, results of arithmetical problems. An algebraic calculus that aims to accommodate all conceivable cases of calculation can therefore make no distinction between zero and one . . ., and the remaining numbers." (p. 139) Their "marginal character" as numbers is retained, however, "in the exceptions they bring with themselves into most types of calculation." (p. 140) The unity of the concept *number* after being broadened to include zero and one is therefore "an extrinsic one established by means of certain relations." Frege is, once again, mistaken in saying that "What does not apply to 0 and 1 cannot be essential to the concept of number." (p. 141)

Next Husserl takes up the distinction between the concepts *unit* and *the number one*. (pp. 141ff.) "Unit" may signify a mere "something," which has nothing of number about it. "One" can also be used in that sense. Thus, "Along with the concept of the multiplicity (or number) the concept of the unit <"something"> is inseparably given." But not the concept of the *number* one. The latter, as we have just seen, "is only a later result of technical developments presupposing the order in the series of cardinal numbers. "Thus we do not speak of the number 'unit'." (p. 141; cp. 193) There follows an interesting – and, for Husserl's further philosophizing, utterly crucial – discussion of how the interaction of abstract and general terms (color/colors, unit/one) leads to the confusion of taking "number" and "multiplicity" as *general terms only*, thus losing sight of the conceptual (abstract) basis which alone enables those terms to serve in the more utilitarian role of general terms. (p. 146)

Next, difficulties having to do with sameness and difference, as they apply to units in a multiplicity, are taken up. (pp. 146ff.) Certainly the members of a multiplicity must be different, and must also be, *as* units or "ones," the 'same.' However, relations of sameness and difference (among others) had already been dismissed by Husserl (in Chapter II) as essential to the content (object) of representations of the totality, and hence as essential to the concept of number. (pp. 150-151; cp. 152-153) Like one-one correlation, they proved to be no part of what we essentially think of or see when we grasp number, in the concrete or the abstract. (e.g., pp. 32-33, 59-61)

Frege's difficulty in reconciling the sameness and difference that must, somehow, be truly ascribable to the units in a group or a number simply shows, Husserl thinks, that he had not adequately grasped the phenomena. "Every intuitive representation of a group of objects the same in kind demonstrates *ad oculos* that sameness and difference stand in no contradiction at all and very well can be given within one unifying act of thought." (p. 155) To take things as the same *as* "somethings" does not blot out their difference, whatever it may be; but neither does it require representing that difference. Husserl supposes it a strength of his analysis that it makes this clear.

Further difficulties about sameness of units derive from misunderstandings about the *applications* of arithmetic. (pp. 156-159) But "arithmetic as a theory of numbers has nothing to do with concrete objects, but rather with numbers in general." (p. 156) Hence the vagueness of object boundaries, and of what is to count in a given case as one or many, that effects the applications of number formulae, do not effect the sense or truth value of those formulae. "But the sameness of the units, as results from our psychological theory, is obviously an absolute sameness. In fact, the mere thought of an approximation is already absurd. For it is a matter of the sameness of contents with respect to the fact *that* they are contents. To deny this sameness therefore is to deny the Evidence of inner perception." (p. 158)

With these discussions as background, Husserl singles out eight equivocations of the term "unit." (pp. 159-162) It is interesting to see here the emergence of what becomes a standard procedure in his phenomenological work: the careful examination and discussion of phenomenologically given distinctions around an issue, and then, on the basis of the results, a listing of the different meanings of terminology.[26]

Finally in Chapter VIII, Husserl discusses the troublesome "fact that we are often in a position to conceive of one and the same object as one and as many, as we may wish." (p. 163) Is it then one, many, both or neither? How can multiplicities be units: a regiment, for example – one thing and yet many? Husserl's elucidation of "one" and "many" makes it clear that in these cases it is never the same thing that is one and many.

A regiment is many *men* and the men are one *regiment*. "The contradiction lies only in the words." "In actual enumeration it is never doubtful or arbitrary what is to be counted as one." (p. 163) A multiplicity can also be one. "Insofar as it functions as an object to be enumerated, it is also thought as a unit, since it is considered with reflexion upon the fact that it is a content, a 'something'." (p. 164)

In this connection Husserl also develops a response to what is now commonly called "Bradley's regress" – actually it is Herbart's: the puzzle about the unity ("oneness") of the thing, given its "many" properties and relations.[27] Herbart "seeks to detect a contradiction in the concept of *one* thing with many characteristics." (p. 165; cp. 166) Now a 'thing' is a "metaphysical whole," in Brentano's language, and it has a more intimate and stronger type of wholeness than what is found in a mere multiplicity. Of course the constituents of a thing will always be of a certain "number." But beyond that there is "a special sense of *unity* of the thing.... The characteristics in the concept of the thing are not raked together in the manner of a mere *collectivum*." (p. 167) But, as Husserl points out, Herbart's argument results from forcing the unity too far, making it out to be a *simplicity*. "If one confounds the concept of the unity of the thing with the concept of the simplicity of the thing, one will necessarily feel it to be contradictory

---

[26] See p. 125 for a rather lengthy statement of the method of investigation of 'indefinables' which Husserl consciously followed at the time.

[27] See Chapters II and VII of F. H. Bradley, *Appearance and Reality*, Oxford: The Clarendon Press, 1962. A rather more accessible exposition of Bradley's points is to be found in Chapter IV of A. E. Taylor, *Elements of Metaphysics*, London: Methuen & Co. LTD, 1921. Also, *Early Writings*, p. XXXIX.

that the concept of the thing does not permit itself to be grasped as a simple concept, to be thought 'without any differentiation of the several properties'." (p. 167) One does not have to choose between simplicity and multiplicity as the only available accounts of unity.

*

In Chapter IX Husserl enters "the debate concerning the true subject of the statement of number." Here, he says, "It is indicative of conceptual confusion when discord can prevail on *this* point." And yet discord surely does prevail. "One can hardly believe how far the opinions of philosophers deviate from one another here." (p. 169) He considers, primarily, the views of Mill, on the one hand, and of Herbart and Frege on the other. He quickly dismisses Mill's view (or at least one of his views), to the effect that "the number is a predicate of the enumerated things." (p. 171) The view of Herbart and Frege, that the statement of number refers to a concept, is more thoroughly examined. Certainly justice must be done to the role played by concepts in our knowledge of number, but, for Husserl, as we now know, "Number is the general form of multiplicity under which a totality of objects *a, b, c* falls." (p. 174) Revealing residual confusion in his own mind, Husserl here continues on to say: "Considered formally, number and concrete group are related as are concept and conceptual object." In fact, this blurs the distinction between the concept and the essence of number. Number and concrete groups are related as are *essence* and conceptual object. He is not yet clear on this matter. But it is nonetheless true to say, in the next sentence: *"Thus the number relates, not to the concept (Begriff) of the enumerated objects, but rather to their totality (Inbegriff)."*

Accordingly, "The relationship between number and the generic concept of the enumerated is ... in a certain manner the *opposite* of what Herbart and Frege maintained. The number does not say something about the concept of the enumerated, but rather the concept says something about the number." (p. 174) The number numbers, not the concept, but the extension of the concept. "It is not to the concept, *horse drawing the king's coach*, that the number four belongs <*it* isn't four>, but rather to the extension of that concept, i.e., to the totality of those horses." (p. 177)

[See Chapter X for further discussion of what the "numbers" dealt with in arithmetic are. They are not the abstract entities – Husserl confusedly calls them "concepts" – which he has identified as the numbers. For then 5 and 5 would be one (number). They are the generally represented objects of the number concepts. That is, 5 etc. occurs in arithmetical statements as general terms, as "a 5" or "any 5." (pp. 191-192)]

*

In the "Appendix to the First Part" (pp. 179-187) Husserl addresses "a Nominalistic interpretation of the number concepts ... based upon the confusion of the concept *one* with the sign '1'." (p. 179; recall p. 119) The struggle against Nominalism was arguably the most primary and pervasive theme of his entire life work. Ironically, the processes of history have brought it about that Husserl himself, seen through the eyes of a "phenomenology" that supposedly emerged from him, is a Nominalist.[28] In this early discussion, as to the end, his position was that Nominalism loses both concept and essence, and has no way of elucidating the relationship of symbols to that which they represent. (Indeed, no way of elucidating the nature of the symbols and symbol systems themselves. See the opening paragraphs of the IInd "Logical Investigation," on the general point.)

Helmholtz, for example, defines numbers as arbitrary symbols. But what, really, are these symbols, and how do they do what they do for knowledge?[29] "What is it, then, that these symbols genuinely *signify*? In the different cases they can designate the most heterogeneous of objects, and yet the designation of the objects is no arbitrary one. Wherever we use the term 'five,' it

---

[28] See J. P. Moreland, "Was Husserl a Nominalist?" *Philosophy and Phenomenological Research*, 49 (June 1989), pp. 661-674; also, the excellent discussions in Chapter 3 of Moreland's *Universals*, Chesham, Bucks, UK.; Acumen, 2001.

[29] For further discussion of these points see the Ist "Logical Investigation" and my paper, "Why Semantic Ascent Fails," *Metaphilosophy*, 14, 1983, pp. 276-290.

occurs *in the same sense*. In what is it therefore grounded that the most dissimilar of representational contents are designated in the same sense by these signs? In short, what is the *concept* which mediates each use of the signs and constitutes the unity of their *signification*." (p. 182) No answer to such questions is available to Nominalism. The number concepts, with the corresponding essences, are simply lost to it, and with them any possibility of understanding the nature and accomplishments of the very symbols on which the Nominalists (Helmholtz and Kronecker in this discussion) rely. With a wave of the hand at the marvelous symbols, nothing is done to clear up what they are and how they do what is claimed for them.

*

Part Two of *Philosophy of Arithmetic* undertakes to answer the third question listed at the outset of this Introduction. That is, the question of how the symbols and symbolic systems used in arithmetical thought enable us to represent, and to arrive at knowledge of, numbers and number relations which are not and cannot be intuitively given to minds such as ours. This is, of course, the very question which Husserl charged the Nominalist with *not* being able to answer. How does he answer it? His answer depends entirely upon the abstract or "Ideal" structures of numbers and number concepts, on the one hand, and of symbols and symbol systems on the other.

He starts from an invocation and extension of Brentano's concept of *inauthentic representation*. There are, as we have seen, operations on numbers themselves by means of authentic concepts. But in arithmetic there is almost nothing at all of this. In arithmetic "operations" are ways of deriving numbers from numbers (whether they can be authentically given or not) by *operations upon symbols* – operations which, of course, are ultimately founded upon a connection with basic (authentic) concepts of numbers and number relations.

What arithmetic calls "operations" are not "actual [*wirkliche*] activities with and upon the numbers themselves," but are "indirect symbolizations of numbers, which characterize the numbers merely by relations, instead of constructing them through operations." (p. 200) And, for the most part, it is by their relations to a specially devised, though naturally developing, symbolism with an internal order that mirrors the order within the domain of numbers. "Actual" operations on numbers are not, somehow, done more concisely or efficiently in arithmetic. They are not done at all, and it is a naïve, though deplorably common, view which does not notice this, and thereby obstructs the way to an accurate understanding of what is really going on in arithmetic. (p. 201)

Accordingly, "If we had authentic representations of all numbers, as we do of the first ones in the series, then there would be no arithmetic." (p. 201) It would meet no epistemic need. "The whole of arithmetic is ... nothing other than a sum of artificial devices for overcoming the essential imperfections of our intellect." (p. 202) It is only these "essential imperfections" that are the occasion for "Symbolic Number Concepts and the Logical Sources of Cardinal Arithmetic," the title of Part Two of the book.

The *logical sources* of arithmetic are the needs which our minds have in representing and knowing the domain of numbers and number relations. The adjective "logical" is applied to three main things in *Philosophy of Arithmetic*: (i) The relation between genus and species. Color, for example, is a "logical" part of red. (p. 144 and elsewhere; cf. *Ideas I*, §12) This usage was taken over from Brentano. (ii) The significations or meanings (and relations thereof) of experiences or words/sentences, in contrast to their parts and properties. (pp. 32, 90-91, 193f., 206, 229, etc. Cf. §21 of the Vth "Logical Investigation.") And (iii) devices which enable us to extend knowledge with absolute assurance, or the needs or conditions of such extension, as well as distinctions bearing thereupon. (pp. 257-258)[30] The term is most often used in the third sense in *Philosophy of Arithmetic*. The three uses are never explicitly integrated in the early writings, nor, perhaps, later. But the "logical" is never simply a fact. It is sense (iii) of "logical" that is intended in the title of the *Logical Investigations*, with

---

[30] On this last application of "logical," the paper "On the Logic of Signs (Semiotic)" is Husserl's clearest statement. (In *Early Works*, pp. 20-51, but especially pp. 44-51.) On the breadth of the "logical" in sense (iii), see *Logic and the Objectivity of Knowledge*, p. 128, note 17.

senses (i) and (ii) playing a subordinate though vital role in that book. If this is not understood, the book will seem to have no overall unity. (On (iii) see also how "logical" is used on pp. 248, 250, 266-267, 287-288, 298f., etc., etc.)

Now in the practice of arithmetic one senses no loss of epistemic status upon abandoning the "numbers themselves," along with actual operations upon them. We speak and act "as though we could continue the series of numbers to infinity, i.e., beyond any limit attained." And the concepts and propositions involved "are even regarded as the logically most perfect ones in the domain of human knowledge." (p. 202) But how is that possible? This is the question which Husserl now attempts to answer by starting from Brentano's descriptive psychology of the inauthentic or symbolic representations. (p. 205n)

On Husserl's view, "if a content <i.e., an object> is not directly given to us as that which it is, but rather only indirectly, *through signs that univocally characterize it*, then we have a symbolic representation of it, instead of an authentic one." (p. 205) Elsewhere he adds the qualification that "the signs are themselves authentically given."[31] But what are these "signs"? He says that actual perception of the "outer appearance" of a house – the parts of it that we see, the surface on "this" side – provides us with an authentic representation of the house. The house itself is, in this case, "bodily present." By contrast, we have a symbolic representation of that house when we know it merely as the corner house on such-and-such side of a certain street. (pp. 205-206)

But "abstract and general concepts" – and what he certainly has in mind are essences or abstract structures generally – can also be symbolically represented. "A determinate species of red is authentically represented when we find it as an abstract Moment of a perception. It is inauthentically represented through the symbolic determination: that color which corresponds to so-and-so billion vibrations of aether per second." (p. 206) And if the authentic concept *triangle* is *closed figure bounded by three straight lines*, then "any other determination that belongs in univocal exclusiveness to triangles can stand in as an adequate sign for

---

[31] *Early Writings*, p. 20.

the authentic concept – e.g., *that figure the sum of whose angles equals two right angles*."

By far the most important cases for our concerns here are what Husserl calls "external" signs – meaning words and other symbols, including letters and numerals. He points out that to the non-musician C^3 is just: that tone which the musician indicates by means of the sign "C^3." "Psychologically considered, external signs mediate every time language comes into play. But so far as *logic* is concerned, such signs come into consideration only in cases where the concept of that which is to be designated by an external sign belongs, as such, to the essential content of the symbolic representation." (p. 206) That is, when what is referred to is referred to *as* what is designated by the external sign.

So *anything* that has a unique relation to an object can function as a "sign" of that object; and the object can be "symbolically" represented and even *known* by means of the sign. And of course the sign can be extremely complicated, as in the case of the decadal system of numerals. The uniqueness of the relation of the sign to the object guarantees that the symbolic representation of the object will be "logically equivalent" to the authentic representation of it: "Two concepts are logically equivalent when each object of the one also is an object of the other, and conversely." (p. 206) This seems closer to what we today might call *material* equivalence, not *logical*; but nonetheless he holds such an equivalence to guarantee that, so far as our interests in knowledge are concerned, "symbolic representations can surrogate to the farthest extent for the corresponding authentic representations." (pp. 206-207)

Intentionality (meaning) seems to resemble electricity in a certain way. An electrical charge will travel a great distance from its point of origin, given suitable connections with that point of origin and suitable conductivity in all the materials involved. The connecting link in the case of intentionality is "logical equivalence" in Husserl's sense. Given this and a suitable "authentic representation" at the point of origin, intentionality can both represent and permit knowledge of vast ranges of objects that can never "themselves" be present to the human mind.[32]

*

Now, once in command of this basic concept of the symbolic representation, there remain three questions for a philosophy of arithmetic to answer:

1. How do symbolic representations of larger multiplicities or groups work? (Chapter XI)

2. How do symbolic representations of numbers and the number series work to make not only thought but knowledge of the domain of numbers possible? (Chapter XII)

3. How does *calculation* (with the symbolisms introduced in response to 2) serve to find numbers from given numbers (known or unknown) "by means of certain known relationships between them"? (p. 271) (Chapter XIII) Husserl is unable to answer this third question, and the Essays, especially Essay III, attempt to deal with the insufficiency of any approach to it in terms of inauthentic or "symbolic" representations (concepts) alone. It is in Chapter XIII of *Philosophy of Arithmetic* that the "logic" he had looked to for help in understanding arithmetic "left him in the lurch" – understanding by "logic" (sense (iii)) an account of how and why the symbolisms used in the formal disciplines are able to accomplish what they do.[33]

*

The problem for Chapter XI, "Symbolic Representations of Multiplicities," is set by the fact that we *do* recognize groups – most obviously, "sensible groups" – as multiplicities (totalities) without observing or in any way singling out all or any appreciable number of the "somethings" and "collective combinations" which, on Husserl's view, constitute them a whole of that particu-

---

[32] For further discussion of the nature of "symbolic" representations, and especially in connection with Bertrand Russell's distinction between "knowledge by acquaintance" and "knowledge by description," see *Logic and the Objectivity of Knowledge*, pp. 89-94.

[33] *Early Writings*, pp. 13, 15-17, 50, 56-57

lar type. (p. 207) Now given the theory of "symbolic representations" just explained, this *must* be in virtue of something we do genuinely ("authentically") grasp, but grasp as a symbol for the multiplicity in question. What is it? How does it work? "Serious and striking difficulties bar the way to the understanding of the Moments that mediate the symbolization" in these cases. (p. 208)

We step into a large room and see "a number of" people – here, and also over there. We subsume the groups under the concept *a number*. But it simply is not possible that we perceive the totality *as such* by "a lightening fast sequence of individual apprehensions and linkings." (p. 209) And certainly we are not aware of any such process. The apprehension of "sensible sets" here in question must, accordingly, be a case of symbolic representation. But what is it that we do grasp "authentically," in these cases, to then serve as a "sign" of the totality in question. Chapter XI attempts to answer this question.

If, as seems clear, the symbolic representation of large sensible groups or sets cannot be some modified authentic representation of them, it must be that "a sense perceptible quality of second order" intervenes. "These quasi-qualitative characters, which in contrast to the element relations conditioning them would be the πρότερον πρὸς ἡμᾶς, could then provide the support for the association with the concept *multiplicity* in each case. They would indirectly guarantee the existence of a relational complex, and therewith that of a multiplicity of relational terms founding it." (p. 214)

We do often obtain a well-founded symbolic representation of larger sets in cases (a box of red marbles, a sizeable group of men) where we take the time to go through the group and to apprehend each member individually, but *not* all in one unified act, as is the case in authentic representations. The group in question may then be symbolically represented as the totality corresponding to the completed process. But in these cases of complete enumeration we may *also* notice that there are certain characteristics "which, arising out of the fusion of the partial contents or their relations, are immediately noticeable in the manner of sense perceptible qualities." (p. 215) An association between these 'fusion' characteristics fully given to consciousness, on the one hand, and the concept of such a successional apprehension of *all* members of the correlative group, on the other, is thus set up in our minds. And then the association is extended onward to the concept of a set or totality corresponding to that full successional process. It may be that in this way there is produced a "bridge to the immediate recognition of what is at first a unified sensible intuition <object>, of the type here considered, *as* a group." (p. 215)

Reflexion on our actual experiences of such groups – "the testimony of experience" (p. 215), as he says – convinces Husserl that this is the correct account of things. We find all around us "quasi-qualitative" properties of the very type described. Ordinary language brings them to expression when, for example, we speak of a *file* of soldiers, a *heap* of apples, a *row* of trees, a *covey* of hens, a *flight* of birds, a *gaggle* of geese, and so on. They are especially prominent with reference to the perceptual organization of space; but we also hear a shift from a major to a minor chord or key immediately, without necessarily being aware of *which* tonal relations have changed. The 'configuration' of flavor and texture in a soup or soufflé is grasped without prior analysis of the complexities upon which it depends. A similar phenomenal fact emerges with ambiguous drawings, such as the well-known duck/rabbit and face/vase designs. Although Husserl calls the qualities that emerge in such cases "figural moments" (pp. 215ff), he is also aware of the use of "*Gestalt*," and there are some rather prickly footnotes about priority of discovery (mainly with reference to Ehrenfels) on pages 217 and 223.

Husserl's main point in these discussions of Chapter XI surely must be conceded. It is probably justifiable to say that it is *chiefly* such immediate apprehensions of unifying qualitative structures, singly or compounded with certain others, that govern or guide human thought and behavior in all of its phases: from architecture and interior decorating to sculpting and painting; from mathematics, cookery and baseball to poetry, conversation, public address, solfeggio, and driving an automobile. Training in all areas consists in large part of learning to pick up on such immediate qualitative structures and to automatically integrate thought and action with them and with their conditioning elements and relations. Of course the question about the ontological status of "figural Moments" or *Gestalt* qualities is by no means settled by that fact.

In any case, this is Husserl's account of "the origin and content" of symbolic representations of large sensible groups as totalities. In such cases there is directly present to perception a pervasive quality of the group, which is known to derive from a relational complex and hence to presuppose a determinate corresponding totality of objects. That corresponding group is, then, *symbolically* represented as *the group whose members are interrelated in such a way as to give rise to this "figural Moment."* With this we have the result of the first phase of his analysis of the role of symbolic representation in the domain of number.

*

Chapter XII is concerned, not with larger totalities, but with larger *numbers*. Since, as we have seen, most of the particular numbers and their relationships are capable of being represented only symbolically, it is necessary to come to an understanding of the type of "signs" and sign systems that function in the symbolic representation (and knowledge) of them.

The signs in question will certainly be determined by the nature of number itself, on the one hand, and the nature of human cognitive capacities on the other. Our cognitive interest in cardinal arithmetic is maximal advancement of our knowledge of the cardinal numbers and of their interrelations. Knowledge of the fundamental nature of number and the basic relations between numbers (greater, less, equal, next) is provided by intuition of those "facts themselves." (Recall Chapters I-V) We also know on the basis of intuition that the numbers form a fixed series of a specific type, and we can simply inspect that series up to a definite point (p. 243) – the number 12, let us say. Drawing upon resources thus provided by intuition, we can form the symbolic representation of the *next number beyond twelve*, of the next beyond that, and so on. Hence, the move beyond the authentically given segment of the number series does not seem to be in general problematic. (p. 239f.) We have the concepts to secure it for rigorous knowledge.

As we move even a short way into the domain of symbolically

represented numbers, however, we begin to find it practically impossible – and further on it becomes utterly so – to keep clear on which numbers we are dealing with, and how they may be related to others. Even our *symbolic* concepts can no longer be rigorously employed. Without some suitable type of support in the intuition of enduring physical entities (pp. 257-267) – the numerals "0" through "9" and their possible arrangements, for example – a relatively small amount of conceptual complexity overwhelms our capacities for clear and ordered thought. One need only try to distinguish 78 from 79, for example, thinking solely in terms of the number 12 and the relation of next-number-after, in order to see the force of this remark. One loses track of the compound relations involved in the symbolic grasp of larger numbers, just as at around the number 12 one ceases to be able to hold authentically in view in intuition the full complexity of the smaller numbers "themselves."

Now, unlike the case of the larger "sensible sets" discussed in the previous chapter, *nature* does not come to our aid here by providing appropriate systems of signs through which the larger numbers can be rigorously specified and interrelated for our thought and knowledge. Also, such sign systems have not originated through outright invention by clever individuals. (pp. 258f; cp. *Early Writings*, pp. 46ff.) Rather, they have emerged in the course of cultural evolution, though interaction between basic human needs and corresponding capacities. Husserl discusses the general character of this cultural development in some detail. (pp. 259ff) The cultural "origin" of systems of inauthentic (symbolic) concepts is, I think, to be regarded as an extension of his basic idea of "the psychological origination" of concepts.

Historical details aside, it has been the system of decadal *numerals* that has been most prominent in providing symbolic representations or concepts of larger numbers and their interrelationships. Certain general requirements of signs adequate to serve in the construction of symbolic representations of the cardinal numbers will be obvious. They must, of course, be themselves clearly distinct from one another; and each must stand in a unique position in a successional structure that runs rigorously parallel to the number series itself and is in principle extendable beyond any given point. (p. 236) Further, the location of each symbol within the sign system must be apparent from the physical, sense-perceptible constitution of the symbol in question. It is, thus, not enough simply that every number have a name. If the names do not wear on their face their place in a system running parallel to the numbers themselves, we will soon be no better off with the names than we were with the numbers alone. (pp. 238, 240f.)

The decadal numerals "2" through "9" are in fact mere names, whose order can only be remembered, along with their correlations to the authentically given numbers 2 through 9. "33" through "47" are, by contrast, not mere names – like "Charles," "Joseph," etc. – but are symbols constructed upon an obvious principle that automatically puts them into relation with one another and with the corresponding numbers. That principle is *understood*, so that one has no need to *remember* the numerals and their order in following it. The familiar place values for units, multiples of tens, and then powers of ten yields a system of classification that enables one easily to handle quite large numbers and their relations. Thus, "121" indicates one unit plus two units of ten plus one unit of ten tens. It is easily seen to designate a smaller number than 133.

It must be understood that, on Husserl's account, the symbol used in a given case is essentially involved in the *concept* of the number represented – though not in the *number* represented. That is because the number is represented *as* something with a certain essential correlation to *this* symbol herewith used. (p. 206) Hence the relation to this symbol enters into the very connotation or sense of the term "235." In their usual employment numerals are therefore, if not token reflexive, at least *type* reflexive.[34] That is, they designate the precise number that they do in virtue of co-reference to the type of physical, sense perceptible structure which they essentially manifest. In this manner, "The composition of the sign is our crutch." (p. 237) We simply cannot succeed in intending or thinking of most numbers without utilizing a carefully constructed symbol in the manner indicated. (pp. 242, 256)

Thus the function of the symbols in arithmetic is not merely a psychological one, but is a *logical* one, in senses (ii) and (iii)

---

[34] On this point see *Logic and the Objectivity of Knowledge*, p. 128, note #19.

explained above. (p. 206) That is, in their normal use they do not merely show up in our consciousness along with or even *causally* supporting thoughts of numbers. What they designate (the various numbers) are designated in virtue of having and *as* having a relationship – a structural analogy or similarity – to the numerals employed. As "morning star" designates Venus in virtue of the relation of that planet to morning, so "235" designates 235 in virtue of 235's relation to "235." And that is precisely what constitutes the "type reflexive" nature of the numerals in their common usage. While we do not, in general, represent or know numbers merely as "whatever" corresponds to a certain symbolism, that is actually the case with nearly all numbers.

What Husserl calls the "systematic numbers" are actually not numbers, but *concepts* of numbers, where the numbers are designated as corresponding to numerals in the system of numerals. Since the parallelism between the three series (of numbers, number concepts, and numerals) is completely rigorous, the arithmetician in the normal case utterly disregards numbers themselves, as well as their explicit conceptualizations, and works almost entirely with the numerals only, "in a wholly external and mechanical manner." (pp. 252-253, 265-266, 272) This mechanical procedure involving numerals is the only thing that allows us to know the numbers of very large groups by enumeration. And it is only because of the structural correspondence of the numeral system to the order in the concepts, and the structural correspondence of the concepts to the order among the numbers themselves, that the mere manipulation of sensible signs, the numerals, serve the epistemic goals they do in arithmetical practice. This is the outcome of Chapter XII, "The Symbolic Representations of Numbers."

*

But representing and knowing numbers by means of the systematic concepts or symbols, e.g. the decadal, is hardly arithmetic. Arithmetic really begins with *calculation*. Arithmetic is to be defined, according to Husserl, as "the science of the relations between numbers," not the science of the numbers themselves. Its "essential task consists in finding from given numbers other numbers, in virtue of certain relationships known to exist between them." (p. 271) The process of "finding" is, of course, nothing other than calculation.

Undoubtedly the most characteristic expressions of simple arithmetic are of the following types: "$85 + 36$," "$92 - 13$," "$7 \cdot 37$," "$92 \div 4$," and so forth, and complications thereof. These may be read as descriptive phrases – as, for example, "the sum of 85 and 36," or "the difference between 92 and 13." Thus, they have a certain similarity to the numerals already discussed. They are clearly distinguished, sense perceptible signs that refer to one and only one of the cardinal numbers.

However, unlike the numerals, these descriptions do not wear upon their faces a precise position in a series of signs running parallel to the number series; and they do not allow an immediate comparison of the numbers designated as to *more, less,* or *equal,* as do the systematic numerals. (Hence they are called "nonsystematic numbers" (p. 276), while simple numerals alone express the "systematic numbers." Of course the numbers themselves are neither.) Each one of the complex expressions in question therefore not only provides a definite description of a particular number, but also poses the task of determining *which* number is specified by it.

In general, the standard arithmetical processes for carrying out this task are calculations that reduce the compound expression given at the outset to a single numeral, corresponding, in turn, to one "systematic number." The numerals "236," "672," and "922" are, for example, placed in the familiar columnar arrangement, and, through well-known processes, the numeral "1830" emerges beneath them. The description, "sum of 236, 672, and 922," is reduced to the numeral "1830," which is the most perspicuous name for classification of the corresponding number within the system of decadal number concepts.

We have then, on the one had, the series of decadal (or other) numerals, and, on the other hand, an indefinitely large group of definite descriptions formed from those numerals, together with the arithmetical signs for addition, subtraction, multiplication, and division – the "four species" of arithmetical operations – and other signs to be introduced. What Husserl calls the *first basic task* of arithmetic has, then, two parts: the classification of all unsystematic (compound) expressions for numbers into their main types, and the statement of the simplest and most certain procedures for the reduction of the expressions of those types to the corresponding systematic expressions (numerals). Indeed, in his view *arithmetical* operations are "nothing other than methods for carrying out this reduction." (p. 278) This reduction must be, if *we* are to be able to do it, a strictly mechanical procedure; but the result is always the same as if it were a fully conceptualized process, or even an intuitive one. And that is precisely because of the rigorous parallelism established between the order of the numbers and operations thereon and the order of the numerals and operations thereon.

But within general arithmetic there are types of indirect or unsystematic characterizations of numbers other than those mentioned up to this point, and they too serve in symbolic representations of numbers. Above all there is the case where a number is indirectly characterized by means of an equation (or even systems of equations) involving one or more unknown numbers. What we know of such a number is that it is the one whose value, *if* it were known, would – given these and those relations – yield the result expressed in the equation. It is symbolically represented, therefore, as *the* number with that relation to *this* equation involving *those* unknowns. In the reductions referred to above we have to do only with the carrying out of simple operations upon known numbers. But "here we have before us a far more difficult problem: namely, that of unraveling complicated number relationships into which the unknown number itself is interwoven." (p. 297)

And this gives rise to *the second basic task* of the science of number: to provide a general theory of operations that will make intelligible all of the types of transformations leading from unsystematic arithmetical expressions in general to the systematic ones (numerals), and especially the transformations of equations found in that peculiar branch of number theory known as "algebra." (p. 298) Such a general theory of operations is what Husserl has in mind when he speaks of a "general arithmetic" (p. 296, and *Early Writings*, pp. 4-5), the "most wondrous mental machine that ever arose." (*Early Writings*, p. 30)

Now at this point Husserl concludes – one must almost say "breaks off" – his first book, believing himself to have considered "all conceivable types of symbolic number determinations" (p. 299) that are to be found in the science of number. He believes himself to have succeeded in delineating the essence of arithmetic as symbolic knowledge of a certain type. Moreover, he has specified the main types of symbolic representations involved in it, and has shows how they depend upon the use of purely formal systems of signs for their construction and transformation. Finally, he has shown that arithmetic, as such a system of symbolic representations, issues from the 'logical' demands made upon our limited cognitive capacities by the domain of number. That is, it is precisely what is required if we are to extend our knowledge over any significant portion of that domain.

*

Still, a note of disappointment hangs over the conclusion of the book. He has no solution to the second basic task of general arithmetic. He has no general theory of arithmetical operations. The purely formal character of those operations is something that cannot be elucidated by a theory of "inauthentic" or symbolic representations. For while the symbols involved in formal calculation are, in a certain respect, symbolic representations, they are not utilized as such in the calculating. In calculating we do not think of what the symbols involved are symbols of. *How, in general, the formal operations of arithmetic work, Husserl cannot say.* And there is a further embarrassment in the fact that within those operations or calculations there show up symbols that – if we did stop to think – could not receive any possible "real" interpretation in the domain of numbers, such as the square root of -1. These are the "imaginary" or "impossible" numbers that long haunted Husserl's efforts to understand arithmetic.

Here we cannot comment at length on his further efforts toward a general theory of operations. But Essays I-VI in this volume arose out of those efforts. Essay I is, as the title supplied by the editor indicates, an attempt to fill out details of a general theory of the totality – of what we today would call "Set Theory." For

Husserl, it was simply an attempt to further theorize his understanding of "totalities" – or of numbers, taken to be abstract categorial structures of reality. He discusses at some lengths the significance of *infinite* totalities and numbers for number theory, and the necessity to take them into consideration in the axioms of a theory of the "totality." Axiom I (p. 360) looks like it might involve him in the now familiar paradoxes of self-reference.

Essay II can be read as an attempt to formulate a strictly formal logic and to understand what happens as you apply an axiomatized formal logic to various specific domains, such as that of number. Of course he is primarily interested in a formal theory of number, and he describes his research here as a study of the "various forms of number determinations that make it possible to determine new numbers by means of arbitrary numbers." (p. 385) A great deal of what he was developing, however, would now be taught as the elements of formal/symbolic logic. But of course at the time it was an area of frontline research. He has numerous interesting observations concerning what "formalization" and the *purely* logical amount to and how they relate to specific domains of knowledge.

The research he did in this area is of considerable significance for his later work, especially for his understanding of *formal* logic (related to sense (ii) of "logical" above) and the corresponding formal ontology.[35]

Essays III-VI continue to record Husserl's research into formalization and how it relates to the axiomatization of arithmetic. However, in Essay III Husserl does resolve one problem that had been bothering him for more than a decade. In a letter to his friend Stumpf, which I think must be from 1891, he remarks: "The opinion by which I was still guided in the elaboration of my *Habilitationsschrift*, to the effect that the concept of cardinal number forms the foundation of general arithmetic, soon proved to be false.... By no clever devices, by no 'inauthentic representing,' can one derive negative, rational, irrational, and the various sorts of complex numbers from the concept of the cardinal number. The same is true of the ordinal concepts, of the concepts

---

[35] See Chapter 11 of the "Prolegomena" to the *Logical Investigations*, and especially its §67, as well as §13, "Generalization and Formalization," in *Ideas I*.

of magnitude, and so on. And these concepts themselves are not logical particularizations of the cardinal concept."[36]

Husserl's gives his conclusion about arithmetic at this point, which I believe was right after *Philosophy of Arithmetic* had gone to press. It is that "no common concept underlies these various applications of arithmetic." "The *arithmetica universalis* is no science, but rather is a segment of formal logic. Formal logic itself I would define as a symbolic technique (etc., etc.), and designate it as a special – and one of the most important – chapters of logic as the technology of knowledge <sense (iii) of 'logical' discussed above>. In general, these investigations appear to push us toward important reforms in logic. I know of no logic that would even do justice to the very possibility of a genuine calculational technique."[37]

In Essay III he is certainly working toward such a "logic," and he does overcome the problem about signs in the algorithm that have no corresponding object in the domain of numbers. The "Double Lecture," from which most of the material in this 'Essay' derives, was given before the Mathematical Society of Göttingen in 1901, with David Hilbert in the audience. The main conclusion reached in these pages is that it does not matter if there show up in the transformations of the algorithm symbols, such as $\sqrt{-1}$ – the "impossible" or "imaginary" in the title assigned (p. 409) – with no possible corresponding entity in the domain of application. What really matters is *completeness*, and this is simply a matter of whether or not every sentence which can be formed within the algorithm from the concepts of the domain is determined as true or false through pure (strictly formal) deductions from the axioms. (pp. 428 & 431) Meaning or reference is not a factor in arithmetical calculation anyway, since it is not a conceptual or intuitive operation. "Completeness" in the sense sought by Husserl was later shown to be impossible by Kurt Gödel.

Does this then mean that Husserl converted to Formalism in the philosophy of logic. Of course it does not. The refutation of Formalism he had painstakingly worked out in his review of Schröder's *Algebra der Logik* and other papers of the early

---

[36] *Early Writings*, p. 13.
[37] *Early Writings*, p. 17

1890s[38] is never retracted. The calculus of logic or of sets works because of its structural parallel with the *categories* of meanings, on one hand, and the categories of objects on the other, not because of merely conventional rules for signs. Thus when he speaks of a "manifold" in the context of these investigations, it is the *Idea* of the manifold (or the Ideas of various types of manifolds) that he is exploring, in conjunction with the *Ideas* of corresponding axiom systems. (pp. 499-500) And the parallel between these Ideas is, of course, what is worked out in some detail in Chapter 11 of the "Prolegomena."

*

Having now emphasized the major issues dealt with in the texts here translated, and indicated something of Husserl's conclusions concerning them, I should note that I have here followed certain conventions of usage and spelling in order to keep some of Husserl's distinctions before us, distinctions which might otherwise be lost sight of:

"Actual" translates *"wirklich,"* which is closely associated with *"eigentlich"* in these texts.

"Authentic" is almost always used to translate the term *"eigentlich,"* which is associated by Husserl with the special type of situation where the object of an act is "itself present," or at least is not represented merely symbolically (*"uneigentlich"*)

"Evidence" and "Evident" with capital "E" translate the word *"Evidenz"* and its variants, the meaning of which – closely related to that of *"eigentlich"* – can only be established from these and other texts

"Idea" with capital "I" always translates *"Idee,"* never *"Vorstellung."* "Ideal" is usually capitalized to emphasize the objective status of what it qualifies, namely, the *"Ideen"* and their laws, which admittedly Husserl had no clear view of at this point.

"Moment" with capital "M" translates *"Moment,"* which Husserl generally uses, even in these early texts, for what will later be thought of as a non-independent individual (or "abstract particu-

---

[38] For details on this refutation see *Logic and the Objectivity of Knowledge*, pp. 136-143.

lar"), in opposition to a "fragment" ("*Stücke*").

"Reflexion" with an "x" indicates reflexivity, self-reference. "Reflection" is reserved for a process of consideration, meditation, thoughtfulness.

"Representation" and its variants, with capital "R," never translate "*Vorstellung*." It translates the special term "*Repräsentation*" and its variants, which were used by Husserl to mark a very important distinction in the analysis of the mental act.

"Representation" and its variants, *without* capital "R," always translate forms of "*Vorstellung*."

I periodically re-introduce the German term in brackets to aid the reader's memory of my usage.

Numbers in "< >" symbols in the text always refer to the pages of the 'Husserliana' volume used in translating. "{LE}" refers to footnote insertions by Lothar Eley, editor of Husserliana XII; "{DW}" to footnote insertions by me.

*

To Lothar Eley, editor of "*Husserliana XII*," and to the Husserl Archives in Louvain, I owe debts I cannot even begin to identify. Dr. Rudolf Bernet, the current Director of the Archives, and others unknown to me, have generously assisted in many ways. Ingrid Lombaerts saved me from literally hundreds of errors. The re-editing of the "Essays," especially "Essay III," by Karl and Elisabeth Schuhmann made an indispensable contribution to the intelligibility of the texts.

I am also indebted to Nora Henry, who worked through all of "*Husserliana XII*" with me, and to Roderick Iverson, who gave painstaking assistance with all of *Philosophy of Arithmetic*. Arthur Szylewicz labored over the "Abhandlungen," both in the "*Husserliana XII*" version and in the revised versions provided by the Schuhmanns. I really could not have done these texts without him. He also gave valuable advice for *Philosophy of Arithmetic*. Professor Stephan Eberhart proved to be a unique source of knowledge without which my translations of numerous passages would have been only the proverbial "shot in the dark." Burt Hopkins also made numerous useful suggestions. To my shame, no doubt, I have not always followed the advice of these fine scholars, and they are not responsible for errors remaining.

I have compared the French translations by Jacques English where these were available.

My immediate family, Jane, John and Becky, make it possible for me to do what I do. Gratitude to them and for them never leaves my mind. In this case, John also computerized the translation and physically incarnated it in the "camera-ready" pages you see before you, and made numerous excellent contributions of language and thought.

The University of Southern California and a succession of Directors of its School of Philosophy – John Dreher, Edwin McCann, and James Higginbotham – have supported and encouraged me in every way possible. My thanks to it and to them.

*

At the opening of the "Translator's Introduction" to his English translation of *Logical Investigations*, John Findlay referred to *Philosophy of Arithmetic* as "a surpassingly excellent but almost entirely forgotten work on the foundations of mathematics." It is all of that and more. The currents and prejudices of the 20th Century went against it, and the peculiar and remarkable impact and reception of Husserl's later works obscured its nature and its contribution.

It contains, however, some of the finest work later characterized by its author as "phenomenological description" to be found anywhere.[39] A huge quantity of it, in fact. The 1887 discussion of "internal time consciousness" and the logical signification of a representation (pp. 325-326; 32-33) is a brilliant case in point. But nothing is gained by giving it that name, and, indeed, in the current context, much may be lost. It is a part of the "spirit" of phenomenology to *not* approach issues or texts under headings

---

[39] On this point see the opening paragraphs of the papers by Walter Biemel and Oskar Becker in R. O. Elveton, ed., *The Phenomenology of Husserl*, Chicago: Quadrangle Books, 1970. See also Roman Ingarden's article, "Główne fazy rozwoju filozofii E. Husserla," in his *Z Badań Nad Filozofią Współczesną*, Warsaw: 1963. This article was first published in 1939. I owe these references to Arthur Szylewicz.

anyway – even the heading "Husserlian" or "phenomenology."

The task for the reader as for the author is simply to give an accurate characterization of the acts and processes of the intellectual work involved in arithmetic. It is to give an adequate description of the essential properties and relations of those acts and processes, and of the corresponding objects as "themselves given," where they *are* given. It is also to correct corresponding misdescriptions. And this is what Husserl relentlessly does, page after page, chapter after chapter.

One of the ironies of philosophical greatness is how the accumulating interpretations of a great philosopher can lead to the obscuring of his own writings – to what, for example, is "Kantian," but very little of Kant.[40] Bernard Williams suggests, I think, that "an expository or subservient" Kantianism is not a particularly good thing. And a similar point would, I suppose, be true of others among the great ones. It is alright for work to be "of basically Kantian <etc.> inspiration," just not merely expository or subservient.[41] Having, by now, seen a great deal of work supposedly in the "spirit" of some great philosopher, one might fairly conclude that a little subservient exposition might not be an altogether bad thing. After all, there really are very few philosophical geniuses, and a sedulous working out of their texts might prove to be one of the better ways toward genuine philosophical enlightenment – better, even, than imaginative constructions under the banner of a great one. Of course it might not seems as "original," "independently minded," or "creative," if *that* is what we are after.

Husserl's early texts invite us to explore the specific subject matters, *number* and *knowledge of number*, in the company of one of the few truly great philosophical minds. As his work unfolds beside us, he calls aspects of these subject matters to our attention, and leaves us to decide for ourselves what is actually there to be found. He suggests ways of going about deciding that. Careful exposition of his texts *may* actually help us to maximally explore the subject matter – even though the exposition is only indirectly

---

[40] I have explored this theme in my "Who Needs Brentano?" in *The Brentano Puzzle*, ed. Roberto Poli, Ashgate Publishers, Brookfield MA.: 1998, pp. 15-43.
[41] Bernard Williams, *Moral Luck*, Cambridge University Press, Cambridge, 1981, p. 1.

focused upon that subject matter. We study number and arithmetic, not Husserl. Or we, momentarily, study Husserl to get a better grip on number and arithmetic. From the very beginning it was "the thing itself" that mattered for him, and that should be our aim as well. No doubt his "spirit" will assist us.[42]

---

[42] For further study of philosophical and historical issues relevant to the texts here translated, see, for both content and bibliography, Claire Ortiz Hill, "Review of Edmund Husserl, *Early Writings in the Philosophy of Logic and Mathematics*," in *Modern Logic*, 8, (January 1998-April 2000), pp. 142-153; and Claire Ortiz Hill and Guillermo E. Rosado Haddock, *Husserl or Frege?*, Chicago: Open Court, 2000. See also Burt Hopkins' Chapter Eleven, "Authentic and Symbolic Numbers in Husserl's *Philosophy of Arithmetic*," in his forthcoming book on Husserl and Klein. Numerous relevant papers by me are available on my web site: www.dwillard.org. José Huertas-Jourda's intriguing study, *On the Threshold of Phenomenology: A Study in Edmund Husserl's Philosophie der Arithmetik*, is, unfortunately, not generally available.

EDMUND HUSSERL

PHILOSOPHY OF ARITHMETIC:

PSYCHOLOGICAL AND LOGICAL INVESTIGATIONS

Volume One

To My Teacher
FRANZ BRENTANO
With Heartfelt Gratitude

FOREWORD

The "philosophy of arithmetic" which I hereby submit to the public does not claim to construct a thoroughgoing system of this boundary discipline, of equal importance to the mathematician and to the philosopher. Rather, in a sequence of "psychological and logical investigations," it claims to prepare the scientific foundations for a future construction of that discipline. In the present state of the science, nothing more than such a 'preparation' could be attempted. I would not know how to indicate even *one* question of consequence where the response could sustain a merely passable harmony among the investigators concerned. This is sufficient proof that in our domain we are as of yet unable to speak even of a merely schematic articulation of truths already secured for knowledge. The task before us here is, rather: through patient investigation of details, to seek reliable foundations, and to test noteworthy theories through painstaking criticism, separating the correct from the erroneous, in order, thus informed, to set in their place new ones which are, if possible, more adequately secured. With that the intention of this work is characterized.

I have not aspired to critical comprehensiveness. It did not seem necessary, in the interest of the subject matter, to take into consideration the innumerable attempts that deal with the fundamental questions of the domain treated and that have come to my attention. It was sufficient to select those which, whether because of their special character, or because of their wide circulation, or because of their inherent significance, seemed to merit preference.

I hope the critical method that I have often followed will not bring me into reproach. Where possible I have endeavored to isolate leading ideas which I found returning again and again in different authors, and which – though not always clearly and consistently pursued – determined their theoretical convictions. Then I have tried to fix those ideas with conceptual rigor and to build upon them <6> a theory that is as consistent as possible.

The critique that follows could then show how far such thought motifs – appearing so plausible at first glance – are really capable of reaching.

In the positive developments I did not allow myself to be exclusively guided by the interests of an epistemological investigation of arithmetic. Where the analysis of the elementary concepts of arithmetic, on the one hand, and of the symbolic methods characteristic of arithmetic, on the other, promised some returns for psychology or for logic, I have engaged in more detailed investigations than a mere "metaphysics of the calculus" would certainly have required. This is true, for example, in the parts of this volume where the psychology of the concepts *multiplicity*, *unity* and *number* is subjected to the most painstaking of treatments – hopefully not a wholly fruitless one. But only the investigations toward a general logic of symbolic methods (i.e., a "Semiotic"), to be made available in an Appendix to Volume Two, fall wholly outside the framework of a philosophy of arithmetic. In them I will brave the attempt to remedy an essential deficiency in logic up to now.

My hope is that such special investigations – precisely in the context in which, as they arose, they also are presented – will not be unwelcome to the philosophical reader. The mathematical reader, on the other hand, can easily pass over what is of less interest to him. Moreover I note, with reference to the latter, that technical philosophical knowledge is not required for an understanding of this work. I have made sparse use of philosophical terminology, which is rather indeterminate in any case. In particular, I have used no terms not sufficiently clarified through definition or illustration. On the other hand, some previous knowledge of mathematics is required for an understanding, if not of this, then of the subsequent volume – as much, perhaps, as is usually provided in a first course of algebra and analysis. This simply could not be avoided. Whoever has not pursued *some* mathematical studies will probably not be able seriously to begin to orient himself in the philosophy of that discipline.

In the first of its two parts, the Volume I before us deals with the questions, chiefly psychological, involved in the analysis of the concepts *multiplicity*, *unity*, and <7> *number*, insofar as they are given to us authentically [*eigentlich*] and not through indirect symbolizations. The second part then considers the symbolic representations of multiplicity and number, and attempts to show how the fact that we are almost totally limited to symbolic concepts of numbers determines the sense and objective of number arithmetic.

The first part of Volume II is to contain the logical investigation of the arithmetical algorithm – still viewed as an arithmetic of cardinal numbers – and the justification of utilizing in calculations the quasi-numbers originating out of the inverse operations: the negative, imaginary, fractional and irrational numbers. Critical reflections on the algorithm repeatedly occasion a closer examination of the question whether it is the domain of cardinal numbers, or some other conceptual domain, that general arithmetic in the primary and original sense governs. To this fundamental question the Second Part of Volume Two is then to be devoted. It turns out that identically the same algorithm, the same *arithmetica universalis*, governs a series of conceptual domains that have to be carefully distinguished, and that by no means does a *single* type of concept – whether that of the cardinal or the ordinal number, or any other – mediate the application of it *in every case*. In the course of these investigations every care will be taken to give a full logical clarification of the true sense of general arithmetic, as well as an analysis of the concepts (such as *sequence, magnitude*, etc.) to be logically mastered by means of it.

Already from these few indications the reader sees that I distance myself by not a little from views currently prevalent. However, I do not fear the charge that I have favored the new for the sake of novelty. I too set out from the dominant views, and it was a full conviction of their untenability – which forced itself upon me in my efforts to give them precise formulation – that first drove me to new theories, the more satisfactory elaboration of which, I hope, is the fruit of reflections over several years. Perhaps my exertions have not been wholly useless. Perhaps I have succeeded in preparing the way, at least on some basic points, for the true philosophy of the calculus, that desideratum of centuries.
<8>

If time and circumstances are favorable, I also propose to develop in Volume Two a new philosophical theory of *Euclid*ean geometry, whose basic conceptualizations stand in close relationship to the questions to be treated there. Perhaps no unfavorable bias against my aspirations will be evoked beforehand if I say that I owe the basic conceptualization of my new theory to the study of Gauss's report – much read, and still only partially exploited – on the biquadratic remainder (II).[1]

In addition I must note that a part of the psychological investigations in the present volume was already included, almost word-for-word, in my *Habilitationsschrift*, from which a booklet four galley sheets in length, titled "On the Concept of Number: Psychological Analyses,"[2] was printed in the fall of 1887 but was never made available in bookstores.

The second volume, now largely complete in concept, is likely to be submitted to the press in the course of a year.

I allow myself to hope that this work, especially in the light of the difficulty of the problems dealt with, will receive that lenient evaluation which the first more extensive work of an author believes it may count on.

Halle a. S., April 1891   E. G. Husserl

---

[1] C. F. Gauss, "Anzeige der Theoria residuorum biquadraticorum, Commentatio secunda," *Göttingsche gelehrte Anzeigen*, 1831, *Werke*, Volume II, Göttingen 1863, pp.174-178.{LE}

[2] Translated below under this title. {DW}

FIRST PART

THE AUTHENTIC CONCEPTS OF MULTIPLICITY, UNITY AND WHOLE NUMBER

INTRODUCTION

There are various concepts of *number*. This is already indicated by the different words for numbers that show up in the language of ordinary life. These are usually listed by grammarians under the following headings: The whole numbers [*Anzahlen*] or cardinal numbers (*numeralia cardinalia*), the ordinal numbers (*n. Ordinalia*), the type numbers (*n. specialia*), the numbers of repetition (*n. iterativa*), the multiplicative numbers (*n. multiplicativa*), and the fractional numbers (*n. partitiva*). That the whole number comes first in this sequence is, as with the other characteristic names which it also bears – the "basic" or "cardinal" number – not founded on mere convention. Linguistically it assumes a privileged position because all of the remaining number terms proceed only through slight modifications from the words for the whole numbers. (E.g., two, second, of two kinds, double, twice, half).[1] The latter are thus genuine terms for basic numbers. In this way language guides us to the thought that the corresponding *concepts* also may all stand in an analogous relation of dependence to those of the whole numbers, and may represent certain ideas, richer in content, where the whole numbers form mere constituents. The simplest consideration appears to confirm this. Thus in the type number ("one kind," "two kinds," etc.) we have a number of differences within a genus. With the repetition number (once, twice, etc.) a number of repetitions. With multiplicative and fractional numbers the number serves to give a more precise determination of the relationship of a whole divided into equal parts to one part or of one part to the whole. If the whole is divided into n parts, then it is called the n-fold of each part and any part an nth (nth part) of the whole. In a similar manner the words "bi-partite," "tri-partite," etc., express the number of the

---

[1] The German is: "zwei, zweiter, zweierlei, zweifach, zweimal, zweitel." The corresponding terms in English show no such common core plus "slight modifications." {DW}

parts of the whole, conceived of as an extrinsic property of that whole. <11> All of these and similar concepts have an obviously secondary character. Even though they are not logical specifications of the concepts of whole numbers, they nonetheless are more restricted formations which presuppose those concepts, in that they link them up in a certain manner with other elementary concepts. This view is contested only in the case of the ordinal numbers. The linguistic dependency of their names upon those of the whole numbers is indeed no less manifest, and the comparison of concepts also seems, as in the other cases, to point toward a corresponding relation of dependency. What we mean is that as whole numbers refer to sets, so ordinal numbers refer to series. But series are ordered sets. Thus one would apriori be inclined to deny an independent standing to the ordinal numbers also. However, leading investigators – no less than *W. Rowan Hamilton*, *H. von Helmholtz* and *L. Kronecker* – are of the opposite view. They go so far as to vindicate the primacy of the ordinal numbers over the whole numbers, in the sense that the latter are supposed to proceed merely from specific applications of the former. But the paradoxical character of this view is made clear by the provable fact that what these investigators call whole numbers and ordinal numbers does not correspond to the concepts that are usually combined with those names.[2] If the common significations of these names are attended to, then it remains correct that the concept of the ordinal number includes, and thus presupposes, that of the whole number, as the manner in which the terminology has developed rightly indicates.

Beyond the types of numbers in practical life there is yet a long succession of others that are peculiar to arithmetical science. This science speaks of positive and negative, rational and irrational, real [*reellen*] and imaginary numbers, of quaternions, alternating and Ideal numbers, etc. But however diverse the arithmetical expressions of all these numbers may be, they always incorporate the signs for whole numbers, 1, 2, 3, . . ., as constituents; and thus within arithmetic too the whole numbers seem to play the role of

---

[2] Cf. with respect to the two investigators last mentioned the Appendix for the First Part, pp. 179-187 below.

basic numbers in a certain manner, provided that the inference from the dependency of the signs to that of the concepts is not totally fallacious. In fact many <12> – and among them highly significant mathematicians such as *Weierstrass*[3] – are even convinced that the whole numbers form the authentic and unique fundamental concepts of arithmetic. Other investigators, certainly, take a different view. For those who regard whole numbers as specifications of the ordinal numbers, the latter are, as a rule, the fundamental arithmetical concepts. Yet others deny both views and take the concept of linear magnitude as the basic concept of arithmetic, etc. In order to have some provisional standpoint to work from, we choose to adopt the first of the views just mentioned as being the most obvious, and accordingly will begin with an analysis of the concept of the whole number that is as accurate as possible. This is in no way to anticipate any definitive resolution of the issue. It may even be that the progress of our developments in Volume II will prove the opinion thus presupposed to be untenable. That would not in the least deprive the following analyses of any of their value, for they are independent of any theory of arithmetic and of value for all of them. In whatever way the various parties may determine the concept domain that is peculiar and primary to arithmetic, they are united, strictly considered, on this: that the concept of whole number plays an exceedingly important role in all arithmetical matters. Whoever, for example, does not regard this concept, but that of linear magnitude, as the concept genuinely fundamental to arithmetic, will still not therefore deny that the presupposed measurements of the magnitudes are always founded on enumerations, i.e., on whole number determinations; or, further, that whole numbers in the form of multipliers and divisors, of exponents for powers and indices of roots, etc., are indispensable means for the formation of arithmetical concepts. A position of the highest

---

[3] *Weierstrass* usually opened his epoch-making lectures on the theory of analytical functions with the sentences: "Pure arithmetic (or pure analysis) is a science based solely and only upon the concept of number [*Zahl*]. It requires no other presupposition whatsoever, no postulates or premises." (Almost identically the same in the summer semester of 1878 and the winter semester of 1880/81.) There then followed the analysis of the number concept in the sense of the whole number.

significance in arithmetic thus remains secure for the whole numbers, even if arithmetic were defined, not as the science of whole numbers, but rather as the science of linear magnitudes, or in some other way. In every <13> case, therefore, an analysis of the concept of whole number is an important prerequisite for a philosophy of arithmetic. And it is its first prerequisite, unless it should turn out that logical priority belongs to the ordinal concepts, as is maintained from another quarter. The possibility of an analysis of the whole number concept that is done in complete disregard of the ordinal will supply the best proof of the inadmissibility of that view. Moreover, such an analysis does by no means merely serve arithmetical purposes. The correlated concepts of unity, multiplicity and number are concepts fundamental to human knowledge on the whole, and, as such, they lay claim to a distinctive philosophical interest, especially since the considerable difficulties that accrue to their understanding have through time occasioned dangerous errors and subtle controversies. These difficulties are closely linked to certain peculiarities of the *psychological constitution* of the concepts mentioned, in the clarification of which psychology also takes a special interest. To satisfy not merely the arithmetical interests, but the logical and psychological interests above all, is what I designate as the task of the following analyses.
<14>

Chapter I

THE ORIGINATION OF THE CONCEPT OF
MULTIPLICITY THROUGH THAT OF
THE COLLECTIVE COMBINATION

*The Analysis of the Concept of the Whole Number
Presupposes that of the Concept of Multiplicity*

The well-known definition of the concept of *number* – as we may more briefly say in place of "whole number," conforming to the common way of speaking – is: The number is a multiplicity of units. Since Euclid[1] used it, this definition has returned again and again. Instead of "multiplicity," the words "plurality," "totality," "aggregate," "collection," "group," etc., are also used, all names of the same or almost the same signification, although not without appreciable nuances.[2]

Certainly very little is accomplished with this definition. What is multiplicity, and what is unity? We have only to raise these questions and we are enveloped in controversies. Some authors have objected against this definition that "multiplicity" means almost the same thing as "number." But that is correct only in that the term "number" can be taken in a *broader sense,* in which it is actually synonymous with "multiplicity." But in the narrower and authentic sense it only stands for some determinate number, such as two, three or four, and it then expresses a richer notion than does the term "multiplicity." Nevertheless, there also persists an intimate connection between the two concepts even in this latter case. Wherever a determinate number is assigned, a multiplicity

---

[1] At the beginning of Book VII of the *Elements*.

[2] Until explicitly indicated, we will refrain from excluding these nuances by using one of the names alone. The reasons why we prefer different names in different expositions (now "totality," now "multiplicity" or "group") will be explained later. {see pp. 100, 146 and 155n. DW}

can always be spoken of; and where we have a multiplicity, there always is a determinate <15> number as well. But the determinate numbers which thereby come into play are different from case to case. The 'number' concept thus encompasses, though only indirectly through the extensions of its species concepts, which are the numbers two, three, four, . . ., the same concrete phenomena as the concept of multiplicity. The close affinity of the corresponding conceptual contents is also clear from the outset, once we consider that the determinate numbers are to be regarded as determinations of the concept of multiplicity, which bears a certain indeterminacy within it. Wherever a multiplicity is given, the question of "How many?" is in order, and it is answered precisely by the appropriate number. So it will be natural first to attempt the analysis of the more general concept of multiplicity, which is more indeterminate in the respect indicated, and only later to characterize those determinations which give rise to the series of determinate numbers and the generic concept of the whole number which presupposes them. But now we leave aside the above definition of number, since it is of no use for our present purposes.

The Concrete Bases of the Abstraction Involved

As to the concrete phenomena that form the foundations for the abstraction of the concepts in question, there is no doubt at all. They are *totalities* [*Inbegriffe*], multiplicities of certain objects. Everyone knows what is meant by this expression. No one hesitates over whether or not we can speak of a multiplicity in the given case. This proves that the relevant concept, in spite of the difficulties in its analysis, is a completely rigorous one, and the range of its application precisely delimited. Therefore we can regard this extension as a given, even though we are still in the dark on the essence and origination of the concept itself. The same holds, for identical reasons, for the number concepts.

We limit ourselves at the outset to multiplicities that are *authentically* [*eigentlich*] represented. We exclude – perhaps the correlative expression seems the clearer to many readers – *symbolically* represented multiplicities. Our domain, accordingly, must be that of the 'totality' <16> of distinct objects, themselves separately given and held together in the manner of a collection. So our analyses will at first concern only the origination and the content of the authentic concepts of multiplicity and number, whereas the detailed analysis of the symbolic formations presupposing them will first be taken up in depth in Part Two.

Independence of the Abstraction from the Nature of the Contents Colligated

We begin with the psychological characterization of that abstraction which leads to the (authentic) concept of the multiplicity, and subsequently to the number concepts. We have already indicated the concreta on which the abstracting activity is based. They are totalities of determinate objects. We now add: "completely arbitrary" objects. For the formation of concrete totalities there actually are no restrictions at all with respect to the particular contents to be embraced. Any imaginable object, whether physical or psychical, abstract or concrete, whether given through sensation or phantasy, can be united with any and arbitrarily many others to form a totality, and accordingly can also be counted. For example, certain trees, the Sun, the Moon, Earth and Mars; or a feeling, an angel, the Moon, and Italy, etc. In these examples we can always speak of a totality, a multiplicity, and of a determinate number. The nature of the particular contents therefore makes no difference at all.

This fact, as rudimentary as it is incontestable, already rules out a certain class of views concerning the origination of the number concepts: namely, the ones which restrict those concepts to special content domains, e.g., that of physical contents. *Leibniz* already found it necessary to combat such errors. "The Scholastics," he remarks, "falsely believed numbers to originate merely from the partitioning of a continuum and to be incapable of application to the incorporeal." But, he continues, number is "an incorporeal figure," so to speak, "originating through the unification of things (*entia*) of any type, e.g., God, an angel, a man and a motion, which together <17> are four." Therefore number is a "transcendental."³ In the same sense *Locke* also calls number the most general of our ideas, "applicable to men, angels, actions, thoughts, to any thing that can be or be thought."⁴ The old error has attained new currency in the Empiricist school of *J. St. Mill*, to which psychology is so indebted in other respects. "The fact expressed in the definition of number," this philosopher supposes, "is a physical fact. Each of the numbers two, three, four, etc., denotes physical phenomena, and denotes *by means of* a physical characteristic of those phenomena. Two, for example, denotes all pairs of things, and twelve all dozens, and denotes them by means of what makes them into pairs and dozens, and that is something physical; for it cannot be denied that two apples are physically distinguishable from three apples; two horses from one, and so forth: that they are visibly and tangibly different phenomena."⁵ This view is so palpably false that one must only wonder how a thinker of *Mill*'s rank could be satisfied with it. No doubt two apples may be physically distinguishable from three apples; but certainly not two judgments from three, or two impossibilities from three, etc. Thus the difference of number as such also cannot be a physical one, visible and tangible. The mere allusion to psychical acts or states, which surely can be counted just as well as physical contents, refutes *Mill*'s theory.

The Origination of the Concept of the Multiplicity through Reflexion on the Collective Mode of Combination

If, however, the general concepts of multiplicity and determinate number do not stand to the respective concreta from which they are abstracted – <18> the totalities of determinate but arbi-

---

[3] *De arte combinatoria* (1666), *Opera philosophica*, Ed. J. E. Erdmann, Berlin 1840, p. 8.

[4] *An Essay Concerning Human Understanding*, Book II, ch. XVI, sect. 1.

[5] John Stuart Mill, *A System of Logic*, London 1970, Book III, ch. XXIV, §5. See also Book II, ch. IV, §7, where the character of number is deemed parallel to the physical characteristics of color, weight and extension.

We frequently find analogous views also among mathematicians. *Frege* cites examples in his *Grundlagen der Arithmetik*, Breslau 1884, pp. 27ff.

trary contents − in the relationship of physical characteristics to physical things, how is their relationship to be understood? The peculiar characteristics of the contents colligated and to be enumerated are of absolutely no significance, as we noted. But if that is so, how are we to arrive at the general concepts desired? How are we to conceive of the abstraction process that yields them? In the abstraction, what is retained as the content of the concept, and what is that from which abstraction is made?

Since, following upon a preliminary observation, we are allowed to take the extensions of our concepts as given, it must after all be possible to approach the contents of the concepts at issue by way of comparison and differentiation of available, appropriately chosen examples from the corresponding extensions. Disregarding the properties that are different, we retain those that are common to all, as those which may belong to the content of the concept in question.

Let us now try to follow up on this indication.

In the first place, it is obvious that the comparison of the particular contents which we find before us in the given totalities would not yield the concept of multiplicity or of determinate number. And, even if that happened, it would be absurd to expect such a thing. Those particular contents are, in fact, not the basis of the abstraction, but rather the concrete totalities *as wholes* are, in which the particular contents are comprised. However, it seems that the desired result is not to arise from comparison of the totalities either. The totalities, after all, consist merely of the particular contents. So how can any common properties of the *wholes* be singled out, if the *parts* that make them up may be utterly heterogeneous?

However, this specious difficulty is easily resolved. It is misleading to say that the totalities consist merely of the particular contents. However easy it is to overlook it, there still is present in them something more than the particular contents: a "something more" which can be noticed, and which is necessarily present in all cases where we speak of a totality or a multiplicity. This is the *combination* [*Verbindung*] of the particular elements into the whole. And things stand here as in the case of many other classes of relations: there can be the greatest of differences between the related contents, and yet there be identity of kind with respect to <19> the combining relations. Hence, similarities, gradations [*Steigerungen*], and combinations involving continua are found in wholly heterogeneous domains; and they can occur between sensuous contents as well as between psychical acts. It is, therefore, quite possible for two wholes to be similar *as wholes*, although the parts constituting the one are completely heterogeneous to those constituting the other.

Those combinations which, always the same in kind, are present in all cases where we speak of multiplicities are then the bases for the formation of the general concept of multiplicity.

As to the sort of abstraction process which yields our concept, we can best characterize it by referring to the way in which concepts of other composites (wholes) originate. If we consider, for example, the cohesion of the points on a line, of the moments of a span of time, of the color nuances of a continuous color spectrum, of the tonal qualities in a "tone progression," and so on, then we acquire the concept of combination-by-continuity, and, from this concept, the concept of the continuum. This latter concept is not contained as a particular, distinguishable, partial content in the image of every concretely given continuum. What we note in the concrete case is, on the one hand, the points or extended parts, and, on the other hand, the peculiar combinations involved. These latter, then, are what is always identically present whenever we speak of continua, however different may be the absolute contents which they connect (places, times, colors, tones, etc.). Then in reflection upon this characteristic sort of combination of contents there arises the concept *continuum*, as that of a *whole* the parts of which are united precisely in the manner of continuous combination. Or, to take another example, consider the quite peculiar way in which, in the case of any arbitrary visual object, spatial extension and color (and color, in turn, and intensity) reciprocally penetrate and connect with each other. With reference to *this* manner of combination[6] we can once again form

---

[6] *Franz Brentano* speaks here of "metaphysical" combination, and Carl Stumpf (*Über den psychologischen Ursprung der Raumvorstellung*, Leipzig 1873, p. 9) of the relationship of "psychological parts."

the concept <20> of a whole the parts of which are united in just such a manner.

We can say quite generally: Wherever we are presented with a particular class of wholes, the concept of that class can only have originated through reflection upon a well-distinguished manner of combining parts, a manner which is identical in each whole belonging to that class.

Just so with the case that concerns us here. We can likewise say that a totality forms a whole. The representation of a totality of given objects is a *unity* in which the representations of single objects are contained as partial representations. Certainly this combination of parts, as found in any arbitrarily selected totality, must be called loose and external when compared to other cases of combination. So much so, in fact, that one would almost hesitate to speak here of any combination at all. But, however that may be, there *is* a peculiar unification there; and it must also have been noticed as such, since otherwise the concept of totality (or multiplicity) could never have originated. So, if our view is correct, the concept of multiplicity has originated by means of reflexion upon the peculiar – and, in its peculiarity, quite noticeable – manner of unification of contents, as it shows up in every concrete totality. And it has arisen in a way analogous to that of the concepts of other sorts of wholes, by reflexion upon the modes of combination peculiar to those wholes.

From here on we shall use the name "*collective combination*" to designate that sort of combination which is characteristic of the totality.

Now before we proceed with the development of our subject, it will be good to deflect an apparent objection: If the multiplicity is defined as a whole the parts of which are united by collective combinations, then one could object that this definition is circular. For in speaking of "parts" we certainly think of a multiplicity; and, since the parts are not individually determinate, we have a general representation of this multiplicity. That means that we are explaining multiplicity in terms of itself.

Nonetheless – and so very plausible as this objection may be – we cannot concede its cogency. First, notice that the intent here is not a *definition* of the concept <21> *multiplicity*, but rather a *psychological characterization* of the phenomena upon which the abstraction of this concept rests. All that can serve this purpose we must therefore regard as welcome. Now the plural term "parts" certainly implies (disregarding its correlation with the concept of the whole) the general representation of a multiplicity; but *that* term does not express what peculiarly characterizes this multiplicity *as* multiplicity. By adding that the parts are collectively combined, we made reference to the point upon which our special interest reposes, and in virtue of which the multiplicity is characterized precisely *as* multiplicity, in contrast to other sorts of wholes.

Chapter II

CRITICAL DEVELOPMENTS

The shortest answer to the question about what kind of unification is present in the totality lies in a direct reference to the
5 phenomena. And here we truly are concerned with ultimate facts. But we are not thereby relieved of the task of considering this kind of combination more carefully, in order to bring into relief its characteristic differences from other kinds, especially since false characterizations and confusions of it with other species of rela-
10 tions have been an all too common occurrence. To this end we shall test a series of possible theories, some of which have actually been advanced. Each of these theories characterizes the collective unification in a different way and, in relation thereto, seeks also to explain in a different way the origin of the concepts *multiplicity*
15 and *number*.

*The Collective Unification and the Unification of
Partial Phenomena in the Total Field of Consciousness
at a Given Moment*

The combination of representations to make up a totality, some-
20 one could say, still hardly deserves the name of a "combination." What then is given when we speak of a totality of arbitrary objects? Nothing further than the co-presence of those objects in our consciousness. The unity of representations of the totality thus consists only in their belonging to the consciousness which
25 encompasses them. Still, this "belonging" is a fact which can be attended to; and with reflexion upon it there then originate those concepts the analysis of which is here in question.

Now this view is obviously wrong. Quite a number of phenomena <23> make up, in each moment, the total state of our consciousness. But it is the role of special interests to lift certain representations out of this plenum and collectively unite them. And that occurs without the disappearance of all of the remaining representations from consciousness. Were this view correct, then in each moment there would be only a single totality, consisting of the whole of the present partial contents of our total consciousness. But at any time, and in any way we choose, we can form various totalities, can expand one already formed by the addition of new contents, or can narrow it down by taking other contents away, without necessarily excluding these contents from consciousness. In short, we are conscious of a spontaneity which otherwise would be inconceivable.

But this view, in its general and indeterminate form, contains in addition an absurdity. In fact, do not continua, with their infinite groups of points, belong to the state of our consciousness? Who has ever actually represented them in the manner of a totality?

It is important to stress that a totality (an authentic representation of a multiplicity) can have as elements only such contents as we are aware of in the manner of things separately and specifically noticed [*für sich bemerkte*]. All other contents, however, which are present only as things incidentally noticed, and which either cannot be separately noticed at all (like the points of continua), or merely are not, for the moment, separately noticed: – all these cannot yield elements out of which a totality is constituted.

Probably all of this will be quickly conceded; and the representative of the view just criticized might forthwith restrict his assertion in such a way that by the "encompassing consciousness" which unites representations into a multiplicity is to be understood a special act of consciousness, and not consciousness in the widest sense, where it takes in the whole of our psychic phenomena. So it would be, accordingly, a matter of a unity in an act of representing that both throws contents into relief and gathers them together, or of a unity of interest, or something similar. We intend to come back later to consider more closely the theory as thus corrected. <24>

The Collective "Together" and the Temporal "Simultaneously"

Let us now turn to consideration of a new theory, which argues as follows:

If a totality of contents is present to us, what else are we to notice but that every content is there *simultaneously* with each other one? Temporal co-existence of contents is indispensable for the representation of their multiplicity. Now, indeed, there is required in any composite act of thought the co-existence of its parts. But whereas in other cases there are present, in addition to simultaneity, distinctive relations or combinations which unify the parts, it is precisely the distinguishing feature of the representation of the totality that it contains *nothing more* than the simultaneous contents. Hence, multiplicity *in abstracto* signifies nothing other than some sort of contents being given simultaneously.

This view, as is easily seen, is subject to precisely the same objections as the previous view, and to many others besides. It would be superfluous to repeat the former objections; and, of the latter, it is sufficient to emphasize that to represent contents simultaneously does not yet mean to represent contents *as simultaneous*. For example, in order for the representation of a melody to come about, the single tones which make it up must be brought into relation with one another. But every relation requires the simultaneous presence of the related contents in one act of consciousness. Thus, the tones of the melody must also be simultaneously represented. But in no wise are they to be represented *as* simultaneous. Quite to the contrary, they appear to us as situated in a certain temporal succession.

It is not otherwise in the case where we represent a multiplicity of objects. That we must simultaneously represent the objects is certain. But that we do not represent them as simultaneous, and that, rather, special acts of reflexion are required in order to notice the simultaneity in the representing of the objects: this is directly proven by reference to inner experience.

In this way it becomes clear to us that the collective "together" must not be described as a temporal "simultaneously." <25>

Collection and Temporal Succession

A third view is likewise based upon time as an insuppressible psychological factor. In direct opposition to the foregoing, it argues as follows:

In virtue of the discursive character of our thinking, several contents which are different from each other absolutely cannot be thought at the same time. Our consciousness can be engaged with only *one* object in each moment. All mental activity of a relational or higher sort becomes possible only in that the objects with which it has to do are given *in temporal succession*. So, then, each complex thought-structure, each whole composed of certain parts, is something that has arisen in stages out of simple factors. In such cases we always have to do with step-by-step processes and operations which, proceeding through time, increasingly intertwine and expand themselves. In its peculiar nature, therefore, each collection [*Kollektion*] presupposes a collecting [*Kolligieren*]; and each number presupposes an enumeration. And herewith there is necessarily given a temporal arrangement of the objects collected, or of the units enumerated. But yet more than this. It is temporal succession and nothing else that characterizes the multiplicity as multiplicity. Certainly a succession is always present wherever contents stand in any sort of relation, simple or more complex, and thus come together to form a representational whole; but then we always can at the same time also speak of a multiplicity. But this cannot happen with respect to just any sort of relations, which may differ from case to case, but rather only with respect to that unique one which shows up in every case: namely, temporal succession. In the case of the weakest combination conceivable, where arbitrary and otherwise unrelated contents – or contents taken in abstraction from all possible relations – are merely thought together, i.e., as multiplicity, there the succession with which the contents came into consciousness is the only relation that still connects them. Therefore, what characterizes the multiplicity, in contrast to other and more intimate forms of assemblage, is the circumstance that in it *mere succession* puts the contents into relationship, while in the other forms of assemblage still other relations are involved. Accordingly, it also follows that *multiplicity in abstracto* is nothing more than *succession*: succession <26> of *any sort of* contents separately and specifically noticed. But the number concepts Represent [*repräsentieren*] the determinate forms of multiplicity or succession *in abstracto*.

Now in order not to dissipate attention through fruitless individual critiques, I have here preferred, instead of criticizing in turn the authors who have propounded such theories or similar ones, to state as plainly and as fully as possible the view which they all more or less clearly have in mind, and to exercise my critique upon that view.

And that view, which is to be combatted here, is based upon essential errors – psychological and logical.

First, it appeals to the psychological fact of the narrow range of consciousness. However, it exaggerates and falsely interprets this narrowness. It is true that the number of separate contents to which we can turn our attention in any one moment is highly restricted. In fact, with maximal concentration of interest the number shrinks to a single one. But it is false that we can *never* be *engaged with* more than *one* content in one and the same moment. Indeed, just the fact that there is thought which relates and connects – as well as, in general, all of the more complicated mental and emotional activities to which this very theory appeals – teaches Evidently the utter absurdity of this conception. If in every instant only *one* content is present to our consciousness, how should we be able to notice even the simplest of relations. If we represent the one term of the relation, then the other either is not yet in our consciousness or it is no longer there. Now we certainly cannot connect a content of which we are not conscious – and which, therefore, does not exist for us at all – with the single content which is at the moment actually given to us. On this interpretation of the narrow range of consciousness, reference to the temporal progression of the representations to be related, instead of grounding the possibility of relational thinking, would, quite to the contrary, bring to Evidence its impossibility.

But then does not experience teach (so, perhaps, our opponent replies) that as a matter of fact we always can have only *one* present representation, and that it is very well possible to bring it into relation with past representations? That a <27> representation is past in no way implies that it ceases to be.

It is easily seen that such an answer would rest upon misinterpretations of experience. One must not confuse temporally present representations with representations of what is temporally present, and past representations with representations of what is past. Not every present representation, as we must emphasize here once again, is a representation of what is present. Precisely all representations directed upon things past constitute an exception; for they all are, in truth, present representations. If we recall a song we heard yesterday, for example, then the memory representation involved is indeed a temporally present representation; only it is referred by us to something in the past. If one keeps this in mind, one will of course no longer find any problem at all in our being able to bring representations of present contents into relation with representations of past contents. In doing so, these representations are all, in fact, simultaneously present in our consciousness. They are *in toto* representations that are temporally present. By contrast, we can relationally connect past representations neither with each other nor with present representations; for, as past, they are irrecoverable and gone forever.

So the alleged fact of experience which our opponent has in view boils down to the claim that, whenever we represent a plurality of contents, there is always one alone which would be a temporally present content, whereas all of the others would exhibit greater or lesser temporal differences. Naturally, then, each total representation composed of distinct (separately and specifically noticed) parts would have to have *originated* through *successive* acts of noticing and relating the particular partial contents, while the total representation itself, as something finished and developed, would contain all of the parts at the same time – only each furnished with a different temporal determination.

Now it is indeed certain that, already with a very modest number of contents, an act of noticing that gathers them together is only possible by apprehending and retaining them successively or in very small groups. But should we not be able with two or three very distinct contents to grasp them in one act and hold them together without requiring <28> a serial progression from one to the other? I would not venture a very confident negative response to this question.[1]

However that may be, we can acknowledge it as a fact that for the origination of representations of groups (some few perhaps excepted), and of all representations of numbers, temporal succession forms an indispensable psychological requirement. One is, therefore, quite justified in designating – if not all, certainly almost all – groups and numbers as results of processes, and, insofar as our will is thereby involved, as results of activities: of *"operations"* of colligating or of enumerating.

But this also is *all* we can agree to. Only this one thing and no more is proven: that succession in time constitutes an insuppressible *psychological precondition* for the formation of by far the most number concepts and concrete multiplicities – and practically all of the more complicated concepts in general. These have a temporal mode of becoming, and thereby each constituent of the completed whole receives a different temporal determination in our representation. But does that also prove that temporal order enters into the *content* of these concepts, or even that it perhaps is the special relationship which characterizes pluralities as such, in contrast to other concepts of composites? In fact, people are often satisfied with such paltry arguments, without considering that time constitutes, in precisely the same manner, the basis for *all* higher thinking, and that, for example, one could with equal right infer the relation of premises to conclusion to be identical with their temporal succession. However, such obvious absurdities have already been removed by the very formulation

---

[1] Perhaps some readers will be surprised at my indecisive language here, where confirming examples for even considerably larger numbers are so easily adduced. As a matter of fact, in playing dominoes, for example, we grasp groups of ten to twelve dots with one glance. Indeed, we even assess their number with total immediacy. It must be observed, however, that in such cases we can speak neither of an actual colligating nor of an actual enumerating. The number name is here directly associated with the characteristic sensuous appearance, and is then recalled on each occasion by means of that appearance without any conceptual mediation. With groups that large, as everyone can test, a direct and authentic collection and enumeration is an impossibility. In the case of very small groups, of two or three objects, the matter stands in doubt because the successive apprehensions of the elements could ensue so quickly that they themselves would escape focused attention. Therefore the cautious manner of speaking in the text.

which we gave the time-theory for our purposes. That formulation <29> asserts only that the case of the totality (of the concrete multiplicity) is distinguished from that of any other composite whole by the fact that in it *mere* succession of partial contents is present, while with the other wholes there is yet *beyond that* some *other* sorts of combinations. So the argument is not simply that, because enumerating requires a temporal succession of representations, number is the comprehending form of the successive *in abstracto*. Rather, it also presumes to be able to show that temporal succession forms the only common element in all cases of multiplicity, which therefore must constitute the foundation for the abstraction of that concept.

But even thus formulated we cannot agree with the theory. If it were correct, then there would fail to be any intelligible distinction between the concepts *multiplicity* and *succession*, which surely no one will seriously take to be identical. For what sense would it then have to speak of a multiplicity of *simultaneous* contents? The origination of the concept of temporal co-existence would, from this point of view, become an incomprehensible enigma.

And in the same way we can, in general, characterize any conceivable attempt to clarify the concepts of multiplicity and cardinal number by reference to temporal succession as hopeless from the outset. The following simple reflection brings this to Evidence. Should temporal succession, in whatsoever form, contribute to the content of the concepts mentioned, then it would not suffice that it be detectable in each of the relevant concrete cases. Rather, it would also actually have to form the object of explicit attention in all cases. But this certainly does not hold true. We do not always attend to the temporal relations, and precisely because of that we are capable of distinguishing between a multiplicity *simpliciter* and a multiplicity of successive (or simultaneous) contents. This mistake is repeatedly committed on one side and censured on the other. To perceive temporally successive contents does not yet mean to perceive contents as temporally successive. This is a point that we also have had to emphasize upon occasion (p. 25) with respect to simultaneous contents. But it is of special importance to consider – something overlooked as a rule – that, even where we <30> notice a temporal sequence of contents, in no way are determinate multiplicities already marked out. That is only brought about by certain psychical acts of collecting. To overlook them means to leave out of account precisely that which forms the true and only source [*Quelle*] of the concept of multiplicity as well as of the concept of number. Some examples may serve to clarify all this.

The clock sounds off with its uniform tick-tock. I hear the particular ticks, but it need not occur to me to attend to their temporal sequence. But even if I do attend to it, that still does not involve singling out some number of ticks, and uniting them into a totality by an inclusive noticing. Or take another example: Our eyes roam about in various directions, fixing now upon this, now upon that object, and evoking manifold representations succeeding one another in a corresponding order. But a special interest is necessary if the temporal sequence involved here is to be separately and specifically noticed. And in order to maintain a grasp on some or all of the noticed objects themselves, to relate them to each other, and to gather them into a totality, here again are required special interests and special acts of noticing directed upon just those contents picked out and no others. That is to say, even if the temporal sequence in which objects are colligated were always attended to, it would still remain incapable of grounding by itself alone the unity of the collective whole.[2] And since

---

[2] By overlooking this circumstance, the most recent analyst of the number concept, W. Brix, has also fallen into errors in his psychologically untenable attempt to conceive of "the number of temporal intuition, or the cardinal" as the second genetic level in the development of the number concept. The first level is supposed to be Represented [*repräsentiert*] through "the number of spatial intuition." (W. Wundt, *Philosophische Studien*, V, Leipzig 1887, pp. 671ff.) From the possibility of continuing the successive positing of units arbitrarily far, Brix draws the highly precarious conclusion (*Ibid.*, p. 675) that in such a way arbitrarily large numbers can be formed. But the mere succession of the repeated positings does not yet guarantee any synthesis, without which the collective unity of the number is inconceivable. It is precisely upon the inability actually to carry out such a synthesis, as we shall later discuss, that every attempt to form an authentic representation of the higher groups and numbers factually runs aground. Through the simplest of experiments Brix could have convinced himself that even nineteen positings of units are not clearly distinguishable from twenty, unless by the indirect means of symbolizations that serve as surrogates for syntheses actually carried out. But of course without the possibility of such distinction there can be no talk of an actual formation of the numbers concerned in terms of sequences of posited units.

we cannot even concede that temporal succession enters into the representation of each concrete totality merely as an invariable constituent always attended to, <31> it is clear that even less can it in any way enter into the corresponding *general concept* (multiplicity, number).

Herbart is completely justified in saying that "number has . . . no more in common with time than do a hundred other sorts of representations which also can be produced only gradually."[3] And *Beneke* expresses the same thought, no less dramatically than cogently: "That time flows over enumerating can prove nothing; for what, indeed, does time not flow over?"[4]

Were it merely a question of describing the phenomenon [*Phänomen*] that is present when we represent a multiplicity, then certainly we would have to make mention of the temporal modifications which the particular contents undergo, although those modifications as a rule are not given any special notice. But disregarding the fact that the same holds true of every composite whole, we have, in general, to distinguish between the phenomenon as such, and that for which it serves, or which it signifies for us. Accordingly, we must also distinguish between the psychological description of a phenomenon and the statement of its *signification*. The phenomenon is the foundation of the signification, but is not identical with it. If a totality of objects, A, B, C, D, is in our representation, then, in light of the sequential process through which the total representation originates, perhaps finally only D will be given as a sense representation, the remaining contents being then given merely as phantasy representations which are modified temporally and also in other aspects of their content. If, conversely, we pass from D to A, then the phenomenon is obviously a different one. But the logical signification sets all such distinctions aside. The modified contents serve as signs, as deputies, for the unmodified ones which were there. In forming the representation of the totality we do not attend to the fact that changes in the contents occur as the colligation progresses. Our aim is to actually maintain them in our grasp and to unite

---

[3] *Psychologie als Wissenschaft*, Königsberg 1825, Part. II, p. 162.

[4] *System der Logik als Kunstlehre des Denkens*, Berlin 1842, Part I, p. 279n.

them. Consequently the *logical content* of that representation is not, perhaps, D, just-passed C, earlier-passed B, up to A, which is the most strongly modified. Rather, it is nothing other than (A, B, C, D). The representation takes in every single one of the contents <32> without regard to the temporal differences and the temporal order grounded in those differences.

Thus we see that time only plays the role of a psychological *precondition* for our concepts, and that in a two-fold manner:

1. It is essential that the partial representations united in the representation of the multiplicity or number be present in our consciousness *simultaneously*.
2. Almost all representations of multiplicities – and, in any case, all representations of numbers – are results of *processes*, are wholes originated *gradually* out of their elements. Insofar as this is so, each element bears in itself a different temporal determination.

But we found that neither simultaneity nor successiveness in time enters in any way into the ⟨logical⟩ *content* of the representation of the multiplicity; and so, likewise, into that of the representation of number.

As is well-known, already in *Aristotle* time and number appear to be brought into intimate connection through his definition: "Time is the number of movement in respect to earlier and later." However, it is only since *Kant* that it has become more generally common to stress the temporal "form of intuition" as the foundation of the number concept. To be sure, this happened much more as a consequence of the authority of his name than as a consequence of the force of his arguments. We do not find in *Kant* a serious attempt at a logical or psychological analysis of the concept of number. Unity, multiplicity and totality constitute the categories of quantity in his metaphysics. Number is the transcendental schema of quantity. *Kant* fully states his view as follows, in the *Critique of Pure Reason*: "But the pure schema of magnitude [*quantitatis*], as a concept of the understanding, is *number*, which is a representation that binds together the successive addition of one thing to another thing (of the same kind). Thus, number is nothing other than the unity of the synthesis of the manifold of a homogeneous intuition in general, a unity which comes about through the fact that I engender time itself in the apprehension of the intuition."⁵

This passage is obscure and is also irreconcilable with the explanations <33> that *Kant* gives of the function of the schema. These themselves certainly are not quite uniform. For example he says: "We wish to call . . . the formal and pure condition of sensibility, to which a concept of the understanding is restricted in its use, the *schema* of that concept of the understanding." On the other hand we read, a few lines later: "This representation . . . of a general procedure of the imagination in giving a concept its model [*Bild*] I call the schema ⟨pertaining⟩ to that concept."⁶

Were we to carry this last definition over to the schema of quantity, then we would have to say that number is the representation of a general procedure of the imagination in giving to the concept of quantity its model. However, by this "procedure" can only be meant the process of enumerating. But is it not clear that "number" and "representation of enumerating" are not the same? Further, it is not very easy to see how, starting out from the category of quantity, we are apriori to attain, by means of the representation of time (as the common schema of all the categories), to the particular, determinate number concepts. Still less intelligible is the necessity which induces us to ascribe to a concrete multiplicity a certain number that is always the same: precisely that number of which we say that it accrues to the concrete multiplicity. The theory of the schematism of the pure concepts of understanding appears here, as elsewhere, to fail in the realization of the goal for which it was specifically created.

We need not consider all the philosophers who, following *Kant*, based the concept of number upon the representation of time. Adherents of extreme Empiricism, such as *Alexander Bain*,⁷ are here in accord with those of *Kant*ian apriorism. From the side of

---

⁵ I. Kant, *Kritik der reinen Vernunft*, *Sämmtliche Werke*, Hartenstein edition, Vol. III, Leipzig 1867, p. 144. {B 180}

⁶ *Ibid*, p. 142. {B 179-180}

⁷ A. Bain, *Logic*, 2nd ed., Part II, London 1873, p. 201f. Cp. the cogent critique of Bain's theory in Chr. Sigwart, *Logik*, Vol. II, Freiburg i. Br. 1878, pp. 39ff.

mathematics two celebrated names must be mentioned here. Sir *William Rowan Hamilton* flatly calls algebra "the science of pure time," as well as "the science of order in progression."[8] In Germany it is *H. von Helmholtz* who, in his *Über Zählen und Messen*,[9] <34> represents the same point of view. We will later find occasion for a thorough discussion of this treatise.[10]

Finally, it should be noted that most of the investigators who took the representation of the *sequence*, instead of that of the group, as a basis for the development of the number concepts, as well as of the axioms of arithmetic, have been essentially influenced by the time-theory.

The Collective Synthesis and the Spatial Synthesis

A. *F. A. Lange's Theory.* Whereas *Kant* places number into relationship with the representation of time, *F. A. Lange* supposes that everything that time achieves for *Kant* can be derived with far greater simplicity and certainty from the *representation of space*. "*Baumann* has already shown," he says, "that number has far greater unison with the representation of space than with that of time.... The oldest phrasings of the words for numbers always designate, so far as we can grasp their sense, spatial objects with determinate properties corresponding to the number in question. Thus, for example, rectangularity [*Viereckiges*] corresponds to the number four [*vier*]. From this we also see that number does not originally arise through systematic addition of one to one, and so on. Rather, each of the smaller numbers, upon which the system arising later is based, is formed through a special act of synthesis of intuitions, so that it is only later, then, that the relations of numbers to one another, the possibility of adding, and so on, are recognized."[11]

---

[8] Cf. H. Hankel, *Vorlesungen über die complexen Zahlen und ihre Functionen*, Part I, Leipzig 1867, p. 17.
[9] *Philosophische Aufsätze. Eduard Zeller zu seinem fünfzig-jährigen Doctor-Jubiläum gewidmet*, Leipzig 1887.
[10] See the Appendix to Part I of the present work.
[11] *Logische Studien*, Iserlohn 1877, pp. 140f.

"It is peculiar to the representation of space that within the great all-inclusive synthesis of that which is manifold there can be singled out, with ease and certainty, smaller units of the most various types. Space is, therefore, the archetype, not only of continuous, but also of discrete magnitudes, to which number belongs; whereas we scarcely can think of time otherwise than as a <35> continuum. To the properties of space belong, further, not only the relations which occur between the lines and surfaces of geometrical figures, but no less so the relations of *order* and *position* of discrete magnitudes. Such discrete magnitudes being considered as homogeneous with each other, and held together in a new act of synthesis, number arises as sum."[12]

"We originally receive each number concept . . .," *Lange* elsewhere says,[13] "in the form of a sensuously determinate image of a group of objects, whether they are only our fingers, or the buttons and spheres of an abacus."

Now our critique certainly will not have to look very far to find a basis. The last quotation is especially shocking; for in it the familiar *general* concept of number appears as an *individual* phenomenon, as the sensuously determinate image of a group of spatial things. However, this may well be only an imprecise mode of expression. Probably the view is that number is something noticeable *in* such groups in the manner of a sensible property, something to be lifted out of them by abstraction. The influence of *J. St. Mill* stands out clearly here. As we saw above,[14] he regards number as a sense perceptible property, on a par with color, weight, etc. But whereas *Mill* explicitly declines to give any further explanation of numerical difference – apparently because he considers it something ultimate and not further definable, just like the difference in color or weight – *Lange* believes that he can demonstrate its source in the nature and properties of the representation of space. The synthesis upon which the concept of number is grounded (the "collective combination," in our language) is, for him, a synthesis of spatial intuitions. In quite

---

[12] *Ibid.*, p. 141.
[13] *Geschichte des Materialismus*, Book II, 3rd edition, Iserlohn 1877, p. 26.
[14] See the quotation on page 18 above.

the same manner as geometry, accordingly, arithmetic is supposed to rest upon spatial intuition. It is to be characteristics peculiar to this latter which ground the obvious truth of the arithmetical axioms and the <36> character of absolute necessity residing in them.

One is startled by the audacity with which this theory defies the clear testimony of inner experience. One immediately asks: Are then only spatially distributed contents enumerable? Do we not speak, for example, of four cardinal virtues, or of the two premises of an inference? Which spatial position or order is the basis of the number designation when we count any sort of psychical acts or states?

Certainly this objection would not alarm *Lange*. After all, he reduces not merely mathematical but all logical thinking to spatial intuition. All of the psychical is, for him, localized. Here is not the place to subject this peculiar view, which seems to me utterly untenable, to a comprehensive critique. We must emphasize only those issues that specifically concern arithmetical problems.

Even if we were to concede *Lange*'s basic outlook, no more would be proven in regard to the representation of space than was earlier admitted in regard to the representation of time. The representation of space would constitute an insuppressible psychological precondition for the origination of the concept of number, but this to no greater extent, and in no other way, than it is for the origination of all other concepts. Even if spatial determinateness did accrue to all contents which we combine in thought, it would always still remain two different things (i) to represent spatially determinate contents and (ii) to represent contents in terms of their spatial determinations. To be sure, this does not yet decide whether the representation of space nevertheless makes a special contribution to the *content* of the number concept. It is easy to see that it does not. Let us represent to ourselves by means of an example how we collectively hold together or count spatial objects. Do we, in doing this, attend constantly and necessarily to the relationships of order and position? Certainly not. There are infinitely many positions and orders for which the number remains unchanged. Two apples remain two apples, whether we set them closer together or further apart, whether we shift them to the right or to the left, up or down. Number has exactly <37> nothing whatsoever to do with spatial location. It *may* be, nevertheless, that relations of order and position are implicitly co-represented in the phenomenon when there is a representation of a multiplicity of spatial objects. But it is certain that in collocating and enumerating they do not constitute the objects of the selective interest determining the content of the number concept.

But *Lange* emphasizes not just the spatial nature and relations of the enumerated contents. In order for the number representation to originate, according to him, it is also necessary that the particular contents be considered as homogeneous amongst themselves and uniformly held together by a peculiar act of synthesis. We leave undecided the first part of this claim. The question of whether in enumeration a comparison is necessarily involved must later be considered with care. The second part is more important for us: namely, the reference to peculiar *acts of synthesis* through which representations of numbers come about. So *Lange* himself seems to have felt that the "all-inclusive synthesis" of space – in other words, spatial relations or combinations – do not suffice to characterize the unification of the enumerated contents in the number. However, he is so far removed from clear insight that he repeatedly designates the representation of space as the "archetype of all synthesis," and especially as the "archetype of discrete magnitudes" (among which number belongs). The unclarity of thought already betrays itself in the vagueness of the language. What is this peculiar expression, "archetype," supposed to convey? Surely some kind of picture-like similarity must be intended. But how is a picture-like similarity to be present between the synthesis of "discrete magnitudes" (thus, of multiplicity and number), which consists in a unifying psychical *act*, and the synthesis in the representation of space, which is the combination of parts of an intuition, and thus a combination in the representation's *content*?

Lange's view becomes somewhat more intelligible to us if we consider the context in which the passages quoted occur. He touches on the questions about the origination and the content of representations of number on the occasion of extended metaphysical reflections concerning the "extremely momentous concept of synthesis" introduced by *Kant*. A few remarks <38> on the *Kant*ian doctrine of synthesis, which *Lange* links up with, may be in place here, especially since it also is of interest for the characterization of the "collective" synthesis advanced by us.

Kant uses the term "synthesis" (combination) in a double sense: first, in the sense of the unity of the parts of a whole, whether segments of an extension, properties of a thing, units in a number, and so on; second, in the sense of the mental activity ("action of the understanding") of combining. Here it is a matter of equivocation by transposition, for the two significations are intimately related for *Kant*. This is because, in his view, every whole, of whatever kind it may be, comes into being from its parts through the spontaneous activity of the mind. "Synthesis" therefore signifies simultaneously, for him, the combining (the act of relating) and the result of combination (the relational content). By running the two significations together he reaches the point of straightforwardly designating combination in a general way – even where only a combination in the sense of a primary content of representation could be intended – as "act of spontaneity" and as "achievement of the understanding."[15] As a result, also *Kant*'s peculiar view on the origination of the representation of combination becomes understandable. If all combination subsists only in virtue of acts that combine, and if its unity lies only in those acts, then obviously the representation of combination can only be acquired by reflexion on such acts. And this indeed is what *Kant* maintains. To be sure, we commonly speak as if the relations and combinations accrue to the objects themselves: as if with relations, just as with absolute contents, it were only a matter of a passive receiving and observing. According to *Kant*, this would be mere illusion. He explicitly states: "Combination does not reside in objects, though, and cannot be derived from them, say, through perception, and be thereby first received into the understanding. Rather, combination is solely an achievement of the understanding, which itself is nothing other than the faculty of combining apriori . . . ."[16] And it is likewise stated in another

---

[15] Cp. the quotation following farther below.

[16] *Kritik der reinen Vernunft*, B, §16 (Hartenstein ed., Vol. III, p. 117. {B 134}

passage "that we can represent nothing to ourselves as combined in the object <39> without having previously combined it ourselves. And among all representations, the *combination* is the only one that cannot be given through objects, but rather can be achieved only from the subject itself, because it is an act of its spontaneity."[17]

Lange starts out from this *Kant*ian theory. He has not eliminated the equivocation in the concept *synthesis* mentioned above, and this no doubt contributed essentially to the frequent obscurities in his own assertions. Thus, on the one hand, he speaks of synthetic concepts, of the synthesis in the representation of space, and the like, and, on the other hand, of synthesis as a "productive act of our mind," of special acts of synthesis of intuitions, which yield the number representations, and so forth.[18] But in other respects he deviates not insignificantly from *Kant*'s view, according to which the concept of synthesis is gained, not from primary contents through analysis and abstraction, but rather only with regard to the understanding's action of combining. *Lange* perhaps proceeded from the observation that this view stands in a certain contradiction with experience. In the case of most composite representations we clearly observe, if they are given to us analyzed, the combination of their partial contents, but nothing at all of a unifying activity that first confers upon those contents their state of being combined. For example, the combination of the color and the extension of a thing is not, and also does not imply, the representation of a psychical activity, but rather is something accruing to the intuition itself and perceptible in it. The same holds for the relation of distance, of direction, etc. Should *Lange* nevertheless wish to hold to the *Kant*ian conception that each combination rests on an activity of combining, his only recourse would be to assume that in such cases the activity of combining, though we perceive nothing of it, is still there: that it therefore is

---

[17] *Ibid.*, B, §15 (Hartenstein, p. 114). {B 130}

[18] Cp., besides the above quotations, the *Geschichte des Materialismus*, Book II, 3rd Edition, pp. 119ff. In the *Logischen Studien*, pp 135f, he explains that we "at first" would have "in this expression ['synthesis'] little more than a formulation of the fact that in all our representations the unity of that which is *manifold* is present . . . ." But a few lines later the fact of synthesis is spoken of as a "process," "through which we, as subject, first arise."

an unconscious one, <40> whereas what it creatively produces, the representation of the combination, falls within our consciousness. In any case this is the course *Lange* adopts, and he pursues it to the end. The representation of the combination is obtained, according to him, just like any other representation, from primary contents by means of analysis and abstraction. But as for the synthesizing acts, exclusively and single-mindedly emphasized by *Kant*, through which the combinations of contents are supposedly first brought about, they are thrust away into the transcendental background of life that is prior to consciousness, which is why they of course can contribute nothing to the origination of the *concept* of combination.

Just like *Kant*, so also *Lange* pursues predominantly metaphysical tendencies in these matters. Consciousness itself, and thereby subjectivity, is supposed first to originate out of unconscious impressions through a process of synthesis at the highest level. Thus is the unity of consciousness explained. Now to this unity there corresponds, as the highest synthesis of the content of consciousness, the "all inclusive synthesis" in the representation of space; for "space is the form of all objects." (p. 148) As, furthermore, that supreme synthesis objectifies itself in the totality of the representation of space, so does each particular synthetic act in intuitive syntheses of spatial images. So all combination is, basically, spatial combination and relation. In the representation of space, *Lange* explicitly says, "we find the intuition for the concepts of connection and separation, of equivalence and of the relationships of a whole to its parts, of a thing to its characteristics."[19] It is because of this that all logical and mathematical thinking is thinking in spatial imagery. It is especially because of this that space pervasively lies at the "origin of all that is apriori."[20] In fact, in the apriori sciences (formal logic and mathematics), what is always at issue is a progression within a chain of interlocking relations. Each axiom passes judgment on relations between relations. If, now, space is the form of all contents also in the sense that, not merely no absolute content, but also

---

[19] *Logische Studien*, p. 148.
[20] *Ibid.*, p. 147.

no relational content, is conceivable which does not have its intuitional basis in the characteristics of the representation of space, in their relations and combinations, <41> then certainly the representation of space intervenes as an indispensable mediator in every case of apriori knowledge.

Given all of this, the expressions peculiar to *Lange* – that space is "the archetype of all synthesis," and in particular is the archetype of number – are understandable. And one understands the significant role assigned to the representation of space, not merely for the origination of the number representations, but also for the entire philosophical theory of arithmetical science as an apriori science, and consequently one grounded in the intuition of space.

I have purposefully gone much more deeply into the exposition of *Lange*'s views than was required by our immediate aims. Now the following critique, although it bears exclusively upon the specific points of interest to us here, also refutes *Lange*'s theory of arithmetical knowledge: a theory that has founded schools, as is well known, and with respect to the further goals of our investigations cannot be passed over.

The theory of synthesis with which we have just become acquainted is untenable and is based upon essential misunderstandings. *Kant* failed to notice that many combinations of content are given to us where no trace of a synthesizing activity that produces connectedness of contents is to be found. *Lange*, again, pays no attention at all to those cases where composite representations owe their unity solely and only to synthesizing acts, while in the primary contents a combination is not present or does not come into consideration. According to him all combination is supposed to occur in the content, and of course in virtue of the form of space encompassing all content. This is false. The very concepts *multiplicity* and *number* resist this view. The combination of the colligated contents in the multiplicity, and of the enumerated ones in the number, is not a spatial combination, just as little as it can be taken for a temporal one – and, we can immediately add, just as little as any other combination within primary contents. If space is the all-inclusive form, then it unifies not merely the contents just enumerated, but rather these along with *all* present contents whatever. But what constitutes, for example, the *special* unity of the collection of five things formed just now, in virtue of which they are, precisely, five? And whatever may be the linkages of these five <42> things, can we not in the next moment pick out from among them merely two, three, or four, through a unifying act of interest, without in the least affecting the factually present combinations of contents (e.g., the distances, physical bonds, or the like)?

One sees that the synthesis in our [numerical] concepts cannot lie in the content, but only in certain synthesizing acts; and also that it can therefore be noticed only through reflexion upon those acts. *Lange* himself surely must have sensed this, since he emphasized special *acts* of synthesis of intuitions. But how is that to be reconciled with his remaining views. These acts surely are not to be considered as something spatial. It would, in any case, be a pure absurdity to suppose them such.

So it is clear that number concepts and relationships between numbers – much more so the whole of arithmetic – have nothing to do with the representation of space. The proposition that space is the archetype of numbers has, however we may twist it and turn it, no signification. It neither makes sense to speak of a picture-like similarity – since there is on the one side a synthesis in the content and on the other a synthesis that is an action – nor does it make sense to seek in space the intuitional basis for number representations and number relations.

Finally, we still have to inquire from whence we are to derive the *concept* of synthetic acts with which *Lange* so often operates – if all synthesizing activity is relegated to the unconscious 'beyond', though its result, the representation of combination, is assigned to the primary content of consciousness, from which we extract it simply by analysis and abstraction. In raising this question it also is to be emphasized that the entire underlying intuition, for *Lange* as for *Kant* – according to which a relational content is the *result* of an act of relating – is psychologically untenable. Inner experience, and it alone is decisive here, shows nothing of such 'creative' processes. Our mental activity does not *make* the relations. They are simply there, and, given an appropriate direction of interest, they are just as noticeable as any other type of content.[21] Strictly speaking, creative acts that produce some new content as a result distinct from them <43> are psychological monstrosities. Certainly one distinguishes in complete generality the relating mental activity from the relation itself (the comparing from the similarity, etc.). But where one speaks of such a type of relating activity, one thereby understands either the grasping of the relational content or the interest that picks out the terms of the relation and embraces them, which is the indispensable precondition for the relations combining those contents becoming observable. But whatever is the case, one will never be able to maintain that the respective act creatively produces its content.

One may perhaps reply to us by pointing precisely to those synthetic acts which we have above verified in representations of number, and which, as we will yet see, are identical with our "collective" combinations. In their case it is indeed the act alone that is supposed to procure the combination. — In a certain sense this is quite correct. The combination of course subsists solely and only in the unifying act itself, and consequently the representation of the combination also in the representation of the act. But there does not exist besides the act a relational content different from the act itself, as its creative result, which the view we are attacking always presupposes.[22]

There is only one argument we have not yet considered: namely, that the most ancient expressions for the number terms allude to objects in space with properties that correspond to number. It would be very risky to conclude from this that in all enumerating the human intellect is necessarily restricted to spatiality, since other explanations so easily present themselves. It is precisely at the more primitive levels of culture that human beings only found occasion to count groups of spatial objects, thus making it possible

---

[21] Cp. C. Stumpf, *Tonpsychologie*, Vol. I, Leipzig 1883, pp. 105ff.

[22] The mistakes here censured led *Lange* to strange results. The origin of the categories is supposed to lie in the representation of space. Its properties form the *"norm for the functions* of our understanding," etc. "Thus," *Lange* finally sums up (*Ibid.*, p. 149), "the representation of space, along with its *properties constitutive of our understanding*, is revealed as the enduring and defining primitive form of our mental essence, as the true objective counterpart of our transcendental ego." As soon as one tries to find a clear sense in these phrases, they disperse into nothingness.

for *their* concept of number to coincide with the concept we can now designate only by the composite term, "number of spatial objects." The more advanced culture <44> carried over the old terms, but their signification in the meantime has, by way of metaphorical transference, been expanded far beyond the spatial domain. As is true of most concepts, the number concepts also have certainly undergone a historical development.

B. *Baumann's Theory.* Analogous errors with respect to the essence of the number concepts are also to be found in *Baumann*, by whom *Lange* was not inconsiderably influenced in his theory of arithmetical concepts and knowledge. *Baumann* repeatedly and energetically affirms the participation of our psychical activity in the formation of the number concepts. He states, for example: "The grasping together of 1, 1, 1 in 3 is a novel act of the mind, incomprehensible to anyone who cannot *do* it. That is, the mere seeing of one and one and one thing still does not yield the number 3, but rather this novel grasping together requires first to be done."[23] Thus arithmetical concepts in general, those of the numbers as well as those of the calculative operations, originate through a "mental action that can be incited and apprehended only in inner intuition."[24] While we thus produce in ourselves "purely mental representations of mathematics," external experience, on the other hand, is supposed to "bear the mathematical in itself, independently of our mind, and to present it in an unmistakable fashion,"[25] a circumstance through which *Baumann* explains the applicability of mathematics to the external world. In particular, it is said with respect to number that it is "contained in space as well as and even more so than in time."[26] "We find number again in the external world, apply it following that world's promptings, and there it proves its worth, i.e., through the success of calculations. It is, thus, joined with space and present throughout it. Therefore

---

[23] J. J. Baumann, *Die Lehre von Raum, Zeit und Mathematik in der neueren Philosophie*, Berlin 1869, Vol. II, p. 571.
[24] *Ibid.*, p. 669
[25] *Ibid.*, p. 675.
[26] *Ibid.*, p. 671.

geometry is also amenable to arithmetical expression."[27] *Lange* has explicitly referred to these latter passages.[28] <45>

Thus, to the "mathematical in us" there corresponds, according to *Baumann*, a mathematical outside of us. The latter is known by the former, entirely in agreement with the ancient formula of *Empedocles*: "Like is known by like." So far as this theory concerns the numbers – and that is all that matters here – it is based on an erroneous conception of the abstraction process that yields these concepts. Certainly *Baumann* sees something true in so strongly emphasizing the inner activity involved in the conceptualization of number. There can be no doubt that in the formation of both numbers and multiplicities, it is *in concreto* not a matter of a passive reception or a mere selective noticing of a content. If anywhere at all, there are here present spontaneous activities which we apply to the contents. Depending upon whim and interest we can grasp together separate contents, and again take away from, or add new ones to, the contents just encompassed. A uniform interest, encompassing all of the contents and linking them together, and – simultaneously with and in that interest, with that reciprocal interpenetration that is peculiar to psychical acts – an act of uniform conceptualization, throws the contents into relief. And the intentional object of this act is precisely the representation of the multiplicity or of the totality of those contents. In this manner the contents are present simultaneously and joined together. They are one. And it is by means of reflexion upon this unification of the separate contents through that complex psychical act that the general concepts *multiplicity* and *determinate number* originate.

Now, if all this corresponds to the truth, then it is clear that *Baumann* starts out with sound remarks, but ends up with untenable ones. On the one hand, numbers are to be, in a certain manner, purely mental creations. And this is actually right, in that the numbers are based upon psychical activities which we exercise upon contents. In this regard we find in *Baumann* various cogent observations, and so in appearance his view stands in perfect

---

[27] *Ibid.*, p. 670.
[28] *Logische Studien*, p. 140.

agreement with ours. But if this were really the case, if *Baumann* wished to maintain no more and no other than that concepts of mental activities enter into the number concepts, how would it be possible for him to speak, on the other hand, of a rediscovery of number in the external world, of number being interwoven with space and in space? <46> In the case of external activities, one certainly distinguishes the activity from the product which it generates and which can externally persist when the activity itself is long since gone. But the psychical activities which ground the number concepts certainly do not produce in them new primary contents which, cut loose from the engendering activities, could then be found again in space or in the external world.

How, in fact, are all of the conceivable numbers – which we can count off through combinatorially arranging spatial contents – to be contained in space? That which is intuitively present, which we can encounter and observe in space, certainly does not consist of numbers in and for themselves, but consists, rather, only of spatial objects and of their spatial relations. But with that no number is yet given. And where a number is given to us, it is not and cannot be the spatial syntheses that bring about the unification of the enumerated as such. The co-existence of objects in space is still not that collective unification in our representation which is essential to number. Which external objects, and how many of them, we colligate and enumerate depends solely upon our interest, and thus the unification of the colligated is exclusively determined and accomplished by a psychical act of the type described above. To seek for numbers in space seems, after all this, no less absurd than to search there for judgments, acts of will, wishes, and the like. Spatial objects are at most the contents of such acts, but they are not these acts themselves. Spatial objects may be at most the objects enumerated, but not the numbers. It is exactly the same error into which *Lange* fell when he explained spatial intuition to be the archetype for the synthesis of numbers, just as, according to him, it is for all synthesis: the error upon which, in general, his theory of synthesis is based.

Comment: We discover number and spatial intuition set into relationship with one another in a wholly new and curious manner in the essay by W. Brix, *Der mathematische Zahlbegriff und seine*

*Entwicklungsformen.*²⁹ "The number involved in spatial intuition" is supposed to represent an original, primitive form of the number concept, and, in comparison with the "number involved in <47> temporal intuition," an "absolutely heterogeneous and independent stage in the genetic development" of it. "To be sure, the number at this level ⟨spatial intuition⟩ is *not yet a concept* . . . . Rather, it is nothing more than a certain *schema of perception*, a *kind of form of intuition* in the *Kant*ian sense. For it still totally adheres to the objects of perception. That is, on this level of development one enumerates not three, four, five, but, perhaps, three houses, four horses, etc., instead. Therefore it also does not yet require *any kind of abstraction*, but consists solely, as *du Bois-Reymond* puts it, 'in the representation of the separateness of perceptual objects'. It *thus nearly coincides with spatial intuition*, since space appears determinate precisely through the individual objects which are grasped together in the representation of number."³⁰

Brix has taken the psychology of the concept of number a bit too lightly. What is really signified in the expressions "three houses," "three horses," "three apples," etc., by the little word "three"? The representation of the separateness of the objects? Very well. Then this separateness, always the same in kind, must have been observed in its own right and as such; and that, specifically, is the abstraction. (A general term without conceptual underpinning would also be a remarkable psychological and logical discovery!) Moreover, one easily sees that the explanation of the collective combination of spatial objects in terms of a spatial separation of them is untenable, for it is subject to an objection we have had to urge repeatedly against the theories considered up to now. When we – willfully and arbitrarily – mentally select from $\underline{n}$ given spatial things 2, 3, . . ., n – 1 (and I say "mentally" because physically we in fact do not move them from their places) and enumerate them, is anything thereby changed in the spatial intuition? Perhaps – what is regarded by *Brix* as a positive Moment of it – the "separateness of the perceptual objects"? The raising of the question suffices to make

---

²⁹ In: W. Wundt, *Philos. Studien*, Vol. V.
³⁰ *Ibid.*, pp. 671f.

clear the untenability of the view. And how are two, three, four spatial contents distinguished from each other *as* two, three and four? Spatial separateness does not, however, admit of a specific differentiation running parallel to that in number. <48> The lovely philosophical terminology, "schema of perception" and "form of intuition," only helps to conceal the unclarity of the thought; and we have no idea what to make of the strange remark that this form of intuition of the spatial number nearly coincides with spatial intuition. How, moreover, this assertion can be reconciled with the one following on the same page of *Brix*'s essay, that "the number here is valid only as a determination of the space of the respective perception," we will leave aside. However, one more quotation may be appropriate here: "The faculty of realizing such ⟨number⟩ representations must therefore be attributed to most higher organisms – perhaps the supposition that it is bound up with the visual faculty ⟨!⟩ would not be wholly groundless – for even animals, as *du Bois-Reymond* emphasizes, battle several attackers differently than they do a single dog. And 'Even ducks count their young', as *Hankel* points out."[31] One sees that these distinguished mathematicians are here involved in a major confusion between the representation of a determinate group of physical individuals and the representation of their number. It is only astonishing that *Brix* is faithful to them in this respect.

Colligating, Enumerating and Distinguishing

Much more scientific and plausible than all of the theories regarding the origination of the concepts of multiplicity and number that have been criticized up to this point is the theory to whose development we now wish to turn. But in order to make completely clear whether or not it does what it promises to do, I shall endeavor to give it as consistent a development as is at all possible, and toward that end I shall decline to tie my exposition and critique directly onto any one of the forms in which this theory has actually been represented by this or that outstanding

---

[31] *Ibid.*, p. 672.

author. The following line of argument should appear to have much to recommend it:

A multiplicity can be spoken of only where objects which *differ* <49> from each other are present. Were all of the objects identical, then we would in fact have no multiplicity of objects, but just *one* object alone. But these differences must also be noticed. Otherwise the different objects would form for our apprehension only one unanalyzed whole, and we would again find no possible way of arriving at the representation of a multiplicity. Hence, representations of differences enter essentially into the representation of any totality. In that we, further, distinguish each single object within the totality from the others, along with the representation of *difference* there also is necessarily given the representation of the *identity* of each object with itself. In the representation of a concrete multiplicity, therefore, each single object is thought of both as an object which is different from all of the others and as an object which is identical with itself.

Given this, the origination of the general concept of multiplicity also, it seems, is evident. What common element could still be present in all cases where we speak of multiplicity other than these representations of *difference* and *identity*, since it is well known that in the abstraction of the concept *multiplicity* absolutely nothing depends upon the peculiarities of the individual contents? Thus, setting out from any one concrete multiplicity, we get the general concept of multiplicity by relating each content to each other one as different – but this completely in abstraction from the peculiar character of the concretely given contents – and by considering each content merely as something identical with itself. In this way there originates the concept of multiplicity as, to a certain extent, the *empty form of difference*.

But along with the concept of multiplicity that of *unity* is now also immediately given, and its content is easily developed from the foregoing considerations. By enumerating in the strict sense of the word – i.e., therefore carrying out number abstraction – we bring each thing to be enumerated under the concept *unit*, we consider it merely as *one*. That means just this: We consider each thing merely as something which is identical with itself and different from everything else. As distinguishing and identifying are reciprocally determining functions which are inseparable from each other, so <50> the general concepts of multiplicity and unity, which are formed through reflexion upon those functions, are also correlative concepts, mutually interdependent.

Out of the concept of the empty form of difference there arise, through the manifold determinations it allows, the *number concepts*. Thus they are nothing but the general forms of difference segregated in such a way as to provide a classificatory system.

We especially find ideas of these and similar kinds at work in the logical writings of *Jevons, Sigwart*, and *Schuppe*.[32] Thus *Schuppe* explains "that the essence of number is indefinable because it flows directly from the principle of identity. In this principle the one thing and another is immediately posited in that the one is distinguished from the other. Therein, then, the plurality or multiplicity is given . . . ." "Red is not green and not blue, a is neither b nor c, and b is neither a nor c, and c is neither a nor b. These judgments of the simplest type form the presupposition of the predication of number; and one can, to express precisely the same sense, assert the number instead of the mere differentiation. Red, green and blue are not, perhaps, one, but are three. One can also more fully say 'three distinct things', or 'three different colors'. But that would be a superfluous clarity. . . . 'Are three colors' says the same thing as 'three different colors'. What I cannot distinguish, I cannot count. It is one."[33] — Number, or the statement of plurality by means of a determinate or indeterminate number term, "only asserts difference without naming the distinction involved."[34]

Jevons seems, in part, to stand even closer to the theory developed above: "Number is but another name for diversity.

---

[32] Among mathematicians, *P. du Bois-Reymond* may be mentioned here. In his *Allgemeinen Functionentheorie*, Part I, Tübingen 1882, p. 16, he comments that "Number is as it were what is left over in our soul when all that distinguished the things has evaporated and there is retained only the representation that the things were separate." To be sure, being distinct in the sense of being *separate* is used here, which is no necessary element in a distinction theory of number. Moreover, *du Bois-Reymond* does not clearly hold to this idea of separateness; and he also seems to restrict himself to spatial objects.

[33] *Erkenntnistheoretische Logik*, Bonn 1878, p. 405.

[34] *Ibid.*, p. 410.

Exact identity is unity, and with difference arises plurality." "Plurality arises when and only when we detect difference."[35] Here, as <51> one sees, "number" is taken in the broader sense as synonymous with "plurality."

With respect to the kind of abstraction which is here present, this same author remarks: "There will now be little difficulty in forming a clear notion of the nature of numerical abstraction. It consists in abstracting the character of the difference from which plurality arises, retaining merely the fact . . . . Abstract number, then, is the *empty form of difference*; the abstract number *three* asserts the existence of marks without specifying their kind." "Three sounds differ from three colours, or three riders from three horses; but they agree in respect of the variety of marks by which they can be discriminated. The symbols $1 + 1 + 1$ are thus the empty marks asserting the existence of discrimination."[36]

However, there is absent in *Jevons* any deeper psychological foundation, which we – chiefly by a free utilization of indications from *Sigwart* – have sought to give above.

At first glance it seems that this type of theory is built upon indubitably secure foundations, so that any critique would primarily question only the indeterminacy of the results. That multiplicity or number, in the broader and more indeterminate sense, is nothing other than "the empty form of difference" is something one could comfortably grant. But this generality does not yet suffice in order to rigorously characterize the content of the sharply distinct number concepts, *two, three, four*, etc., in relation to one another. They, indeed, are all "empty forms of difference." What differentiates *three* from *two, four* from *three*, and so on? Are we perhaps to give the dubious answer: With two we notice *one* relation of difference, with three, *two*, and with four, *three* such relations, and so on?

The information which the last of the passages quoted gives us is obviously very meager. That phrase, "variety of marks," either signifies the same thing again as "number," or it signifies the same as "form of difference." But what characterizes these

---

[35] *The Principles of Science*, 2nd Edition, London 1883, p. 156.
[36] *Ibid.*, pp. 158 and 159.

"forms" psychologically in contrast to each other, so that they can be grasped through their peculiar determinations, clearly distinguished from each other and, accordingly, also designated by different names? <52> Here lies an essential defect of the theory. Let us see whether it can be eliminated.

For the sake of simplicity we will consider only a totality of three objects A, B, C. Into the representation of this totality there must enter, according to the view in question, these relations of difference:

$$\overgroup{AB}, \overgroup{BC}, \overgroup{CD}$$

(where the ties indicate the relations). They are given together in our consciousness, and they effect the unification of the objects into the collective whole. However one may now replace A, B, and C with contents of any type, these differences constantly remain present as somehow determinate. They thus constitute the "form" of difference which is characteristic of the number three.

However, certain objections to this present themselves: If these relations of difference are together in our representation, then, if the basic viewpoint of the theory is correct, each of the differences represented must also be in turn perceived as identical with itself and as different from any other. For were $\overgroup{AB}$ and $\overgroup{BC}$, for example, not recognized as different, then they would just blend together as undifferentiated; and then, as one immediately sees, their terms [*Fundamente*] also could not show up in the representation of the totality as distinct from each other. So the sum total of the differences of differences would also have to be in our representation. That is:

$$\overgroup{\overgroup{AB}\ \overgroup{BC}};\ \overgroup{\overgroup{BC}\ \overgroup{CA}};\ \overgroup{\overgroup{CA}\ \overgroup{AB}}$$

But the same would also be true with respect to them. And so on.

Hence, in order to get hold of the "form of difference" we would fall into a lovely *regressus in infinitum*.

But there is still a way of avoiding this consequence. One could reply that if we pass in our distinguishing from A to B, and from there to C, then a new distinction of C from A is no longer required. That is to say, in relating the two differences $\overgroup{AB}$ and

BC (which are connected by the one term B) to one another in a higher act of differentiation, the possibility of C and A blending into one is *eo ipso* excluded. So the true schematization would be: <53>

A B C

Then, whatever A, B and C may signify, this schematic figure refers us to a process which is everywhere the same. If we therefore abstract from the peculiarities of the particular contents, retaining each only as *somehow* determinate, then we have here the desired form which is common to all multiplicities with three contents, and in virtue of which we also ascribe the number three to such multiplicities.

In this way one could set up all of the forms of difference which are to form the basis of the numerical denominations. Thus, for example, AB would be the schema of the simplest number, two. It would express the fact that in all cases where there is a duo one object is there, and in addition an object *different* from it. If we assign to a concrete totality the number two, that means: we direct our attention *merely* toward the fact that one content, and still one other content, is present. Our attention would not focus upon the specific nature of the difference, but rather upon the mere fact of it.

The schematic form for the number four would be:

A B C D

And one now easily grasps how the forms are further complicated. In all cases, the distinctions are ones which border on one another (i.e., have a term in common), thus making it possible for all of them ultimately to be grasped together in a single act, by means of higher-order acts of distinguishing.

These schemata would have to be regarded as diagrams of those mental processes as they occur in the representation of any totality of, respectively, two, three, four, etc., contents. And in reflexion upon those mental processes, whose well-characterized difference would have to be given to inner perception, the number concepts would arise.

So it turns out that the defects which we criticized above can be completely eliminated. Yet more, in fact. The more refined elaboration that the theory has now undergone seems to imply, as a direct consequence, the solution to many pressing questions regarding the number concepts. <54> So, for example, the extremely rapid expansion in those forms would make it intelligible why we attain authentic [*eigentlich*] representations only of the smaller numbers, while of the larger ones we are able to think only symbolically – indirectly, as it were.

The independence of the number from the order of the enumerated objects would be immediately brought to Evidence by a glance at the schematic form. Obviously the form remains unchanged regardless of how the A, B, C, . . ., may be rearranged.

In the same way, yet many other points could be cited that are favorable to the theory. But linguistic usage appears to offer an especially significant confirmation of it. With the same sense we say that A and B are *different* things and that they are *two* things. Red, green and blue are three. That means they are not perhaps *one*, but rather three different colors. But this "is superfluous clarification and emphasis. That there are three colors says *the same* as that there are three different colors." Thus *Schuppe* – for it is his example – also understands linguistic usage as harmonizing completely with the "difference" interpretation of the number concept.

After all this it seems we have here a well-founded and consistently elaborated theory that lays claim to our acceptance. Nevertheless, a thorough investigation reveals the theory to be untenable. Even if a plausible objection can no longer be brought against the consistency of its structure, its psychological foundation still does not withstand a rigorous critique.

It is correct that we can speak of a totality only where there are contents present that are different from each other. But the assertion annexed to this truth is incorrect: That these differences must be represented *as such*, because otherwise there would be in our representation only an undifferentiated unity, and no multiplicity. It is important to keep distinct: "to notice two different contents," and "to notice two contents *as different from one another*." In the former case we have – presupposing the simultaneous, unitary grasping of the contents – a representation of a totality; in the latter case there is a representation of a difference. Wherever a totality <55> is given, our apprehension primarily lays hold of *absolute* contents alone (namely, those which compose the totality). By contrast, wherever a representation of a difference is given (or a complex of such representations), our apprehension lays hold of *relations* between contents. This much alone is correct: Where a plurality of objects is perceived, we are *always justified*, on the basis of the individual contents, in making Evident judgments to the effect that every one of the contents is different from each other one. But it is not true that we *must* make these judgments.

With regard to the concepts *distinguishing* and *distinction* various obscurities generally prevail. These have arisen from certain equivocations and have substantially contributed to the errors that I here touch upon.

(1) "Distinction" or "difference" signifies *the result* of a comparison. A comparison can yield either of two results: that the contents considered are the same, or that they are different, i.e., *not* the same. Thus, difference here signifies something negative, the mere absence of a sameness. In this sense one speaks of comparing and distinguishing as correlative, intimately connected activities. Indeed, in any case where we have an arbitrary act of comparison, both sorts of results may occur: affirmative judgments which acknowledge samenesses are made, or, on the other hand, negative judgments which reject them. To this affirming of samenesses, then, the term "comparing" refers, while the term "distinguishing" refers to the denial of samenesses, whenever we use the combination "comparing and distinguishing."

In the case where comparison of contents in a certain respect leads to the result *non-sameness*, it can, nonetheless, happen that at least a similarity, or "gradation," etc., is noted. These are well-characterized classes of relations, in the case of which, quite as in the case of sameness, the representation of the relation Represents [*repräsentiert*] a real [*reellen*], positive content of the representation. Now these relations, too, are called relations of difference; and, in particular, the names "distinction" or "difference" are customary for *intervals* in continua. We speak in this sense of distinctions of place, distinctions of time, <56> distinctions of intensity and quality (between two colors, tones, odors, etc.).[37] But, then, this narrower signification of those terms led again, conversely, to cases of *mere* non-sameness (since such cases, too, were called distinctions) being thought of as if they were primary relations (cf. pp. 71-72) – i.e., as if in their case too the relation could lie in the content of the representation – whereas, in fact, nothing further is given than an Evident negative judgment which denies the presence of such a relation (*viz.*, the relation of sameness).

From the *practical* point of view, it may still be useful to classify all of the results to which comparison can lead under the two headings, "sameness" and "difference." It must not, however, be overlooked that classes of relations are then grouped together under the latter heading which, as to their phenomenal character, are foreign to each other, while, moreover, a sub-group of them is closely related to the sameness relations that have been brought under the other main heading. From the psychological point of view, the relations of similarity, sameness, metaphysical combination, etc. – in short, all relations that have the character of representation phenomena in the narrower sense (hence, primary contents but not represented psychical acts) – belong in one class.[38] But difference in the broadest sense does not belong in that class; for it is *not* a representation content that is immediately observable at the same time as the terms are. Rather, it is a negative judgment made, or represented as made, upon the basis of those terms.

---

[37] This also is probably connected with the fact that from time to time, especially in the case of physical contents, one uses "being distinct" as synonymous with "being separate" (in intuition), expressions which do not always coincide, as is to be seen from the relation of whole and part. We therefore do not take this to be an essentially new kind of equivocation.

[38] We will call them "primary relations." The further details on the division here indicated, which will prove to be important for purposes of a characterization of the collective combination, are given in Chapter III.

(2) But the term "distinguishing" is used in yet another signification, which is connected with *analysis*. According to this signification the "distinguished" is that which has been thrown into relief and especially noticed through analysis; and "to distinguish" means the same as "*to segregate*," "to analyze."

By investigating the conditions which favor analysis, it is found that a plurality of partial contents <57> is all the more easily and surely segregated the greater, in number and degree (or interval), are their distinctions amongst themselves and over against the environs. Such reflections, which consisted of comparisons and distinctions dealing with contents that were already analyzed, frequently misled people into believing that the process of distinguishing (in the sense of analyzing) is also such a *judgmental* activity of distinguishing (in the sense of distinguishing compared contents). Then one reasoned: To be able to retain several contents in consciousness *as segregated* – i.e., as analyzed and separately and specifically noticed – they must be thought of as *distinguished* from each other, i.e., as compared and specifically characterized in terms of their distinctions. But this is false. In fact, it is obviously absurd. The *judgmental* activity of distinguishing Evidently presupposes contents that are already segregated and separately and specifically noticed. Hence, these contents cannot have *first* become noticeable through their being distinguished from one another [in judgment].

Now it is this error which the theory we are contesting commits by arguing: "The differences between objects of a multiplicity must have been noticed *as such*. Otherwise, in our representation we would never get beyond an unanalyzed unity, and there would be no talk of any multiplicity. Hence, the representations of difference must be explicitly contained in the representation of the multiplicity."

What is true is that, if the contents were not different from each other, then there would be no multiplicity. Further, it is true that the distinctions, if they were intervals, would have had to exceed a certain measure. Otherwise just no analysis at all would occur. But the supposition is not true that every content first becomes a distinct content, i.e., one which is separately and specifically noticed, by means of the apprehension of its distinctions from other contents; whereas it is surely Evident that every representation of such a distinction presupposes, as its terms, contents which are already separately and specifically noticed and which, in *that* sense, are distinguished.

In order for a concrete representation of a totality to originate, all that is necessary is that each of the contents involved therein should be a content which is noticed separately and specifically, a segregated one. However, there is no absolute necessity that the distinctions of the contents be attended to, although this will happen often enough. <58>

What has been explained of the representation of [mere] distinction also holds true quite analogously of the representation of *identity*. Both are concepts that issue out of reflexion on certain judgmental activities: judgmental activities that have far-reaching importance in practical life, and that also may proceed in parallel with the representing of a plurality. But that they always and necessarily do this, that they in general Represent [*repräsentieren*] "ever-present activities, which repeat themselves in each act of thought," activities in which the "self-consciousness, being one and the same in all acts . . . is actualized";[39] and above all that the concepts of unity, multiplicity, and determinate number are obtained with reference to them: — All of this we cannot regard as established by the lines of argument presented to us. Here as elsewhere, in the analysis of an elemental concept one has all too readily succumbed to the temptation to regard the results of subsequent reflections on its content as something originally contained within it or as a necessary Moment of the concept's development.

If, now, the line of argument – so evident at first sight – which seemed to force us with inescapable necessity to accept the distinction theory also proves to be untenable, that still does not settle the question of whether, in spite of all this, representations of differences constitute essential elements in the number concepts. To this point all that has been refuted is the view that the representation of a plurality *in concreto* cannot come about unless the particular objects are first held in separation from each other

---

[39] Chr. Sigwart, *Logik*, Vol. II, p. 37.

by activities judging them to be distinct. From this it follows only that there is on no account a, so to speak, apriori necessity to appeal to representations of differences in the development of the number concepts, and, in consequence of that, to identify precisely these concepts with those acts of distinguishing of higher order stacked on top of each other in the manner of a pyramid. It would still be conceivable, in spite of all, that the direction of interest in the formation of number is turned, rather, upon the distinctions between the objects to be enumerated.

But the reference to inner experience is decisive. It shows with total clarity that neither the representation of a <59> concrete multiplicity nor that of the corresponding number necessarily includes explicit representations of the distinctions between the particular contents enumerated. Distinctions in the sense of intervals would not come into consideration here, in any case, since totally disparate contents can be collected and enumerated, while between disparate contents interval relationships certainly cannot subsist. There remains, accordingly, only those representations of distinction that originate from comparisons and presuppose reflexion upon negative judgments. But of such judgmental activities, through which the particular contents are supposed to be grasped as distinct from each other when we enumerate them, inner experience discloses nothing. Much less then does it show a trace of those layered, higher order acts of distinguishing, grounded one on another, which had to be assumed for the consistent elaboration of the distinction theory. Certainly it is true that we at any moment can make the particular contents into substrates for acts of distinguishing. But it is no less certain that the latter are not *what is meant* in enumerating. Distinguishing, on the one hand, and collecting and enumerating on the other, are totally different mental activities. Only this *one* thing and no more is required: that the contents to be enumerated be segregated (i.e., separately and specifically noticed), but not in any sense distinguished *from each other*. Friends of unconscious psychical activities may always transport the distinguishing acts, which we here reject, into the nebulous region of the unconscious. Then the "forms of difference" described would also have their place there. But this much is clear, I think: that those unconscious psychical mechanisms neither could have contributed anything to the *content* of our conscious representation of number, nor would they be capable of explaining the least thing about the *origination* of such representations.

It only remains to unravel the last argument, which seemed to give an especially favorable impulse toward the distinction theory: namely, the corroborating reference to certain equivalences in linguistic usage. "Red, green and blue are *three* colors" says *the same* thing as "They are three different colors." In fact, *Schuppe*[40] thinks, the latter form of assertion involves a superfluous clarification and emphasis. He surely is not quite <60> right about this. Just consider the special emphasis placed on the number words, which is indispensable for the sense of the two assertions to become genuinely the same. "Red, green and blue are three *colors*" already expresses a quite different sense. In order to prevent an impending confusion of several contents we can lay emphasis precisely upon their number, for without difference there is no number. That is a transposed function of this concept of number, adapted to a quite special purpose. If we say that red and green are two contrasting colors, then number no longer has this special function. Difference is still involved in a certain manner, but to express it is not the specific intention. And the same thing is shown by other examples, as many as you wish. "Saturn has eight moons and three rings." "This rod has a length of ten meters," etc. It is just not right to say "that the number only asserts difference, without mentioning it." Besides, the [German] language possesses particular forms of number words where this purpose forms an essential part of the signification: the so-called "genus" number terms, "one kind" [*einerlei*], "two kinds" [*zweierlei*], etc. It is obvious that one cannot substitute them everywhere for the basic number terms, "one," "two," etc.

Critical Supplement: Above we mentioned *Sigwart* also, along with *Jevons* and *Schuppe*, among the main Representatives [*Repräsentanten*] of the distinction theory. And in fact the basic conceptualizations in this theory have been repeatedly defended

---

[40] Cp. the quotation above.

by him. But that is not to assert that his view on the content of the number concepts fully agrees with the developments given above. When he says that "all of the number concepts are, consequently, only developments – realized in ever higher syntheses – of the formal functions that we exercise in every act of thought in general by positing a unity and distinguishing,"[41] then one would want, in any case, to be reminded of the forms of difference previously deduced. The same holds true for another passage: "For what is posited as identical and distinguished from another is *precisely thereby* – just as is that other – posited as *one*. And by raising these correlative functions to consciousness *in relation to each other* <61> there originates with the concept of *one* also that of *two*, and therewith the foundation of all number concepts."[42] Other passages, however, speak of a "distinguishing and collecting of the individual *acts of progression* from one object to another,"[43] of a consciousness of the "transitions of consciousness."[44] And enumerating itself is designated as "the general form of conscious *progression* from one unity to another."[45] Thus in *Sigwart* yet other elements besides acts of identification and distinguishing, elements not taken into consideration by us above, come into play in the number concepts. It seems that for him those syntheses or 'combinings' do not consist merely in differentiating acts of higher level, as we presupposed in our consistently worked out structure of the forms of difference. Nevertheless, in many passages from his *Logic* he advocates precisely the essential conceptualizations of the distinction theory. Distinguishing and positing as identical, according to him, must be activities that we carry out in each representation of objects. They are "simple, interconnected acts" . . . "through which alone the 'many' and the 'distinct' first gain access to our consciousness."[46] The same opinion is expressed no less clearly in another passage: "With the

---

[41] *Logik*, Vol. II, pp. 38f.
[42] *Logik*, Vol. II, p. 37.
[43] *Ibid.*, p. 38.
[44] *Ibid.*, p. 41.
[45] *Ibid.*, p. 43.
[46] *Ibid.*, p. 36.

fact that *several different* objects are in our consciousness, distinguishing is certainly presupposed. But first only the *result of this function* comes to consciousness, consisting in several objects being 'together', each being also retained in its own right."[47] Here we find entirely the same view we have just refuted. It is impossible for distinguishing and positing as identical to have the function *Sigwart* ascribes to them. Where the terms of the relation are not already separate, where no 'many' and 'different' are already present, there no kind of relating activity – and consequently also not that of distinguishing and comparison – has any possibility of taking hold. Distinguishing and positing as identical are judgmental activities the practical point of which in the context <62> of our thinking seems to me to lie in a wholly different direction. That A is identical to itself means that A is not B, C, D, ..., but precisely A. Such a reflection is directed toward preventing confusion of A with other contents, a goal that is attained by seeking out and displaying the "distinctions" of A from B, C, D, ... (that is, the characterizing properties that belong to it and not to the others). But while this process is developing, A, B, C, D, ... are already present to consciousness as contents separate from each other; and it is absolutely not the task of that process first to separate what originally is an identical 'one', and through the segregation of the units to make the multiplicity possible for the first time. The equivocation residing in the terms "distinguishing" and "distinction," which links together two concepts that must be clearly separated (those of analyzing and of distinguishing in judgment), is precisely what has misled *Sigwart* into his way of thinking. This may be sufficiently clear from the passages cited.

It surely is strange enough that *Sigwart* himself incidentally takes note of the proper relationship, in that he charges *Ulrici* with almost the same error.[48] In the light of this, one will be inclined to

---

[47] *Logik*, Vol. I, p. 36

[48] *Ibid.*, p. 279n. "The opinion that it is first through distinguishing that a representation becomes determinate forgets that the distinguishing itself is only possible between different representations already present, and that the distinction therefore does not produce the different contents." Here *Sigwart* is referring to H. Ulrici, *Compendium der Logik*, 2nd Edition, Leipzig 1872, p. 60.

assume that *Sigwart*, in speaking of distinguishing, does not in the final analysis have a judging activity in mind, but rather the activity we have contrasted to distinguishing in judgment as the analytical mode of segregation. But I do not find it possible to sustain this interpretation of him. Distinguishing is repeatedly associated with *comparison* by *Sigwart*, as psychical activities in reflexion upon which the concepts of sameness and distinction would be obtained. So it is said, for example, precisely in the continuation of the last passage cited on p. 63: "The representation of distinction . . . first develops when distinguishing is consciously carried out and *that activity* is *reflected upon*." It is inconceivable that psychical activities are meant here other than the acts of judgment in which we grasp differences or samenesses. Certainly analyzing would not come into consideration, <63> for it is after all no psychical activity in the true sense of the word, i.e., one which would fall within the domain of reflexion. One distinguishes between a psychical *event* and a psychical *act*. Representing, affirming, denying, loving, hating, willing, and so forth, are psychical acts, which are intimated to us by inner perception (*Locke*'s "reflection"). It is totally otherwise with analyzing. No one can inwardly perceive an analyzing activity. We can have an experience where a content unanalyzed *at first* then becomes an analyzed one, and where earlier there was *one* content, now a multiplicity is observed. But nothing more than this can be substantiated *post hoc* from within. Of a psychical activity through which the multiplicity first arises out of the unanalyzed unity, inner perception teaches nothing.[49] But we become cognizant of the fact of analysis having occurred by comparing the memory representation of the unanalyzed whole with the present representation of the analyzed one. Such acts of comparing and distinguishing do occur, which, however, presuppose that analysis has been carried out. If all of this is right, then a view that would derive the concepts of distinction, plurality and number through reflexion upon activities of distinguishing – in the sense of analyzing – is deprived of any basis. Consequently, this also is no way

---

[49] Very appropriately, therefore, in his *Tonpsychologie* (Vol. I, p. 96) *Stumpf* defines analysis as the noting of a plurality.

to establish *Sigwart*'s statements by giving them a reinterpretation, and to construct, in place of the one developed above, a new and more defensible distinction theory of number.[50]

---

[50] Errors with regard to the function of distinguishing in the representation of several objects are so easy to fall into that we are not surprised to find them already present in older writers. Cf. J. Locke, *An Essay*, Book II, ch. XI, sect. 1, and Book IV, ch. I, sect. 4, and ch. VII, sect. 4, and elsewhere. Further, James Mill, *Analysis of the Phenomena of the Human Mind*, ed. by *J. St. Mill*, Vol. II, London 1879, p. 15: "As having a sensation, and a sensation, and knowing them, that is, distinguishing them, are the same thing; . . . ." He also repeats this assertion many times elsewhere.

Chapter III

THE PSYCHOLOGICAL[1] NATURE OF
THE COLLECTIVE COMBINATION

Review

Let us now review our deliberations up to this point and their results.

We undertook to bring to light the origin of the concepts *multiplicity* and *number*. To that end it was necessary to obtain a precise view of the concrete phenomena from which they are abstracted. These phenomena lay clearly before us as concrete multiplicities or totalities. However, special difficulties seemed to obstruct the transition from them to the general concepts. It was clear to begin with that the specific nature of the particular objects which are gathered in the form of a multiplicity could contribute nothing to the content of the respective general concept. The only thing that could come into consideration in the formation of these concepts was the *combination* of the objects in the unitary representation of their totality. It was then a question of a more precise characterization of this mode of combination. But that was exactly what seemed to be no easy task. In fact, we examined a sequence of theories concerning the origin and content of the concepts *multiplicity* and *number*, which all ran aground on misunderstandings with respect to the syntheses here present. The first of them characterized the collective combination as merely a

---

[1] Because of the obvious suggestion of this terminology to the contrary, it is necessary to point out immediately that Husserl's position is *not* that the collective combination is psychological or "mental" in any usual sense of the word. Rather, it is 'psychological' only in the sense that it is a member of a unique class of relations the defining features of which are paradigmatically exemplified by *intentionality*, which Brentano had used to characterize the psychological or psychical over against the physical. {DW}

matter of belonging to *one* consciousness. It was obviously inadequate, but it did call our attention to an important psychological precondition: each of the contents colligated must be one that is separately and specifically noticed. It was also previously suggested to us that the unification of the contents is to be regarded as mediated by special acts of consciousness. We were repeatedly confirmed on these points by means of the critique of the three subsequent theories. <65> These presumed to be able to find the basis for our concepts in the "forms of intuition," of time and of space. Here we came to see time as a psychological precondition of number. The last theory we considered – and the only one that approaches genuine scientific status – was the distinction theory. It set out directly from certain psychical acts. But they were acts of distinguishing, which a more deeply penetrating critique could not acknowledge to be the synthetic acts that are relevant for *collectivity* and *number*.

The Collection as a Special Type of Combination

What, then, are the possibilities that still remain? We have not yet investigated all classes of relations. Is collective combination to find its place among those yet remaining? For obvious reasons, however, we are exempted from a detailed examination of all the particular species of relations. Since we know that the most heterogeneous of contents can be united in the collective manner, all relations with a range of applicability restricted by the nature of specific contents fall aside without examination; in other words, relations like similarity, gradation, continuous combination, etc. In fact, it appears that none whatsoever of the familiar sorts of relations can satisfy the required conditions, once temporal relations and relations of distinction are excluded. At most one could still think of relations of identity in this connection; for, however much two contents may deviate from each other, it will always be possible to indicate a respect in which they are identical to one another. In fact it is often enough thought (indeed, most investigators hold this view) that with regard to the origination of the number concepts we must have recourse to relations of identity.

We must take this up later on. (Cf. Chapter VIII.) But here it suffices to note briefly that the identities which we might at any time discover between colligated contents certainly cannot form the 'glue' that achieves the synthesis of objects held together in the representation of the totality. For the collecting presupposes no kind of comparison at all. <66> In thinking the totality of clock, ink and pen, for example, we need not first compare these contents. To the contrary, in order to compare them we must already have colligated them.

So there seems nothing left to do but claim for the collective combination a novel class of relations, well-distinguished from all others. Inner experience also speaks in favor of this. When we think particular contents "together" in the manner of a totality, this "together" does not permit itself to be resolved into any other relations, so as to be defined by means of them.

That would seem to find its confirmation in the following reflections, which seek to characterize the collective combination yet more precisely in its peculiar nature over against other relations.

On The Theory of Relations

Since I am not in a position to rely upon a firmly established and generally acknowledged theory of relations, I find myself forced to insert here a few general observations concerning this very dark chapter of descriptive psychology.

First, it will be useful to come to agreement on the term "relation." What is the element common to all cases where we speak of a "relation," in virtue of which precisely that name is used? To this question *J. St. Mill* gives us – in a note to his father's work on psychology – an intelligible and essentially adequate answer: "Any objects, whether physical or mental, are related, or are in relation, to one another, in virtue of any complex state of consciousness into which they both enter; even if it be a no more complex state of consciousness than that of merely thinking of them together. And they are related to each other in as many different ways, or in other words, they stand in as many distinct relations to one another, as there are specifically distinct states of consciousness of which they both form parts."[2] <67>

Let us add a few supplementary observations. The expression "state of consciousness" ("state of mind") is not to be understood here as "psychical act," but must instead be taken in the widest of senses, in such a way that the range of its signification coincides precisely with that of "phenomenon." — Upon close examination *Mill*, in the above sentences, has really only defined the concept "standing-in-relation." But what, then, is to be understood by "relation"? That this is no idle question already follows from the fact that *Mill* himself vacillates in his terminology. He repeatedly designates that complex state of consciousness mentioned above as the "term [*Fundament*] of the relation," while by "relation" pure and simple he understands the relative attributes to be formed by reflexion on the term.[3] But it also happens that he explains that "term" itself as constituting the relation. In this way the name "relation" certainly becomes equivocal.[4] In order, now, to stabilize our usage we stipulate that by "relation" is to be understood that complex phenomenon which forms the basis for the formation of the relative attributes, and that by "term of the relation" (in agreement with the usage now generally prevalent) we mean any one of the related contents.

I further note that only in one respect is the definition a little too narrow, inasmuch as it speaks of relations between only *two* terms. For there also are relations between several terms, and indeed even *simple* relations – as we will show in what immediately follows.

For the purpose of a classification of relations, one might at first use as a guideline the characteristics of the contents which they interrelate (thus, that of the "terms"). However, such a classification would remain superficial. In the most diverse of domains we find relations that have one and the same character. Thus, identities, similarities, etc., occur both in the domain of primary

---

[2] James Mill, *Analysis of the Phenomena of the Human Mind*, ed. J. St. Mill, London 1897, Vol. II, pp. 7ff. {DW}

[3] *Op., cit.*, p. 9n. Cp. further J. St. Mill, *Logic*, Book I, ch. 3, § 10.

[4] *Logic*, Book I, ch. 2, § 7.

contents ("physical phenomena") and in that of psychical acts ("psychical phenomena") as well.[5] <68>

But one can also classify relations in terms of their peculiar phenomenal character. This is the more penetrating principle of division, and from this perspective several ways of dividing up relations result – among others one which divides them into the following two main classes:

1. Relations which possess the character of primary contents (of "physical phenomena", in the sense defined by *Fr. Brentano*).

Every relation rests upon "terms." It is a complex phenomenon which comprises – in a certain way which cannot be more closely described – partial phenomena. But in nowise does every relation comprise its terms intentionally,[6] i.e., in that specifically determinate manner in which a "psychical phenomenon" (an act of noticing, of willing, etc.) comprises its content (what is noticed, willed, etc.). Compare, for example, the way in which the representation called the "similarity" of two contents includes these contents themselves, with any case of "intentional inexistence," and it will have to be acknowledged that we have here two wholly different kinds of inclusion. It is precisely for this reason that similarity must not be subsumed under the concept of "psychical phenomena." In this regard it belongs, rather, among the primary contents. The same is true of other relations as well, e.g., identity, gradation, continuous combination (the combination of the parts of a continuum), "metaphysical" combination (the combination of properties, as in the case of color and spatial extension), logical inclusion (as color is included in red), and so on. Each of these relations Represents a particular type of "primary content" (in the signification of this term here assumed); and in that respect each belongs in the same main class.

I would, in addition, expressly point out that it makes no difference at all here whether the terms are themselves primary contents

---

[5] In regard to the signification of the terms "physical" phenomena and "psychical" phenomena, and the fundamental distinction underlying them, which is indispensable for the following reflections, cp. Franz Brentano, *Psychology from an Empirical Standpoint*, Vol. I, Book 2, ch. 1, (translated by A. C. Rancurello, D. B. Terrell and Linda L. McAlister, New York: Humanities Press, 1973 {DW}).

[6] *Ibid.*, pp. 88ff.

or are some type of psychical phenomena (represented psychical states). Such identities, similarities, etc., as we perceive to hold between psychical acts or states (judgments, acts of will, and so on) also have in the respect under consideration <69> the character of primary contents. But they are occasioned by those psychical phenomena and are grounded in them.

As belonging among the primary contents, relations of this class could, not inappropriately, be simply designated by the name *"primary relations."* But one would have to guard against the misunderstanding that thereby we always have to do with relations *between* primary contents, whereas, as was just emphasized, it is not a matter of that at all.

2. On the other hand there is a second main class of relations, characterized by the fact that here the relational phenomenon is a "psychical one." If a unitary psychical act is directed upon several contents, then, with regard to it, the contents are combined or are related to each other.

Were we to carry out such an act, then we would of course seek in vain among the contents of the representation which it encompasses for a relation or combination (unless *in addition* there were also a primary relation in it). The contents are, in this case, unified precisely by the act alone; and the unification, therefore, can only be noticed by means of a special reflexion upon the act.

As an example, any arbitrary act of representation, judgment, or emotion and will that is directed upon a plurality of contents will do. Of any of these psychical acts we can say, in agreement with *Mill*'s definition, that it sets the contents into relation with each other. Especially relevant here, e.g., is the relation of "distinctness" in that widest of senses discussed earlier, in which case two contents are brought into relation by means of an Evident negative judgment.

The characteristic difference between the two classes of relations can also be marked by saying that primary relations belong in a certain sense among the representational contents of the same level as their terms [*Fundamente*], which cannot, however, be said of the psychical relations. In the first case, the relation is immediately given along with representing the terms, as a Moment of the same representational content. But in the second case, that of the psychical relation, in order to represent the relation there is first required a reflexive act of representing bearing upon the relating act. The immediate content of this latter is the act instituting the relation, and only through that does the representation bear upon the terms. <70> The related contents and the relation thus form, as it were, contents of distinct levels.[7]

Another division of relations that comes into consideration for us is the familiar one between *simple* and *composite* relations. Often an erroneous principle of differentiation is taken to govern these cases, namely that relations between two terms were supposed to be simple, while those between more than two terms are composite.[8] However, mere number alone surely grounds no distinction in essence. Rather it can at most only indirectly refer us to such a distinction, which here would have to be precisely the distinction between the simplicity and compositeness of relations, in the genuine sense of those terms. Actually, the deeper reason for citing number as the criterion for the distinction resides in the opinion, taken to be self-evident, that just as every composite relation, as a relational complex, necessarily includes more than two terms, so also, conversely, every relation with more than two terms would necessarily be a complex of relations – and indeed of relations which hold between any two of the terms.

I cannot see any of this to be correct. On the one hand there are cases of composite relations between two terms. Simply consider the relation between the two terminal members of a sequence. On the other hand there are simple relations involving more than two terms. And precisely to this latter group we will later have to assign the collective combination among arbitrarily many elements.

---

[7] In the foregoing discussions I have avoided the expression "physical phenomenon," which in *Brentano* is paired with "psychical phenomenon," because it is somewhat awkward to designate a similarity, gradation, and the like, as a "physical phenomenon." Also, *Brentano* himself had in mind with that phrase only the non-relational [*absoluten*] primary contents – and, indeed, individual phenomena, not abstract Moments in an intuition. However, it is clear from the above expositions that the property of intentional inexistence – which in *Brentano* functions as the first and most penetrating mark distinguishing psychical from physical phenomena – also leads to an essential division in the classification of relations.

[8] Cp. for example, M. W. Drobisch, *Neue Darstellung der Logik*, 4th edition, Leipzig 1875, p. 34.

Furthermore, the case of sensuous identity can also serve to make Evident at least the possibility of relations of the latter kind. Under favorable and frequently realized circumstances we can grasp in one glance an identity amongst more than two sense perceptible objects without taking the least notice of the vast <71> manifold of simple relations that can be established between any two of the objects.[9] Already in the case of small groups (still authentically representable), the number of such relations would no longer be manageable within a uniting act. With six objects there are fifteen of these relations, with seven twenty-one, and so onward. To be sure, this does not yet rule out the possibility that such an identity is, nevertheless, a composite relation. Even if it is not composite in the manner of a relational complex in which the elemental relations are contained as separate and specifically noticed constituents, it still could be composite in the manner of a "fusion" of relations,[10] in the unanalyzed unity of which the elemental relations would at first be present as unnoticed factors. But this is not self-evident. That the relation ⟨of identity⟩ at first *appears* as a simple one sufficiently proves, in any case, the possibility of simple relations with more than two terms. At the same time one also sees that the view according to which any relation can *directly* link only two terms, and no more, is not justified.

The most appropriate definitions here might be the following: Relations that are themselves in turn composed of relations are *composite*. Relations of which this is not true are *simple*.

Psychological Characterization of the Collective Combination

After this digression into the theory of relations we turn back now, once again, to those specific relations the characterization of which is our goal. And we first raise the question: Are the relations that unify the objects in the totality, and which we have named "collective combinations," primary relations in the sense

---

[9] Cp. also C. Stumpf, *Tonpsychologie*, Vol. II, p. 310.
[10] Concerning the concept of the fusion of relations see Chapter XI below.

made precise above – as, for example, metaphysical and continuous combinations are? Or must we perhaps assign them to the class of psychical relations? More precisely stated: Are the "collective" combinations intuitively contained in, and to be separately noticed among, the contents of the representation of the totality as partial phenomena – as are, say, metaphysical <72> combinations in the metaphysical whole? Or can nothing of a combination be noticed *in* the representation contents themselves, but rather only in the psychical act which encompasses the parts in a unifying manner?

To resolve this question let us first compare the totality with any arbitrary primary representational whole.

In order to observe the combining relations in such a whole, analysis is necessary. If, for example, we are dealing with the representational whole which we call "a rose," we arrive at its various parts successively by means of analysis: the leaves, the stem, etc. (the physical parts); then the color, its intensity, the scent, etc. (the properties). Each part is picked out by a distinct act of noticing, and is held *together* with those parts already segregated. As the immediate consequence of the analysis there results, as we see, a totality: namely, the totality of the separately and specifically noticed parts of the whole. But with regard to the unification of the parts in the intuitive whole, there are still to be added the combining relations – as distinct and specifically determinate primary contents that are relational. In our example we have the continuous combinations within the leaves; or the combinations of the properties such as redness and spatial extension, combinations characterized in a wholly different way again from the continuous. Thus these combining relations present themselves as, so to speak, a certain "more," in contrast to the mere totality, which appears merely to hold its parts together, but not ⟨really⟩ to combine them. What, then, distinguishes the case of these primary combinations from that of the collective combinations? Obviously it is this: that in the first case a unification is intuitively noticeable *among* the representational contents, while this is not so in the latter case.

The same thing is also shown by a comparison of the collective combination with the relations of identity, similarity, gradation, etc. (which within the class of primary relations constitute, like the relations of combination, a group of relations that are psychologically well-characterized). Although they do not "combine" the contents upon which, as terms, they are based, they still constitute primary contents; and in contrast to them, once again, the collective combination appears almost as a case of *un*relatedness, so to speak. And so one also speaks of "disjoined" or "unrelated" contents when it is a matter <73> of emphasizing the absence of any primary content relation whatsoever, or of any upon which the current governing interest is directly focused. In such cases the contents are just simply thought "together," i.e., as a totality. But in no wise are they truly disjoined or unrelated. To the contrary, they are joined by means of the psychical act holding them together. It is only that *within* the content of that act all perceptible unification is lacking.[11]

The following circumstance also shows that between the collective combination and all of the primary content relations there is an essential distinction, which can only find its explanation in the fact that the former absolutely is not to be counted among the primary relations. Every relation rests upon terms and, in a certain manner, is dependent upon them. But while with all content relations there is a limit on the range of variability of terms that is admissible without altering the species of the relation, with the collective combination any term can be varied completely without restriction and arbitrarily and the relation still remains. The same also holds true of the relation of distinctness in the broadest sense. Not every content can be conceived of as similar, continuously joined, etc., to every other. But they *can* always be conceived of as different, or as collectively united. Precisely in these two cases

---

[11] Therefore *Mill* is quite right in strongly emphasizing that objects already stand in relation to each other even if we only think of them together. Precisely with respect to the psychical act which thinks them together they form parts of a psychical whole; and by means of reflexion upon that act they also can at any time be recognized as combined. That constitutes their "relation." And only if one were to restrict this term to what we have called "primary relations" could there, of course, be no more talk of relation in the case of a psychical combination. On the one hand this certainly is a terminological matter. But, on the other hand, there is *de facto* so much in common between the primary relation and the psychical relation, as to their essential Moment, that I fail to see why a common term would not be justified here.

the relation does not reside immediately in the phenomena themselves, but is, so to speak, external to them.

So testimony from many sources – and, above all, from inner experience itself – tells us that we must decide in favor of the second view mentioned, according to which collective unification is not intuitively given *in* the representation content, but instead has its subsistence only in certain psychical acts that embrace the contents in a unifying manner. This is a result which also repeatedly pressed itself upon us <74> in the critical discussions of the previous chapter.

And obviously these acts can only be those elemental acts that are capable of taking in any and all contents, however unlike they may be. So, then, a careful examination of the phenomena teaches the following:

A totality originates in that a unitary interest – and, simultaneously with and in it, a unitary noticing – distinctly picks out and encompasses various contents. Hence, the collective combination also can only be grasped by means of reflexion upon the psychical act through which the totality comes about.

Again, the fullest possible confirmation of our view is offered by inner experience. If we inquire what the combination consists in when we, for example, think a plurality of such disparate things as redness, the moon and Napoleon, we obtain the answer that it consists merely in the fact that we think these contents together, that is, think them in one act.

Let the following serve as a further characterization of the collective combination. For the apprehension of each one of the colligated contents there is required a distinct psychical act. Grasping them together then requires a new act, which obviously includes those distinct acts, and thus forms a psychical act of *second order*. If a multiplicity is represented in terms of subgroups, as when we represent a multiplicity of six objects in the form of 3 + 3 or 2 + 2 + 2, then the formation of each one of the sub-groups requires a psychical act of second order. Consequently, the collective unity that embraces all the sub-groups must be established by a psychical act of *third* order. How it comes about that we frequently prefer representations of this type, in spite of such a complication of acts directed upon one another, shall occupy us in the course of Chapter XI.

Since the collective combination presents us with a distinctive type of relation, it is obvious that at least the two-member collections have the character of *simple* relations. But how do matters stand in the case of collections with *several* members? Are they, perhaps, complexes or networks made up of collective combinations <75> between any two of its members? I think not. The colligating act encompasses all members without collectively linking them one by one; and, where we think we observe such an individual linking, closer consideration always shows that what is involved are other types of linkages that compete with the collecting act. So it is, for example, when we successively undertake the separate apprehensions whereby individual elements are sequentially linked. We must abstract from this temporal linkage if we wish to attain purely to the collective combination. — One also will not be able here to defend the view that the unity of the collection as a whole presents us with a fusion in which the elemental collections at first form indistinct Moments. For even in subsequent analysis we do not find such elemental collections present, unless we then actualize them anew. Therefore we look upon the collective combination of an arbitrary number of terms as likewise a *simple* relation.

Collective combination plays a highly significant role in our mental life as a whole. Every complex phenomenon which presupposes parts that are separately and specifically noticed, every higher mental and emotional activity, requires, in order to be able to arise at all, collective combinations of partial phenomena. There could never even be a representation of one of the more simple relations (e.g., identity, similarity, etc.) if a unitary interest and, simultaneously with it, an act of noticing did not pick out the terms of the relation and hold them together as unified. This 'psychical' relation is, thus, an indispensable psychological precondition of every relation and combination whatsoever.

Given the elemental character of the collective combination it is natural that it also must have found its imprint in ordinary language. In this regard the little syncategorematic word "and" satisfied all practical requirements. In and of itself it is without

signification. But where it links two or more names, it indicates the collective combination of the contents named. That ordinary language possesses no categorematic [*selbständigen*] name for the concept of collective combination is nothing to be wondered at.
5 That concept is only of an occasional, scientific interest. The usual aims of thinking and speaking require only the <76> linguistic formulation of the circumstance that given contents are combined in the collective manner, and this is accomplished in our language in a perfectly adequate manner by the conjunction
10 "and."[12]

---

[12] *Locke* describes the activity of colligating in *Essay*, Book II, ch. XI, sect. 6, under the heading "Compounding." He also observes that it links the units in a number. Nevertheless, he has not recognized the role which this activity plays in the abstraction of the concept of number.

Chapter IV

ANALYSIS OF THE CONCEPT OF NUMBER
IN TERMS OF ITS ORIGIN AND CONTENT

Completion of the Analysis of the Concept of Multiplicity

Now that we have established the 'psychological' nature of the collective combination, we can bring to completion our task of exhibiting the origin and content of the concepts *multiplicity* and *cardinal number*, and those of the individual numbers as well.

Already at the outset of our investigations we have laid the essential basis for an understanding of the origination of the concept of multiplicity (Chapter 1, pp. 18-22). By means of reflexion upon the psychical act that effects the unification of the contents combined to form the totality, we obtain the abstract representation of the collective combination. And by means of this latter we form the concept of multiplicity as that of a whole which combines parts solely in the collective manner.

Since the expressions "whole" and "part" are commonly used with a narrower extension than is here the case, and could easily lead to the thought of some more intimate combination lying within the primary contents themselves – which is here absolutely not intended – we would like to express our result in yet another way. A representation falls under the concept of the multiplicity, we may say, provided it combines any separately and specifically noticed contents whatsoever in the collective manner. Since we have now sought out the source from which the collective combination originates, and have identified it with certain psychical acts, there can hardly remain any obscurity concerning our concept.

Doubts might arise on only one point, which, however, is largely a matter of terminology. To what possible end, one might ask, are the terms "multiplicity" and "collective combination" – and thus the concepts as well – still held to be distinct? <78> Since in all cases where one speaks of a multiplicity there is nothing else in common than the collective combination, surely the two concepts are therefore identical. We will quickly come to agreement here if we only attend to the equivocation contained in the term *concept*. If we understand by "concept" the abstractum underlying the name, then we actually do have here an identity of the respective concepts. But this does not then demonstrate that the names involved also share a single sense. If, namely, we understand by the "concepts" the thought-correlates of the names, then the senses ⟨of the concepts⟩ are in fact different here, and the differentiating terminology is justified. For wherever we use the name "multiplicity" the object of interest is, occasional exceptions aside, not the collective combination as such, but rather the collective whole. "Collective combination" Represents the abstractum that underlies the general concept *multiplicity* or *collective whole*, and consequently the "signification" of the name "multiplicity" in the sense of logic. But this "signification" still does not constitute the total logical content [*Gehalt*] of the name. The corresponding total concept is that of a "something which possesses this abstract Moment of collective combination." So understood, the concept of the collective combination forms the most essential constituent of the concept of the multiplicity, without the two being identical. And that is indeed how things usually stand with general names. If we speak simply of *a man*, then we have the concept of a something that possesses such and such abstract properties. If we wish to direct specific interest upon the union of those properties in their own right, then we must say: the abstractum "man." The expression, "the concept *man*," is equivocal. It can signify both the general and the abstract concept. — By way of exception, moreover, the name "multiplicity" is also, if used by itself, understood in the abstract sense. Thus, outstanding psychologists such as *Lotze*[1] and *Stumpf*[2] speak of the multiplicity as of a relationship, by which, obviously, nothing other can be meant than the collective combination. By introducing the term "collective combination" we have eliminated this equivocation. <79>

---

[1] *Metaphysik*, Leipzig 1879, pp. 530/31
[2] *Tonpsychologie*, Book I, p. 96

We can express the content of the concept of the multiplicity in yet another way that is important for our further analyses. For this purpose we must analyze somewhat more closely the peculiar abstraction process that yields that concept.

No concept can be thought without foundation in a concrete intuition. Hence, even when we represent the general concept of the multiplicity we always have in consciousness the intuition of some concrete multiplicity by means of which we abstract the general concept. In what way, then, does this abstraction proceed? As we have established, total abstraction from the peculiarities of the individual contents colligated must be effected, retaining, however, their combination. This appears to present a difficulty, if not a psychological impossibility. If that abstraction is performed in all seriousness, then of course the collective combination disappears along with the individual contents, instead of remaining behind as a conceptual distillate.

The solution is clear. To disregard or abstract from something means merely to give it no special notice. The satisfaction of the requirement wholly to abstract from the peculiarities of the contents thus absolutely does not have the effect of making those contents, and therewith their combination, disappear from our consciousness. The grasp of the contents, and the collection of them, is of course the precondition of the abstraction. But in that abstraction the isolating interest is not directed upon the contents, but rather exclusively upon their linkage in thought – and that linkage is all that is intended.

The abstraction to be carried out can now be described in the following manner: Determinate individual contents of some sort are given in collective combination. In abstractively passing over, then, to the general concept, we do not attend to them as contents determined thus and so. Rather, the main interest is concentrated upon their collective combination, whereas they themselves are considered and attended to only as some contents in general, each one as a *certain something*, a *certain one*.

We will capitalize upon this result by combining it with an earlier observation, according to which the collective combination can be indicated in language by the conjunction "and" in a completely clear and intelligible <80> manner. Multiplicity in general – as we can now express ourselves quite simply and without any circumlocution – is nothing other than: a certain something and a certain something and a certain something, etc.; or, some one and some one and some one thing, etc.; or, more briefly, *one and one and one*, etc.

The Concept 'Something'

This makes it clear that the concept of the multiplicity also contains that of *something* along with and in the concept of the collective combination. It is therefore our task to characterize this concept more precisely as to content and origination.

"Something" is a name which is suited to any conceivable content. Any real or conceptual entity is a "something." But we can also give this name to a judgment, an act of will, a concept, an impossibility, a contradiction, and so on. Of course the concept *something* is not to be obtained by any conceivable *comparison* of contents which takes in all objects, physical and psychical. Such a comparison would simply remain without a result. The "something" is quite certainly no abstract partial content. That wherein all objects – actual or possible, real or non-real, physical or psychical, etc. – agree, is this alone: They either are contents of representations, or are surrogated in our consciousness by means of contents of representations. Obviously the concept *something* owes its origination to reflexion upon the psychical act of representing, for which precisely any determinate object may be given as the content. Hence, the "something" belongs to the content of any concrete object only in that external and non-literal fashion common to any sort of relative or negative attribute. In fact, it itself must be designated as a relative determination. Of course the concept *something* can never be thought unless some sort of content is present, on the basis of which the reflexion mentioned is carried out. Yet for this purpose any content is as well suited as another: even the mere name "something."

The crucial role of the concept *something* or *a* in the origination of the general concept of the multiplicity consists in the fact that each particular one from among the determinate contents encompassed by the concrete <81> representation of multiplicity is thought under the mediation of the concept *something*, and is attended to only insofar as it falls under that concept. In this way there comes about that utter depletion [*Entschränkung*] of content which confers upon the concept of the multiplicity its generality.

The Cardinal Numbers and the Generic Concept of Number

The expression "one and one and one, etc." which, according to the elucidations given, clearly expresses the content of the concept of multiplicity, refers by means of the "etc." to a certain indeterminateness that is essential to the concept in its broad sense. Not as if the collection of 'ones' which the concept of the multiplicity Represents to us were endless. Much less, then, that the form of multiplicity which, starting from a determinately given totality, we attain to through the abstraction process described above would be without closure. Rather, nothing else is meant than that no determination is set with respect to an upper bound, or else that the factually present limitation is to be regarded as something that does not matter.

But if we want to do away with this indeterminateness, then there are many possibilities; and it is clear that, corresponding to these, the concept of multiplicity immediately splinters into a manifold of determinate concepts that are most sharply bounded off from each other: the *numbers*. There arise concepts such as: *one and one; one, one and one; one, one, one and one*, and so forth. These, by virtue of their extremely primitive character and importance for practical life, at least within a limited range – so far, namely, as they can be easily distinguished – have already been formed at the lowest level of human mental development. Accordingly the names two, three, four, etc., belong to the earliest inventions of all languages.

It is of course not necessary to assume the general and indeterminate concept of multiplicity as mediator in the derivation of the number concepts. We come by them directly, setting out from arbitrary concrete multiplicities; for each of these falls under one, and indeed a determinate one, of those concepts. The abstraction process which yields the determinate number accruing to a given concrete multiplicity <82> is, in terms of the above analyses, completely clear. Disregarding the specific character of the particular contents grasped together, one considers and retains each content only insofar as it is a 'something' or 'one'. And thus one obtains, in taking account of the collective combination of them, the general form of multiplicity appertaining to the multiplicity at hand: one and one, . . ., and one – a form with which a definite number term is associated.

All concepts that arise in this way are clearly akin to each other. Their similarity rests upon the sameness of the partial representations that make them up (the 'ones' or units), as well as upon the elemental similarity of the psychical acts combining those representations. Their similarity suffices to delimit the number concepts as a well-characterized class of concepts and to serve as the basis for a general name. The name *cardinal number* [*Anzahl*] performs this function. "Cardinal number" is a common name for the concepts two, three, four, etc.[3] Now we certainly do also speak of a general concept and not merely of a general name, "cardinal number." But we cannot explain this concept otherwise than by pointing to the similarity which all of the number concepts have to each other. There is no cardinal number in general, understood as a separately noticeable partial representation – in the manner of a physical or even just a metaphysical part (e.g., the color or shape of an external thing) – which might be isolated within the representation of each of the cardinal number concepts. Rather, the relationship between the logical part and logical whole (e.g., between the ⟨genus⟩ color and the difference red) seems to correspond to the one between cardinal number in general and determinate number (two, three, etc.).

---

[3] *James Mill* assigns the name "cardinal number" to that category of names which he calls "names of names." (Cp. *Analysis*, Vol. II, p. 4) This is a highly inappropriate manner of expression. "Cardinal number" did not arise as a general name for the *names* "two," "three," etc., but rather for the concepts designated by means of them, whose inherent kinship offered the reason and the occasion for a common term.

Relationship Between the Concepts
'Cardinal Number' and 'Multiplicity'

How, then, are the concepts *cardinal number* and *multiplicity* related to one another? That they coincide as to their more essential content [*Gehalte*] <83> is obvious from the start. The distinction consists only in this, that the concept of the cardinal number already presupposes a differentiation of the abstract forms of multiplicity from each other, but that of the multiplicity does not. The former is to be taken as the generic concept which originates from the comparison of the determinate multiplicity forms or numbers, as species concepts, already differentiated from each other. The concept of multiplicity, by contrast, arises directly out of the comparison of concrete totalities. Certainly the abstraction procedure described does not, in being applied to different totalities, always lead to the same multiplicity form. But at the level of abstraction to which the concept of multiplicity belongs, a differentiation and classification of the various multiplicity forms either has not yet been carried out or falls outside of the governing interest. Observation discovers only the obvious sameness of kind in those psychical acts of reflexion which always come into play when we rise from concrete multiplicities to the corresponding concept, whereas the distinctions which lead to the rigorous segregation of that unbounded sequence of species concepts remain still unnoticed or are deliberately disregarded. In consequence of this the concept of the multiplicity carries within itself precisely that vague indeterminateness which we have characterized above. What it lacks is that which first gives to numbers their full and distinctive character: the sharply defined *how many*.

That the indeterminate multiplicity concept actually Represents a considerably lower level of concept formation is confirmed in experiences with children and with people who are still in a state of childhood. It is well known with what difficulty children who have already long been in possession of the name and concept of the 'many' are to be led to a clear differentiation of the number concepts. As to primitive people, even those of the lowest cultural level, where terms for the determinate number concepts do not extend past three or five, all still possess the name and concept of the indeterminate *many*.⁴ <84>

One and Something

The relationship between the concepts *one* and *something* still requires elucidation. In our view *one* essentially coincides in its very concept with *anything, any one thing*, or simply *a thing*, where "a" signifies the indefinite article. And all these names have, in turn, essentially the same signification as "something." In counting we bring each of the concretely given things under this concept. But then there obtains between the multiplicity as a whole and the particular objects as its parts a relationship of correlation; and thus also between the multiplicity *in abstracto* and the unit as individual element of the multiplicity thought under the mediation of the concept *something*. Now, since the name "one" came into use exclusively in the process of enumeration, a certain difference in signification arose between "one" and "a thing" or "something," because "one" preserved a correlation to multiplicity as a connotation. This resulted from the mode of usage entirely of itself. Thus "one" came to mean the same as "enumerated thing," or "*one* thing" in opposition to many things, whereas "one thing" pure and simple (with the "one" unaccentuated), and the equivalent "something," remained free of this relation to the concept of the multiplicity. In the characterization of number abstraction I purposely said: We bring each content under the concept *something*, and not: We bring each content under the concept *one*. For the correlation with the concept of the multiplicity, which alone marks out the concept of *one* over against that of the *something*, is not a point that in any way comes into consideration for number abstraction. In that each object of the multiplicity is thought merely as a "something," the "something" is already "one." It stands, as a "something," in the multiplicity and thereby possesses *eo ipso* that correlation to it.

---

⁴ Cp. E. B. Tylor, *Primitive Culture*, Vol. I, ch. 7, "The Art of Counting," London 1871, p. 240, and J. Lubbock, *The Origin of Civilization and the Primitive Condition of Man*, London 1870. {LE}

We can with full justification designate the concepts *something* and *one*, *multiplicity* and *cardinal number* – these most general of all concepts, and most empty of content – as form concepts or *categories*. What characterizes them as such is the circumstance
5 that they are not concepts of contents of a determinate genus, but rather in a certain manner take in any and every content. There are also still other relational concepts <85> of a similar character, e.g., the concepts of *difference* and *identity*. The all-encompassing character of these concepts finds its simple explanation in the fact
10 that they are concepts of attributes which originate in reflexion upon psychical acts. And such acts can be brought to bear upon all contents without exception.

Critical Supplement: Our analyses have led to an important result, firmly established by both the critical and the positive routes. It is
15 impossible to explain the origination of the number concepts in the same way as, say, that of the concepts *color*, *shape*, etc., which, as positive Moments in the primary content, are isolated through mere analysis thereof. Therefore it was not only *Aristotle* who was in error, when he attributed the numbers and the *one* to the
20 ἀισθητὰ κοινά, to the objects common to all the senses[5], but also *Locke*, when he assigned the *one* to the concepts that have their source simultaneously in the domain of sensation and in that of reflexion.[6] The enumerated contents certainly can be physical as well as psychical, but the number concepts and the *one* belong
25 exclusively to the domain of reflexion. And accordingly it is also absurd from the outset when *Locke* (like so many after him) considers the represented numbers to be "primary qualities," as perfect copies of original qualities, which have their subsistence in the things themselves and independently of our mind.[7]
30 In this most general of forms our result is not new. *Sigwart* has

---

[5] *De Anima*, Book II, ch. 6, p. 418, a 16; Book III, ch. 1, p. 425, a 13; etc. (Becker edition, Berlin 1831). Cp. Fr. Brentano, *Die Psychologie des Aristoteles*, Mainz 1867, p. 83. *Leibniz*, among others, also followed *Aristotle*. See his *Meditationes de cogitatione, veritate et ideis* (1684), *Opera philosophica*, ed. Erdmann, p. 79.

[6] *Essay*, Book II, ch. VII, sect. 7.

[7] *Essay*, Book II, ch. VIII, sects. 11, 17, etc. Cp. the critique of *Baumann's* views in chapter two of this present work.

already stated it in his *Logic* – the first to do so, so far as I know – and thereby has shown the correct path toward the analysis of the number concept. However, there is lacking in *Sigwart* any soundly worked out and sustainable theory. His expositions are successful and to the point only in all those places where he critically demonstrates the untenability of the physical abstraction theory in the teachings of *J. St. Mill* and *Bain*.[8]

Let it be here stated that it was the critical study <86> of *Sigwart*'s investigation that first led me to the theory developed above.

Wundt's expositions of the concept of number in his large work on logic stand under the influence of *Sigwart*, if I am not mistaken. He too finds that concept to emerge from reflexion on psychical acts – in doing which, however, he knows to distance himself from the distinction theory. I have to admit that I have not succeeded in arriving at a completely clear picture of his view. What I do understand I am unable to accept in the sense of a clear and consistent view. "The starting point for the development of the number concept," says *Wundt*, "is the *unit*. It appears in the primitive activity of the function of enumerating as an abstraction from the particular object. A common view therefore regards the number as a replica of the particular enumerable things, in which their distinguishing characteristics are disregarded. Now it is clear, however, that the things can only become *enumerable* by thought grasping them as units. Certainly the motivation for this logical transaction resides in the representations of the things, in their status as closed off and independent over against other representations. But it would be completely unintelligible how this motivation is to become effective if our thinking did not have the ability to grasp the individual object as a unit. *So the genuine bearer of the concept of the unit is the individual act of thinking.* Therefore that alone is enumerable which can always be separated into individual acts of thinking bound up with one another."[9]

I search in vain for the *nervus probandi* of this line of argument. Could one not show, in precisely the same way, that the individual

---

[8] Chr. Sigwart, *Logik*, First edition, Vol. II, Freiburg i. Br. 1878, p. 39 ff.
[9] *Logik*, Vol. I, Stuttgart 1880, p. 468.

acts of thought are "bearers" of *every* concept, and for example argue: Color is an abstraction from the colored object? The grasping of the colored thing as such is a logical transaction, motivation to which can be present in the things themselves. But it would be unintelligible how this motivation is to be effective if our thought did not have the ability to grasp the colored object as colored. The genuine bearer of the concept *color* is therefore the individual <87> act of thought (that grasps the colored object). — Taken in a certain sense, no doubt, the proposition is valid here and in all cases, but not in the quite specific sense it has as applied to the unit. Two quite different assertions are here confused, unless I am totally misled:

1). The concept of the unit cannot originate without an act of thought that bears it – namely, one abstracting it. This assertion is valid for *every* concept, and not merely for that of the unit. That the above line of argument applies to this assertion alone is clear. Precisely for this reason it has the possibility of being carried over to every other concept. For the rest, there surely is no need to take the trouble of demonstrating something that is completely obvious.

2). The abstract concept of the unit cannot originate without an act of thought that bears it – namely, a certain act that belongs to its *content*. This assertion certainly is not at all obvious, but it becomes confused with the one above. The proof of the first comes to be regarded as valid for the second assertion as well.

Therefore we also cannot concede the further propositions, which *Wundt* thinks follow, to be what they profess, namely, logical consequences of the foregoing line of argument: "Therefore everything is enumerable that can be separated into individual acts of thought bound up with one another. Thus, not merely objects, but rather properties and events are enumerable as well . . . . The function of enumeration, whatever it may have reference to, always consists in a combining of individual acts of thought into composite units . . . . It originates from the combining of acts of thought that follow each other, when complete abstraction from the content of those acts is effected. As the 'one' designates any sort of thing that can be given as an individual act of thought, so any number composed of units presents us with a sequence of acts of thought of arbitrary content . . . ."[10]

Although some expressions are not wholly clear to me, I would still have been inclined to interpret these and similar propositions which recur in other places in the sense of the theory proven correct in our investigations above, if so much did not follow for *Wundt* that absolutely cannot be reconciled with that theory. <88> I do not understand how, following the proposition last cited, according to which "any number composed of units" is to present us with "a sequence of acts of thought of arbitrary content," we can speak of the "development of the formations composed of units into the *persisting* sequence of number concepts," and of a "conceptual development" which is to form "the source of all the abundant *transformations*" which the number concept has undergone. Here is meant the origination of the negative, fractional, irrational and imaginary numbers.[11] The fractional number, for example, is to derive from the problem of determining the number $\underline{a}$ which originates when a number $\underline{c}$ is divided by another number $\underline{b}$. Now it can happen "that such a number $\underline{a}$ does not exist in the sequence of the positive whole numbers, but rather that $\underline{a}$ corresponds to the concept of a number positioned between two adjacent whole numbers."[12] Certainly that problem could not be formulated as such were we to put the above explication by *Wundt* in place of "number." If, now, one first considers the fact that "positive whole number" here means nothing more than "number,"[13] then it becomes wholly unintelligible how this $\underline{a}$, which does not exist in the number sequence, can correspond to the concept of a number which is supposed to subsist *between* two numbers, thus supposedly Representing something which is not composed of $\underline{n}$ and not of $n+1$ thought acts, but "is positioned between them." "Since the quotient $\frac{c}{b} = a$," *Wundt* continues, "can assume all possible intermediary values between two whole

---

[10] *Ibid.*

[11] *Op. cit.*, p. 469. {LE}

[12] *Op. cit.*, p. 470. {LE}

[13] "The number as a combination of units is first of all the positive number and . . . whole number." *Op. cit.*, p. 469.

numbers, there arises from this the need to expand the number concept in such a way that it encompasses everything which comes from adding and subtracting in the two mutually opposed directions. Those numbers which, in order to meet this demand, one must interpolate between the positive and negative whole and fractional numbers as intermediary values, are the irrational numbers."[14] How the concept of a sequence of acts of thought can provide for the possibility of "expansions" or "transformations" which would have the capability of meeting this or that demand <89> remains just as puzzling as the ability to think – and the concept – of the "intermediary values."

One further comment: *Wundt* calls number "the most abstract form in which the law of discursive thought, according to which each composite thought consists of individual acts of thought, comes to expression . . . ."[15] It does not seem to me that number acquires a special characterization in this way. How then do matters stand in general with this *law* of discursive thought? I am surely correct in saying that if it is true, then it is a tautology, and if it is no tautology, then it is not true. If we understand by a composite thought one that consists of individual acts of thought, then the tautology is obvious. But if we understand by a composite thought any composite content in general, then the proposition is incorrect, precisely because compositeness of content by no means ever consists of (represented) acts of thought. Only this is right, that the originally undifferentiated unity of a composite phenomenon passes over into a plurality that requires a plurality of acts of thought to articulate it. But also for this elemental fact of our psychical life, it is not made clear how it is supposed to come to expression in the concept of number, since this concept can also subsist without it. Given the collective combination, the number concept is possible, no matter the source of that combination. <90>

---

[14] *Op. cit.*, pp. 470f. {LE}
[15] *Op. cit.*, p. 468.

Chapter V

THE RELATIONS "MORE" AND "LESS"

According to the analyses of the last chapter, the number concepts have presented themselves as an indeterminate sequence of concepts, apparently continuing even to infinity, whose clarity and easy mutual distinguishability seemed to be beyond question and to render superfluous further investigations for the purpose of precise reciprocal delimitation. One and one is sharply distinct from one and one and one, and this in turn from one, one, one and one, etc. One sees, however, that the ease of differentiation noticeably diminishes the further we advance in the sequence of numbers. Nineteen is much less easy to distinguish from twenty than nine is from ten, and the latter less easy than three from four. That this circumstance does no harm, that in spite of it we consider number determinations and number distinctions to be the most rigorous in the domain of our knowledge, has its basis in certain instrumentalities [*Hilfsmitteln*]. Through these we are in position – in cases where direct intuition is either totally denied or could easily err – to attain the goal of rigorous differentiation indirectly and to confine error to a very narrow range. The instrumentalities are those of enumerating and calculating, i.e., certain 'mechanical' operations, as it were, the true basis for which lies in the elemental relations between the numbers. The psychological analysis of these relations of *more* and *less* is the goal which we now set before us. <91>

The Psychological Origin of These Relations

Let us turn back once again to the sphere of concrete phenomena. Imagine a given group [*Menge*], perhaps of balls. Add, now, one or several balls to that group. Then we say that the new group has *more* balls by those added. But if balls are taken away, then we say they are *less* by those taken away. In this case we are dealing with physical objects and with a physical operation upon them. But also in cases where we collectively *think* contents together – and not just external contents – such an adding to and taking away is present. What is meant thereby certainly can only be shown and not defined. It is an elemental fact [*Tatsache*], to be described in no other way than by reference to the phenomena, that while certain contents are thought 'together' by us, still other contents can then be added and grasped together with the ones already present. The original act is expanded by the taking in of new contents. But the opposite can also occur. Some of the contents already brought together are omitted, while the unifying act retains and includes only the remaining ones.

But these facts of the expanding and narrowing of totalities do not yet just by themselves suffice to ground the relational concepts of the 'more' and the 'less.' Of course one can attach to these facts the correct explanation that the expanded totality contains more, by the elements newly taken in, and likewise that the narrowed totality contains less, by the elements omitted, than the one originally represented. One can, further, draw the correct conclusion that *more* and *less* are correlative concepts, in that the return from the expanded totality to the original requires a diminishment, and the return of the diminished totality to the original requires an augmentation. All of this is correct, but it does after all presuppose a *novel* fact [*Factum*] of inner experience in order to be possible. As any relation requires that the terms be together in a single act of consciousness, so also with our relations of more and of less. It therefore presupposes for its realization that the original and the expanded totality be present to us simultaneously and in *one* act. <92> And even that does not yet suffice, for the latter totality must even appear as the "sum" of two totalities, one of which is recognized as identical with the original totality, while the other Represents the totality of the newly added contents. If, for example, I expand the totality (A, B, C) to form (A, B, C, D, E), then the judgment that the second is more by D and E requires the simultaneous representation of:

(A, B, C), (A, B, C, D, E) and (A, B, C; D, E),

and indeed in *one* act.

Consequently it is a fact that we have the capability of representing several totalities together as unified into *one* totality, without thereby their separate unifications being lost. We represent totalities whose elements are in turn totalities. In fact, even totalities of totalities of totalities are thinkable, etc. That the limits of authentic representing are imposed very early on, and all the rest consists only in inauthentic (symbolic) representing, requires no wide-ranging discussions. It is certain, however, that authentic representing is present at least in the first stages. For otherwise the very idea of such composition would be absurd, and there would be no question of a comparison of multiplicities as to more and less.

As to the psychological foundation of these more intricate modes of formation, one recognizes that there are here present *psychical acts of higher order*, i.e., psychical acts which are directed in turn upon psychical acts and bear upon primary contents only through mediation of these latter. If in one act we represent several totalities, there is required for the formation of each particular totality a unifying act of the type described above. And if each of them is to be consciously held fast in its unity, and thought as unified with the others, then a psychical act of *second* order must be directed onto the acts of *first* order – upon which the specific unifications of the partial totalities rest – and only through them onto the primary contents. In the case of totalities of totalities of totalities, we would arrive at psychical acts of the *third* order, etc. Indeed, more closely considered, these numerical levels are to be raised by one degree further; for already in the simple totalities, i.e., <93> those whose elements are unitary contents not further analyzed, acts of the second order are present, to the extent, that is, that the particular contents are thrown into relief by special acts and only then are encompassed by a common act which unites them all. (Cp. p. 77)

Comparison of Arbitrary Multiplicities, as well as of Numbers, in Terms of More and Less

To this point we have merely considered totalities which proceed from one another through augmentation or diminishment. One easily understands, however, that the comparison of wholly arbitrary totalities in terms of more and less gives rise to no new difficulties in the psychological respect. To be sure, such a comparison can occur *in concreto* only under one specific condition. The totalities to be compared must consist, wholly or in part, of contents that are the same on both sides. And this in such a way that all of the contents of the one are represented by like contents in the other. If for example the first is called M and the latter N, then we can think of this latter as split into two partial totalities: M' and N',

$$N = M' + N',$$

of which M' encompasses the same elements as M and therefore is identical with M. Now since the totality N contains more by the elements N' than its partial totality M', and since this latter is *identical* with the totality M, we also say that the totality N contains more, by the elements N' (or by the totality N'), than the totality M.

The presupposed condition for a possible comparison is of course always fulfilled when the two totalities consist of contents of one and the same genus. Two such groups therefore always find themselves in the relationships of more and less. But if there are totalities that consist of heterogeneous contents, or if that condition is not fulfilled, then one can only compare their *numbers* in terms of more and less. Now this happens exactly in the manner just described. For in consequence of the complete depletion of content which occurs with the abstraction of numbers <94>, these most general of totality forms, considered extrinsically, do themselves, in turn, Represent totalities of like contents, namely, of units. Since each concretely present content is considered and thought together with the others only insofar as it is a 'something', each one becomes like the others – precisely as a 'something'. That is why numbers can likewise be compared with one another with respect to more and less, just like totalities of homogeneous concrete elements. At this high and vacuous level of abstraction all distinctions disappear precisely *eo ipso*.

It results from our last inquiry that we have with specific justification utilized the expression "*comparison* of totalities or numbers in terms of more and less." Any recognition of such a relationship includes the knowledge of an identity. If we are dealing with concrete totalities, then the comparing act consists of recognizing the identity of the one with a part-totality of the other, and grasping the surplus in its own right. And it is just the same with numbers. If we judge that five is more by two than three, then we represent five split into the two partial numbers, two and three. And in establishing the identity of the three represented as a partial number with the three represented in its own right, the surplus comes before consciousness as the partial number two. Now since this 'more' (or, from the standpoint of the second multiplicity, the 'less') is recognized as the ground of the distinctiveness of the two multiplicities, one simply calls it their "difference." And in the same way we speak of a difference of two numbers in cases where a number is understood that is thought as the surplus of one number in relation to another.

So this is the characteristic manner in which the mental activity of comparing and distinguishing comes into play for numbers. We will continue our examination of equality of number, in and for itself a simple matter, in the next chapter. This is because it has occasioned, in the Modern period, some noteworthy discussions, and has even served as the point of departure for peculiar analyses and definitions of the concept of number. <95>

The Segregation of the Number Species Conditioned upon the Knowledge of More and Less

The multiplicity relations of equal, more and less essentially condition the origination of the number concepts. Certainly *Herbart* goes too far in saying: "The truly scientific concept of number" is "none other than that of the more and the less."[1] But it is correct that the determinate numbers, two, three, etc., presuppose a comparison and differentiation of delimited multiplicities, thought *in abstracto*, in terms of more and less. In order to be able to rise above the concept of the "multiplicity of units" and form the series of numbers, two, three, etc., we must classify multiplicities of units. And this requires judgments of identity and non-identity. But an exact judgment of non-identity is in this case not possible without recognition of more or less. Two numbers are non-identical if the one is identical with a part of the other.

Note: Since in this chapter it is above all the psychical activities essential to the multiplicity concept that have come into consideration, I have, as elsewhere in similar cases, preferred to use the term *totality* as opposed to the terms "multiplicity," "plurality," etc. It expresses in an especially clear manner the way in which colligated contents are grasped-together-in-one[2]. <96>

---

[1] *Psychologie als Wissenschaft*, Part 3, p. 163.

[2] "*In-eins-zusammenbegreifen.*" The point, of course, does not come through in English. "*Inbegriff*," which we translate as "totality," involves *Begriff* or "concept," which in turn involves *greifen* or "to grasp." With the meaning of "in" which is common to both languages, Husserl's point about the usage is clearer. {DW}

Chapter VI

THE DEFINITION OF NUMBER-EQUALITY THROUGH THE CONCEPT OF RECIPROCAL ONE-TO-ONE CORRELATION

Ever since *Euclid*'s "Elements" attained the status of the model of scientific exposition, mathematicians have followed the principle of not regarding mathematical concepts as fully legitimized until they are well-distinguished by means of rigorous definitions. But this principle, undoubtedly quite useful, has not infrequently and without justification been carried too far. In over zealousness for a presumed rigor, attempts were also made to define concepts that, because of their elemental character, are neither capable of definition nor in need of it. Of this sort are the so-called "definitions" of equality and difference with respect to number whose refutation will now engage us. And they have indeed a special claim on our interest precisely because they have led to a class of definitions of the number concepts themselves. These definitions, baseless and scientifically useless, have nevertheless, in virtue of a certain formal character, found favor among mathematicians and among the philosophers influenced by them.

Leibniz's Definition of the General Concept of Equality

The most extreme in this respect are those investigators who, following the lead of the brilliant *Hermann Grassmann*, think it necessary to define even the general concept of equality in order to be able to apply it to groups and to numbers. Here is the definition by the philosophical mathematician mentioned: "Two things are said to be equal if the one can replace the other in any assertion."[1] In essentials <97>, *Leibniz* had already offered the

---
[1] H. Grassmann, *Lehrbuch der Arithmetik*, Berlin 1861, p. 1

same definition: "Eadem sunt quorum unum potest substitui alteri salva veritate,"[2] which *Frege* adopted as the basis for his construction of the concept of number.[3]

We are unable to convince ourselves that this definition is of value. First, it is obvious that instead of equality it defines identity. So long as there is a trace of difference there will be judgments in which the things concerned cannot be substituted "*salva veritate.*"

Secondly, it is clear that the definition stands the true state of affairs precisely on its head. Assuming we had come to believe, with respect to two contents, that they satisfy the definition (and even that would have its difficulties!), the quite justifiable question still arises: What is the basis of the fact that one may substitute the one content for the other in some or in all true judgments? The only appropriate response is: The equality or else the identity of the two contents. Every equality of property [*Merkmal*] founds judgments of equal truth-value, but equal judgments do not found equal properties. So if one were to pose the converse question, "Why are the two contents equal to one another?" the response, "Because they admit of substitution in true judgments," would be obviously wrong.

The following simple reflection shows to what absurd consequences the view we are opposing leads. If the basis for the knowledge of equality of two contents lay in the interchangeability required above, then in every case our acknowledgment of the equality would have to be preceded by that of the substitutability. But the latter act itself, after all, consists in nothing other than an (even endless!) number of acts, each one of which implies the acknowledgment of an equality. Namely, the equality of each particular true judgment bearing upon the first content with "the same" judgment bearing upon the second content. But in order to acknowledge all of these equalities there would be required, in turn, the knowledge that with respect to each of these pairs of

---

[2] "Non inelegans specimen demonstrandi in abstractis," *Opera philosophica*, J. E. Erdmann, ed., p. 94. Substitutability in all judgments is expressly stipulated in the elucidation adjoined.

[3] G. Frege, *Die Grundlagen der Arithmetik*, Breslau 1884, p. 76.

judgments <98> "the same" true judgments are valid. And so forth. We therefore fall into a genuine labyrinth of infinite acknowledgments [*Regressen*].[4]

The Definition of Number-Equality

In other cases one is satisfied to define equality, along with more and less, specifically for multiplicities, with respect to their number. In particular, mathematicians currently have a preference for the following definition, which I cite exactly in the formulation I find for instance in *Stolz*: "Two multiplicities are said to be *equal* to one another ⟨or, more correctly: are said to be *equally many* or *equal in number*⟩, provided that each single thing in the first can be correlated with one in the latter, and none of the latter remains uncorrelated."[5] This is followed by the supplementary definition: "The *greater* of two multiplicities is said to be that one in which, after each thing in the other (the *smaller*) is correlated to one in it, some things (a *remainder*) are left uncorrelated.[6]

One may indulge no illusions as to what these definitions accomplish either in the psychological or the logical respect. More closely examined, it turns out that the representation of the more

---

[4] *Helmholtz* too, although under the influence of *Grassmann* in other respects, has rejected his definition of equality in his treatise, *Über Zählen und Messen*, in Philosophische Aufsätze z. Zellers Jubil., p. 38.

[5] O. Stolz, *Vorlesungen über allgemeine Arithmetik*, Part I, Leipzig 1885, p. 9. *Stolz*'s definition would be quite misleading without the correction inserted above, especially since he immediately before gave this definition: "By multiplicity is understood a group of objects *equal among themselves*, i.e., the totality of disjoined things whose differences are nevertheless not considered, with no regard to their order." But then for the equality of two multiplicities that reciprocal, one-to-one correlation without remainder would contribute nothing. Equality with respect to the genus of the things in the totality would surely be required over all else. In the above definition *Stolz* would precisely have to say, instead of *equal, equally many*, or, as other mathematicians prefer, *equal in number*.

[6] In agreement with the more common linguistic usage in mathematics, *Stolz* says "greater" and "smaller" where we, in order to avoid a needless involvement of the concept of magnitude, would say "more" and "less." — *Stolz* further joins to these definitions a demonstration which, in its essential concepts, originates in Schröder (*Lehrbuch der Arithmetik und Algebra*, Leipzig 1873, p. 14), to the effect that the result of those correlation operations is independent of the manner of linkage. From this it follows that an equality or difference which is established under one manner of linkage remains valid for every other.

and less is already included in this definition of equality, whereas that representation itself, as we concluded above <99>, cannot be conceived without representations of equality. When we say that the reciprocal correlation must leave no element uncorrelated, this is only another way of saying that on neither side may there be one element more or less. Thus the circularity is obvious.

One could respond that with this line of argument the futility of the definitions is not yet proven. Perhaps one must view them as merely explanations of terms of the sort that are of value as aids to understanding in all cases where a term lacks a rigorous and unequivocal signification. — However, also from this perspective we are unable to convince ourselves that the definitions are of value. On the one hand the expression "equality of two multiplicities" requires no further explanation. On the other hand, by the above definition the obvious and intrinsically familiar is rendered obscure by what is distant and strange. And in fact we have to designate as distant and strange that reciprocal correlation[7] which is invoked to explain equality. It is not true that that definition is a mere explanation of terms, although it presents itself as such. When *Stolz* says, "Two multiplicities *are said to be* equal to one another ⟨namely, with regard to their number⟩ provided that each thing in the first can be correlated with each one in the latter . . .," there is surely nothing more certain than that the definiendum and the definiens are not conceptually equivalent. To represent to oneself two equinumerous multiplicities, and to represent two multiplicities reciprocally correlated term-by-term, is not one and the same thing. The definition states a true proposition, but not one of identity. It may well happen that, in order to verify *in concreto* the equality of two groups as to their multiplicity, we place pairs of elements alongside one another or link them up in some other way. But neither can we regard this operation as necessary in every case, nor does there reside in it alone, where this happens, the essence of the comparative act.

In order to clarify the true sense of the definition of equality here offered to us, we will begin by making some general pre-

---

[7] Instead of "correlation" [*Zuordnung*] other expressions are also used, among others, "combination," "linkage," "pairing," "association." Cp. E. Schröder, *op. cit.*, pp. 7-9.

liminary <100> remarks concerning the sense and purpose of definitions of equality for special cases.

Concerning Definitions of Equality for Special Cases

What is it for two relatively simple, i.e., not further analyzed, contents to be equal to one another neither permits of nor requires an explanation. So let us immediately turn to composite contents. In relation to these, linguistic usage seems at first glance to vacillate. If contents of this type are equal, then some property must be equally present in both of them (thus, perhaps physical parts as well, for the possession of such parts can also serve as a property). This is obvious. But the converse seems not to hold. Sometimes properties are equally present on both sides, and nevertheless we do not speak of the equality of the objects. In fact, it can even happen that *the same* objects at one time are called equal by us, and at another time unequal. So for example in Geometry two straight lines are said to be equal if they have the same length. But sometimes such lines are said to be unequal, in that those are understood to be equal which are equally long and moreover are parallel and have the same "sense." And it is similar with the concept of the equality of figures and solids, which now covers the content measured, and then again the position also.

This apparent vacillation in the terminology is cleared up very simply. We say *purely and simply [schlechthin]*, of whatever contents, that they are *equal* to one another *if there is equality with respect to the* (intrinsic or extrinsic) *properties which directly form the central focus of our interest*. In metric geometry (as in *Euclid*) the interest is directed precisely on the quantity of the structures (length, surface, volume). In topology [*Lagengeometrie*] we are also interested in the position, and the comparison is oriented toward it. "Two structures are equal" – for the geometer this is only a practically abbreviated expression of the fact that those structures are equal in a determinate respect of governing interest in the respective field of investigation.

Such abbreviated ways of speaking are of course only utilized where misunderstandings are precluded. Otherwise <101> the

Moments with respect to which equality obtains will either be expressly stated – the objects are, we say, equal with respect to size, relative position, color, etc. – or else *specific concepts of equality* are designated by special names and thereby rigorously distinguished. This is the source [*Quelle*] and purpose of the "definitions of equality" for special cases. They will prove themselves useful and necessary in science everywhere that different types of comparison cases stand in close proximity. And it is natural, where these cases frequently show up alongside one another, to seek to facilitate their differentiation by means of special terminology. Thus for example, metric geometry distinguishes concerning figures an equality pure and simple, as equality with respect to surface areas, from congruence, as equality where the corresponding parts of the figure have equal measure; and both of these again from similarity, as equality with respect to the form quantitatively determined by certain lengths or angle measures, etc.[8]

Application to the Equality of Arbitrary Multiplicities

Now let us consider multiplicities in particular. If, first of all, two groups of determinately given objects are set before us for comparison, then equality can be noticeable in several respects. This or that element on the one side is perhaps equal to this or that on the other. Perhaps there is even for *each* element in the one group a correspondingly equal one in the other, in which case, the converse may or may not hold true. Which among the many possible cases of equality is the one that causes us to claim the two groups to be equal to one another stands out immediately in accordance with the principle just formulated. In forming the representation of the group, a unitary interest distributes itself uniformly over the particular contents encompassed. <102> Thus, we

---

[8] It is a mark of the origin [*Ursprung*] of geometry in praxis that equality pure and simple has been utilized from the beginning precisely in the sense mentioned, whereas the present day science of geometry would be far removed from finding equality as such, or κατ' ἐξοχήν, in quantitative equality (i.e., that of the numbers used in measuring).

will only speak of the equality of two groups where equality in terms of all parts occurs. That is, where to any element of the one set there always corresponds an equal one in the other, and conversely. This requirement we would further explain in brief: If as is customarily done we consider the act of comparison to be a sequential process, then it must be the case that:

1) To each element from the group with which we begin, there corresponds an equal one, and indeed step by step a new one, in the other group.

2) In the second group there can be no element left over. For that would mean nothing other than that it would have parts which the first group did not possess. There would thus be no equality of the two collective wholes in terms of all their parts.

So it is only upon the fulfillment of these conditions that two multiplicities as such are to be called equal.

This does not mean, however, that the act of comparison in every case actually implies such a sequence of individual comparisons of the elements on the two sides. In the case of groups of a few elements (perhaps of two up to four) the recognition of the equality of the collections as wholes may, under certain especially favorable conditions, set in with one glance, as it were, without there being any need of sequential comparisons of the individuals. If, however, the number of elements is larger, or the nature of the remaining circumstances less favorable, then, given our limited mental powers, step-by-step operations are indispensable. Consequently, the resolution of the comparison into a series of comparisons of elements can plainly be regarded as the typical case.

We can also express our result as follows, utilizing a not inappropriate image: Two collective wholes are compared when one tries to bring them element-by-element into reciprocal *coincidence*. In fact, the procedure described is analogous to that the geometer uses to prove the congruence or incongruence of spatial formations. Indeed it is, one recognizes, the same in all cases where an analyzed whole is compared with another one of the same type. For each whole thereby appears represented in the form of a collection, namely, that of its analyzed parts. And the comparison process proceeds in such a way that <103>, part by part, one sets the two sides over against one another and tries to bring them into coincidence.[9]

Comparison of Multiplicities of One Genus

We now turn to the special case – but one of singular importance in practice – where the two comparison groups consist of
5 things of a single genus. If we know this beforehand, then the comparison process undergoes an essential simplification. We no longer have to start by investigating whether to each element in the one set there corresponds an *equal* one in the other. It suffices that, in general, to each element on the one side there corresponds
10 one on the other, and conversely. To have two elements in the representation, and to have two *equal* elements in the representation, is in this case one and the same thing. If, therefore, we grasp any element of the first group in unison with any one from the second, then the representation thus originating of a sequence
15 of pairs of elements (of the equality of which nothing further need be thought) holds as a fully valid Representative of that otherwise necessary act which would embrace just as many relations of equality. Instead of comparing pair-by-pair we therefore merely juxtapose the elements in our representation. We form a collec-
20 tion of the collections of each two corresponding elements.

Comparison of Multiplicities with Respect to their Number

Up to now we have considered the comparison of concrete groups as such. How, then, do matters stand with the comparison

---

[9] The spatial superposition in geometry, the bringing-into-coincidence in the narrowest of senses, is a procedure of the imagination, which certainly involves somewhat more than a common case of comparison. Through a gradual depletion of the Moments which are irrelevant to the comparison, the ones that bring about the spatial differentiation of the formations (their differences of position) are continuously transposed into each other to the point where they become identical. Thus there comes to *Evidenz* the equality with respect to the Moments which remain unaltered during the process. We do not adopt a similar procedure in every case because the usefulness is not so striking in all cases. As a rule we concentrate attention upon the constituents to be compared, abstracting as far as possible from the properties that differ.

of given groups with respect to their number? <104>

We obtain the abstract multiplicity form belonging to a group by diminishing each of its elements to a mere "one" and collectively grasping together the units thus originating. And we obtain the corresponding number by classifying the multiplicity form thus constructed as a two, a three, etc. If it is, now, merely a matter of confirming the *numerical equality* (or else a more or a less) and *not the number itself* (the "how many" or the "how many" more or less), then we do not first need the classification, and the act of comparison seems to simplify itself by amounting to no more than the comparison of the multiplicity forms corresponding to the given groups. But these can be regarded as groups of identical content. Two groups are, therefore, equal in number, if *their units* can be put by thought into a reciprocal, one-to-one correspondence.

One easily sees, however, that yet another and much simpler manner of comparison is possible. Instead of first rising to the general multiplicity forms, and then comparing these with one another, we also can establish equality by operating directly upon the concrete groups. Since we know from the start that each particular thing is to come into consideration only as a "one," we may think of both groups as totally consisting of identical things (identical as units). The groups will, therefore, be of equal number if the elements themselves on the two sides can be mutually correlated in the manner repeatedly indicated.

But for the confirmation of equality of number yet a third method is open to us, and one which is by far the most preferred. With the first method, upon which the second is also grounded, we had recourse merely to the confirmation of the "equally many," but avoided the determination of the numbers themselves – with the idea of economizing on mental labor. But this is a mistake. Indeed, in virtue of the ease of the familiar symbolic process of enumerating, determination of number is the resource most immediately available to us. And it is a totally mechanical process. We follow it without thinking the concepts themselves. But we nonetheless are certain that the resulting number term, when we bring its signification to consciousness, really presents the correct number concept. What is simpler than <105> comparing the two multiplicities in terms of their number by counting them both in the symbolic sense? And we obtain in this way not merely conviction of the equality (or inequality) of the numbers, but also *these numbers themselves*. That, already with groups of a relatively small number, the mechanical process of enumerating will proceed with incomparably greater speed and certainty than the seemingly so simple process of reciprocal correlation surely requires no demonstration.

The True Sense of the Equality Definition under Discussion

Our last analyses set in the clearest of lights the sense and significance of the definition of equality in question. The possibility of the reciprocal one-to-one correlation of two multiplicities *is not* the same thing as their number-equality, but rather only *guarantees* it. The knowledge that the numbers are equal absolutely does not require the knowledge of the possibility of their correlation, much less, then, that the two would be the identical. The definition contested is, therefore, far removed from providing a nominal definition that fixes the signification of the expression, "equality of two multiplicities with respect to number." All that we can grant is that it formulates a *necessary and sufficient criterion in the logical sense*, valid for all cases, for the obtaining of equality. Although it is not necessary to undertake the comparison by means of one-to-one correlation, we indeed can undertake it in all cases, and then the condition expressed in the definition must always be satisfied. In this consists, accordingly, the only serviceable sense and the achievement of the "definition."

As to the benefit of the criterion, we will hardly be able to see it as amounting to very much. Only for that mental level at which the classification of multiplicities (the differentiation and the naming of the numbers), and the mechanical enumeration procedure founded thereon, is still in a backward condition will we have to concede a significant value to it. For those primitive peoples who experience insurmountable difficulty, as is reported, in counting beyond five, the comparison of the numbers of two groups will be greatly facilitated <106>, if not first made possible, through the carrying out of the reciprocal correlation – which itself may be in turn facilitated by means of physical linkages and replaced thereby. Where, on the other hand, there stands available a procedure as rapid and secure as our symbolic enumerating, there, as already observed, the simplest criterion for the equality of the number is precisely the result, *the same number*, upon the enumeration of the groups compared.

Reciprocal Correlation and Collective Combination

Above we have thought of the reciprocal one-to-one correlation as a collective combination of pairs, combined with the understanding that in each pair each element belongs to the one multiplicity and the other element to the other. Precisely in this way the two multiplicities attain distinctiveness from each other, which grounds a certain contrast between them. With reference thereto we may speak of them "being contrasted" in our representation.
— This conceptualization stands in absolute contradiction to that of our opponents, who, with a certain justification, could invoke here the witness of experience. When we – so they will object – confirm the equality of two groups of physical objects through reciprocal correlation, we then do something more than juxtapose each thing of the one *only in thought* with each one from the other. We situate them in pairs with and upon one another, perhaps we also link them up together, etc. What in such cases brings about the reciprocal one-to-one correlation or connection is thus not mere collection, but rather a spatial, physical or other type of combination. Moreover, more closely considered, any arbitrary relation seems equally well suited to effect the correlation. If we compare, for example, the tone multiplicities (c, d, e) and (C, D, E), then we can also use the octave interval and correlate c/C, d/D, e/E. This is a manner of correlating one-to-one, and from its possibility results the number-equality of the two multiplicities. Such conceptualizations we find in *E. Schröder*[10] and *Frege*, both of whom define number-equality <107> in the

---

[10] E. Schröder, *Lehrbuch der Arithmetik*, pp. 7-8.

manner we reject. Thus for example, the latter, for his definition, requires that there be an arbitrary relation ϕ which achieves the one-to-one correlation. A stands in relation ϕ to B, and A is correlated to B, is regarded by him as the same thing.[11]

It will not be difficult for us to prove the soundness of our view in the face of these objections. We have seen that the comparison of groups as to their number, when we do not utilize the technical expedient of enumeration, is as it were the faint shadow of the comparison process which we must follow in the comparison of arbitrary concrete groups. Here, namely, the necessity falls away of still comparing separately each pair of juxtaposed elements. The equality with respect to being a "one" is there *eo ipso*. The mere "juxtaposing" within our representation stands in for the comparison still necessary in other contexts. One sees immediately that this "juxtaposing," irrespective of the opposing view mentioned, is identical with what we have called collective combining. Now it is correct that in the comparison of external things we in every case utilize yet other, e.g., physical, relations in order to establish a correlation. But one must not therefore confuse those relations with the correlation itself. The correlation as such is, under all circumstances, a collective combination. Those juxtapositions or superpositions or other types of manipulations pursue certain *concomitant* goals. They are meant to facilitate the formation of the collection consisting of pairs, and chiefly to give it a more stable configuration. When we compare a pile of apples and a pile of walnuts in order to confirm whether equally many are present on each side, or to which side the "more" accrues, then it often would be hard to avoid errors in the reciprocal correlating <108>, to not overlook individual elements, to not count others

---

[11] *Die Grundlagen der Arithmetik*, p. 83. Accordingly it thus seems to follow that there would have to be just as many types of number-equality, and therefore of number concepts also, as there are conceivable conceptually different types of the one-to-one correlational relations. I said: "of number concepts also," for, for *Frege*, the concept of number is above all to be defined by means of number-equality. The unity of the number concept rests for him only upon the indeterminacy of the relation ϕ that mediates the correlation. Once we established determinate relations of correlation, we would obtain determinate species of numbers, and therefore distinct twos, threes, etc. – a result which *Frege* certainly has not intended. For more details on his purely logical construction of the number concept, see Chapter VII to follow.

twice. But if we always place one apple and one walnut together, and the pairs thus originating again in a series or in a clearly arranged pattern, then the risk of error is enormously diminished. The distinction between an individual element and a pair is one that is palpable. We therefore survey in a single glance whether only pairs are before us or not. Each pair detaches itself from the surroundings as a unitary representation – thus saving us the mental labor of constantly holding fast the collection of pairs as such. The intuitive, relationally unitary representation, which we have produced in the way stated, is precisely of such a kind that it yields the collection intended through a very easy analysis, while it otherwise would have to be produced by the much more laborious route of successive synthesis. But it is not to be overlooked that we, in order to obtain the representation of the "equally many," must also actually undertake the analysis mentioned (or, in the case of a symbolic representation, at least strive for it) following that process. A representation must result which contains each pair segregated off to itself, and in each pair each thing segregated off to itself. It is not that external perception, but rather a collective representation to be formed on the basis of it, which guarantees the equality of the numbers. — The psychical process requisite for the formation of this collective representation can certainly be vastly simplified. For the specific character of the intuition which has been produced through those external manipulations that link up the pairs paves the way to the symbolic representing which abbreviates the process without essentially impairing the value of the result for knowledge. The intuitive unity of the two-fold group – comprehensible in one glance, perhaps in the form of a sequence of pairs – resolves itself, immediately and at will, into a multiplicity each element of which (i.e., each pair) will first be given as an unanalyzed intuitive unity. If, now, interest is directed upon any one of them, then there immediately originates the representation of a collection of two elements, of which we recognize that the one had belonged to the one group, and the other to the other group. The intuitive equality of the pairs makes it superfluous to undertake the same deliberations on each pair. Therefore the representation of the series, together with the thought that these analyses <109> could be undertaken, suffices as the symbolic place-holder for the full-fledged representation of the collection of collections.

In such a way, then, depending on the circumstances, at one time this, and at another time that type of linkage can serve as the instrument for the correlation and number comparison. But the auxiliary status of such types of linkage is certain. If one keeps this in mind, then it also becomes clear that not just any relation can, as is said, be used as correlational, but rather the collective alone, while all others can only come into consideration precisely insofar as they are suited to stand in symbolically for the collective relation. Assume ϕ, an arbitrary but determinate relation which sets the elements of two groups into a reciprocal, one-to-one relation. Then I ask: What is it that this relation is supposed to accomplish for us? It produces, one says, the desired correlation. But what it does any other possible relation between the elements of the two sides also does. But if the essence [*das Was*] of the correlating relation does not matter, then it also cannot be essential that a relation be there at all. The general idea that an arbitrary relation $\underline{x}$ can be assumed is, after all, of no benefit here. The essential thing is precisely that the corresponding elements are linked in our thought, that they are colligated. For this is of value to us as an absolutely sure sign of the fact that to every unit of the one group there always corresponds a single unit of the second. If by chance we come across a linkage of content which unifies the elements of the two sets in pairs, it is a welcome discovery, because it serves as a convenient guide to the linking up we enact in thought, or even symbolically stands in for it. But from it abstraction must again be made if the purpose of the process in general is to be realized. To introduce a relation as a technique only intending a one-to-one correlation means to forget and to miss its very purpose. For this would result in the directing of attention away from what is at issue here: namely, away from comparison as to number.

The Independence of Number-Equality from the Type of Linkage

After these explications there also can be no doubt what our attitude must be toward a related proposition <110> to which mathematicians attribute great importance. It states that the number-equality of two multiplicities is independent of the type of linkage. Or, more precisely, it states that two multiplicities which prove equal in number by the pairing up of each determinate element in the one with a single determinate element of the other, also remain equal in number when we dissolve those pairings and combine anew each element of the first multiplicity with an element – and indeed now a *different* element – of the second. "To fully explain why we are compelled by our understanding" to affirm this proposition, is seen by *E. Schröder*[12] as "a task for psychology." And *v. Helmholtz* so far agrees with him on this point as to attribute a special merit to him for recognition of this situation.[13] I think that everything psychological which can come into consideration here at all is brought to light by our analyses, and it would be tiresome to repeat in new phraseology what has been discussed in detail. As to the frequently reproduced proof of the proposition, we do not wish to contest it. Obviously it is necessary only for the case where equality with respect to number is actually defined by means of reciprocal correlation, i.e., where the two are regarded as meaning the same thing. But if one starts out from the true and authentic concept of equality, then the demand for proof implies an absurdity. We have acknowledged that the possibility of some particular – or any arbitrary – reciprocal, one-to-one correlation could serve as a logically necessary and sufficient criterion of number-equality (in the genuine sense of the word). The thought that variation in the type of linkage would lead to an uncorrelated remainder is equivalent to thinking that the two multiplicities which we are comparing fall simultaneously under the same and under different number concepts – which is absurd.

---

[12] *Op cit.*, p. 14.

[13] Apparently *v. Helmholtz* refers, in the passage here considered (*Op. cit.*, p. 19), to something else, namely to "the fact that the number of a group of objects is to be found independently of the sequential order in which one enumerates them." However (as is also occasionally the case with *Schröder*), number is here understood as the number-sign, and the enumeration of the group regarded as the successive correlation of the number-signs, in their natural sequence, to the elements of the group.

Chapter VII

DEFINITIONS OF NUMBER IN TERMS OF EQUIVALENCE

Structure of the Equivalence Theory

It is not without reason that we devoted so much attention, in
the foregoing chapter, to the elucidation of the misunderstandings
that are always linked to the definition of number-equality in
terms of reciprocal, one-to-one correlation. Those misunderstandings have in fact entailed unfortunate consequences by leading to a
total misconstrual of the concept of number itself. It will perhaps
not be inappropriate if at first, without taking into consideration
theories that have actually been advanced, we consider the following line of thought, which draws together ideas strewn here and
there into the form of a maximally coherent theory.

The definitions of the "equally many," the "more" and the
"less," as here set forth as our basis, are independent of the
concept of number. They only require that one correlate, element
for element, the groups to be compared with each other, and then
check back to see whether or not there are elements left over from
that process. Thus without counting the groups, indeed without
even having to know what counting is, one is nevertheless in
position to make a firm judgment on whether they are equinumerous or not. In this one only has to watch that the word
"equinumerous" not be understood otherwise than as the definition
specifies it. We therefore prefer to say *"equivalent"* instead of
"equinumerous," since the latter way of speaking bears in its
connotation the concept of number, while the definition is independent of that concept. If, now, we start with an arbitrary concrete group M, then we can place all other given or conceivable
groups into correspondence with it, and in this way single out the
entirety of the groups equivalent to the given group M. <112> In

this sense we will speak of the *class of groups* K belonging to the group M. Now we add an arbitrary new element to M and form the corresponding class of equivalent groups. Then we once more add a new element and form the corresponding class, and so on. The process goes on *in infinitum*, as one sees, since there is no conceivable group to which we could not add a further element. Likewise we proceed in the opposite direction: we remove some one element from M and form the corresponding class, then again an element, and so on until all of the available elements in M are removed. The classification of all conceivable groups thus achieved is the most rigorous imaginable. One group can never belong simultaneously to two different classes. By application of the definition of equivalence previously given, each given group is subsumed under one definite class, and under that one only. And conversely, each class is totally determined by any arbitrary one of the groups belonging to it. Any one of its groups can, thus, with equal right be used as the basis for the construction of the class and be regarded as Representative [*Repräsentant*] of the class. It is also obvious that out of one group the entire class originates by running the individual elements through all conceivable qualitative transformations (thus, no partitions).

The entirety of the classes is to this point given to us as an unordered totality. We easily discover a principle of ordering. It is the same one that already guided us in the successive construction of the classes. We start with any class K. The group M serves as its Representative. If we now think of some element in M as removed, then we will call the class K', of which the group M' thus originating is the Representative, the class *next lower* to K. It is easily shown that the class K' remains always the same whichever of the elements of the group M we may remove, so that K' is univocally determined. Further, if from M we form a new group M'' by adding any arbitrary thing to it, then the class K'' belonging to M'' would be called the *next higher* class relative to K. It too is a completely determinate one.

These procedures obviously suffice to order all classes <113> into a sequence in a univocal manner, where each one obtains a wholly determinate position.

DEFINITIONS OF NUMBERS IN TERMS OF EQUIVALENCE   119

And now from this point the following simple line of thought leads to the number concepts. Each class encompasses the entirety of conceivable groups of the same cardinal number. To different classes correspond different numbers. That we assign one and the same number to all groups of one class can only result from there being a characteristic which is common to all groups in that class. But what they all have in common, and what distinguishes them from all remaining conceivable groups, is surely nothing other than the circumstance that they belong precisely to the same class, i.e., that they stand in the relationship of mutual equivalence. In order to express this property for some given group M, a uniform mode of designation is required that mirrors the classes in their natural relationships, in their sequential order. A class can be univocally Represented through any one of its groups. Although it is a matter of total indifference which one we select for this purpose, we still must choose a definite one in order to attain uniform designations suited to the scientific use of language. We select the concrete groups, 11, 111, 1111, ... – originating through repetitions of the mark "1," or through repetition of the tonal complex "one" – or (in order to prevent confusion with certain composite signs of the decimal number system) the groups $1+1, 1+1+1, 1+1+1+1, \ldots$ as Representatives of the classes, and name them in sequence by 2, 3, 4, ... These groups formed out of marks are the natural numbers, in that as Representatives of the class they are also Representatives of the number concepts.

A group concretely before us is counted when the natural number that is its equivalent is sought, and thereby it is also assigned to the class to which it belongs. We often find the number corresponding to the group by "modelling" each element by means of a mark. Thus there results a group of marks equivalent to the group, and this is the natural number. The numbers form an ordered sequence corresponding to the sequence of classes.

This may amply suffice as a characterization of a singular <114> attempt to derive the concept of number originally from that of number-equality, and, by passing all psychological analyses (which are always somewhat precarious), to gain insight into the foundational concepts of arithmetic.

120   PHILOSOPHY OF ARITHMETIC

Illustrations

In order to make clear that this theory is not merely the product of phantasy run wild, we will now quote an illustrative passage from a mathematical work recently published, namely, Stolz's
5   *Allgemeinen Arithmetik*, already cited above. After laying down the definitions of "multiplicity" and of the relations of "equal," "greater" and "smaller" between multiplicities with which we are now familiar, *Stolz* gives the following explanation of the concept of number, or, as he puts it, of the "natural number"[1]:
10   "The common property of all multiplicities that are equal to a determinate one is expressed by the name of a cardinal number. We compare the multiplicities with the groups originating by continuous repetition of a mark, 1 (a *one*, a *unit*): 11, 111, . . . (only later are these signs to be introduced for the numbers eleven, one
15 hundred eleven, . . .). Anything capable of being repeatedly posited is called a **concrete unit (*benannte Einheit*)**, but only 1 is called the *unit pure and simple. The natural number is a multiplicity of units*, i.e., of ones. Every other multiplicity is called a *concrete **number**.* To each such multiplicity there corre-
20 sponds, namely, a natural number equal to it, which is to be found by selecting from that multiplicity the units belonging to it, one after the other, modelling each unit with the mark 1, and successively laying the units aside. To the multiplicities equal among themselves there correspond equal numbers, to the larger multi-
25 plicity the larger number." — "We can speak of equal natural numbers only insofar as <115> such a number, like every concept, can be thought of as posited arbitrarily often."[2]

At first it certainly may seem as if *Stolz* has defined number

---

[1] *E. Schröder* introduced this term. Cp. *op. cit.*, p. 2 and pp. 5ff. It is probably supposed to serve to mark the distinction of the cardinal numbers over against the other forms of number which come into play in arithmetic: the rational and irrational, the positive, negative and imaginary numbers. Moreover, the word "number" is not totally univocal, since it has sometimes been used to designate the concepts of numbers in series. Cp. G. Cantor, *Grundlagen einer allgemeinen Mannichfaltigkeitslehre*, Leipzig 1883, p. 5, and F. Meyer, *Elemente der Arithmetik und Algebra*, Halle 1885, p. 3. Nevertheless, we have thought it most suitable in this work to adhere to the older and almost universally customary use of language.

[2] *Vorlesungen über allgemeine Arithmetik*, p. 10. {LE}

merely as a group of "1"-marks. But the first sentence does state that by a name for a cardinal number is expressed the common property of all multiplicities equal (i.e., in our terminology "equivalent") to a determinate one. Since neither earlier nor later on do we hear the least mention of such a property, we must assume that the addition, "which are equal ⟨equivalent⟩ to a determinate one," expresses this property itself, which shows the agreement of the view with the theory developed above.³

Critique: And now let us turn to the critique. The fallacies committed in this extreme relationalistic theory stand in the most intimate of connections with the misunderstanding of the essence of one-to-one correlation and of the role it plays in the knowledge of the equality of two groups. The definition of equivalence is, as we have shown, nothing more than a mere criterion for the existence of equality of number in two groups, whereas here it is taken to be a nominal definition of it. But it is not true that "equivalence" and "equal in number" are concepts with the same content. Only this much is true, that their *extensions* are the same. If one identifies equivalence with equality of number <116>, then it is of course natural to regard equivalence as also the source of the concept of number itself, and to conclude: the entirety of the groups equinumerous to one another (i.e., equivalent, belonging to one "class") can surely have nothing else in common than number-equality defined in the manner indicated. Belonging to the class would therefore be what is essential for the number

---

³ Even *G. Cantor* himself formulates definitions in his *Grundlagen einer allgemeinen Mannichfaltigkeitslehre* (1883) which have quite the same ring to them. For example, on p. 3: "To each well-defined group . . . accrues a determinate power, the same power being ascribed to two groups if they can be reciprocally correlated one-to-one with each other, element by element." Cp. also the corresponding formulation of the definition of "number," *ibid.*, p. 5. ("Power" ["*Mächtigkeit*"] in *Cantor*'s terminology means the same as cardinal number, and "number" ["*Anzahl*"] the same as ordinal number.) Nevertheless, this mathematical genius in no way belongs to the tendency to be criticized above, as is apparent from all his later publications. Already in his letter to *Lasswitz* (dated Feb. 15, 1884, and published in the "Communications on the Theory of Transfinites," *Zeitschrift für Philosophie und philosophische Kritik*, Vol. 91, 1887, p. 13) the first definition appears in a deepened version which gives it a quite different character. And in another passage from the "Communications" (p. 55n) he very cogently says: "For the formation of the general concept 'five' there is required *only one* group . . . to which that cardinal number accrues."

concept in question. To ascribe a number to a concrete group would consequently mean nothing other than to classify it in this sense.

Of course we cannot accept this type of reasoning. What the equivalent groups have in common is not merely the "number-equality" or, more clearly expressed, the equivalence, but rather the same number in the true and authentic sense of the word.

We say: "number in the true and authentic sense of the word." For between that which we call numbers in agreement with the general use of language in life and science, and that which is to be so called according to this theory, there is, as we can easily show, absolutely nothing in common. If numbers are defined as those relational concepts grounded in equivalence, then surely every numerical assertion, instead of being directed upon the concretely present group as such, would always be directed upon its relationships to other groups. To ascribe a determinate number to that group would mean to classify it within a determinate cluster of groups equivalent to one another. But this is absolutely not the sense of a numerical assertion. Let us consider a specific case. Do we call a group of nuts lying before us "four" because of the fact that it belongs to a certain class of infinitely many groups that can be mutually put into a univocal, one-to-one correspondence? Very likely no one has ever thought of such a thing in this context, and we would be hard put to find any practical occasion whatever that would make it of interest. What does in truth interest us is the fact that here is one nut and one nut and one nut and one nut. We immediately give this awkward and complicated representation (and it all the more deserves this characterization once we come to larger groups!) a form more convenient to thinking and speaking by thinking it under mediation of the general group form one-and-one-and-one-and-one, which has the name "four." <117> The indeterminate "one" attains its determination through the species term added to the name of the number – a determination reaching exactly as far as our logical interest does. We are here interested precisely in the concrete individual as "one nut," and not in it as *this* nut with such and such characteristics. In this practice, already present in the most common employments of thought, is grounded the interest in extracting the general form of the group, or the number. But those equivalence relations of the given group to other groups, in which the theory under examination seeks the origin and sense of the concept of number, seem to us completely useless and uninteresting.

The outlook of course becomes no more favorable for this theory by having recourse to those groups of marks, 11, 111, . . ., which serve as standardized Representatives (as standard measures, so to speak) of the classes, and with the aid of which the subsumption of the group to be counted under the respective class is carried out. We cannot but find it totally absurd if these groups of marks are designated as "natural numbers," and the names "two," "three," etc., are thought of as *their* names; and not less so if the concept of unit is identified with that of a single such "1" mark. We certainly do not ascribe the number four to a group of nuts, and the number one to each individual instance of those nuts, just because this group can be "modelled" by 1111, and each individual nut by 1!

What, then, we want to ask, is the basis of the fact that we may designate all individual contents which we count – and nothing is conceivable that could not also be counted – by means of such a mark? If this designation is to have a genuine basis, then it must reside in a characteristic common to each and every one of those contents. But there is only one all-encompassing concept: that of the *something*. The mark "1" can therefore only designate of each content that it is a something, and number is accordingly something and something and . . . so forth. It may therefore seem that the slightest deliberation would have to lead from error to truth. It may seem that merely to raise the above question already suffices to put us on the right path. However, this response is just too obvious, and rings too trivial at first glance. And so, in order to avoid it, many fall into those remote and artificial constructions which, with the <118> intention of building up the elemental arithmetical concepts from their ultimate definitional properties, distort and dismiss them to such an extent that, finally, totally strange conceptual formations result, equally useless for praxis and for science.

Frege's Attempt: The cogency of the last observations are also strikingly illustrated by the often cited and ingenious book by *Frege*, which is devoted exclusively to the analysis and definition of the concept of number. In fact, he raises the question of why it is we can designate all things with the name "one," and devotes lengthy discussions to it.[4] He also occasionally brushes up against the correct answer, but then subsequently distances himself only that much further from the truth. This is the place to discuss *Frege*'s noteworthy effort, for the view which he ultimately achieves stands, if we consider its essential points, in close relation to the equivalence theory criticized above.

What *Frege* has aimed at is absolutely not a *psychological* analysis of the concept of number. It is not from such an analysis that he hopes for an illumination of the foundations of arithmetic. ". . . [P]sychology should not imagine that it could contribute anything to the grounding of arithmetic."[5] And also in other passages he is unsparing of resolute protests against the presumed incursions of psychology into our domain.[6] One already sees the direction *Frege* is taking. "However much . . . mathematics must refuse all assistance from psychology, it just as little can deny its relationship with *logic*."[7] A founding of arithmetic on a sequence of formal definitions, out of which all the theorems of that science could be deduced purely syllogistically, is *Frege*'s ideal.

Surely no extensive discussion is necessary to show why I cannot share this view, especially since all of the <119> investigations which I have carried out to this point present nothing but arguments in refutation of it. After all, one can only define what is logically composite. As soon as we come upon the ultimate, elemental concepts, all defining comes to an end. Concepts such as quality, intensity, place, time, and the like, no one can define.

---

[4] *Op. cit.*, pp. 40ff.

[5] *Op. cit.*, p. vi {LE}

[6] "One must not take the description of how a representation originates for a definition . . ." (*Ibid.*) For the "number is as little a subject of psychology, or a product of psychical processes, as is, say, the North Sea." (*Op. cit.* p. 34) In one passage *Frege* bemoans the fact that even in mathematical textbooks psychological phraseology occurs. "When one feels obliged to give a definition without being able to do it, then one wants at least to describe the route by which one arrives at the objects or concepts in question." (*Op. cit.*, p. viii)

[7] *Op. cit.*, p. iv. {LE}

And the same is true of elemental relations and the concepts grounded on them. Equality, similarity, gradation, whole and part, multiplicity and unity, and so on, are concepts that are totally incapable of a formal logical definition. What one can do in such cases consists only in pointing to the concrete phenomena from or through which the concepts are abstracted, and laying clear the nature of the abstraction process involved. One can, where it proves necessary, rigorously mark off the relevant concepts by means of various paraphrases, and thus prevent the confusion of them with related concepts. What can reasonably be required of the presentation of such a concept in language (e.g., in the exposition of a science that is based upon it) would accordingly be this: It must be well-suited to place us in the correct attitude for picking out, in inner or outer intuition, those abstract Moments themselves which are intended, and for reproducing in ourselves those psychical processes that are requisite for the formation of the concept. Such a procedure certainly will be useful and necessary only when the name designating the concept does not by itself suffice for understanding, whether because of existing equivocations, or because of some misinterpretations occasioned by the concept. We have precisely such a case before us in the number concepts, and therefore we can find absolutely nothing inherently blameworthy when mathematicians, at the apex of their system, "describe the route by which one arrives at the number concepts" instead of giving a logical definition of those concepts. It is only required that those descriptions be correct, and ones which also achieve their goal.

As for the rest, it results from our analyses, with incontestable clarity, that the concepts of multiplicity and of unity rest directly upon ultimate, elemental psychical data, and consequently belong among the concepts that are indefinable in the sense indicated. But the concept of number is so closely joined to them that also in its case one can scarcely speak of any "defining."[8] <120> The goal *Frege* sets for himself must therefore be termed chimerical. It is therefore also no wonder if his work, in spite of all ingenuity, gets lost in unfruitful super-subtleties and concludes without

---

[8] Concerning the usual definitions of number, compare Chapter VIII of this Part.

positive results. It would take us too far from our task were we to follow his expositions step by step. Here it will suffice to select and examine a few of his more important definitions. In order for them to be intelligible it must be understood beforehand that, according to *Frege*, the statement of number involves an assertion about a concept.⁹ The number accrues neither to an individual object nor to a group of objects, but rather to the concept under which the objects enumerated fall. When we judge: Jupiter has four moons, then the number four is assigned to the concept *moon of Jupiter*.

The basic conceptualization in the *Fregian* exposition agrees with that of the equivalence theory above inasmuch as he too wants to obtain the concept of number by setting out from the definition of "number-equality." The method which he forges is considered by him to be a special case of a *general logical method* that is supposed to make it possible to obtain from a familiar concept of equality the definition of what is to be considered equal. "That certainly does seem to be a very unusual type of definition, and one insufficiently attended to by logicians as of yet. But that it is not totally unheard of, a few examples may show. The judgment, 'The straight line a is parallel to the straight line b,' or in symbols:

a // b,

can be understood as an equation. When we do this we obtain the concept of *direction* and say: 'The direction of the straight line a is identical to the direction of the straight line b.' We thus replace the sign '//' by the more general '=', by distributing the specific content of the former to a and b. We split up the content in a manner different than before and obtain thereby a new concept."¹⁰

And *Frege* gives yet a second example: "The concept of shape [*Gestalt*] issues from geometrical <121> similarity in such a way that instead of saying, for example, 'The two triangles are similar,' one says, 'The two triangles are identical in shape,' or 'The shape

---

⁹ *Op. cit.* p. 59. Compare on this the discussions in Chapter IX of this Part.
¹⁰ *Op. cit.*, pp. 74-75.

of the one triangle is identical to the shape of the other'."[11]

Parallelism and geometrical similarity supply, in these examples, "the familiar equality concepts." Let us now look at how, by means of them, *Frege* intends to obtain the definition of what is to be considered equal, i.e., the definitions of the direction of a straight line and of the shape of a triangle. What follows is the result of a longer discussion:

"If the straight line a is parallel to the straight line b, then the extension of the concept 'straight line parallel to the straight line a' is equal to the extension of the concept 'straight line parallel to the straight line b'; and conversely, if the extensions of the concepts referred to are equal, then a is parallel to b. So let us try the explications:

> The direction of the straight line a is the extension of the concept 'parallel to the straight line a';
> The shape of the triangle d is the extension of the concept 'similar to the triangle d'."[12]

One now sees immediately how these ideas and definitions can be utilized with the concept of cardinal number. As direction accrues to straight lines and shape accrues to the triangle, so number accrues to concepts. We thus have to put concepts in place of the straight lines and triangles. Further, the place of parallelism and similarity is taken by the equality concept that applies here: the "number-equality" of concepts. The concept F is said to be equinumerous to the concept G, if there exists the possibility of a reciprocal one-to-one correlation of the objects falling under the one concept and those falling under the other. In this way we obtain the definition:

> "The number that accrues to the concept F is the extension of the concept 'equinumerous to the concept F',"

which, together with the foregoing definition, forms the point of

---

[11] *Op. cit.*, p. 75. {LE}
[12] *Op. cit.*, p. 79.

departure for a long sequence of further definitions and subtle analyses connected therewith.[13] <122>

I am unable to find that this method represents an enrichment of logic. Its results are of a type that can only make us wonder how anyone could even provisionally take them to be correct. In fact, what this method allows us to define are not the contents of the concepts *direction, shape* and *number*, but rather their *extensions*. Thus it yields: "The direction of the straight line a is the extension of the concept 'parallel to the straight line a'." By the extension of a concept, however, we understand the totality of the objects falling under it. The direction of the straight line a would therefore be the totality of the straight lines parallel to a. Similarly we get: "The shape of the triangle d is the extension of the concept 'similar to the triangle d'," i.e., the totality of all the triangles similar to d. And thus "the number that accrues to the concept F" is then also defined as the extension of the concept "equinumerous to the concept F." In other words: the concept of this number is the entirety of the concepts equinumerous to F, thus an entirety of infinitely many "equivalent" groups. Further commentary is surely pointless. We note, however, that all the definitions become correct statements if the concepts to be defined are replaced by their extensions. Correct, but certainly entirely obvious and useless statements as well.[14] <123>

---

[13] {*Ibid.*, LE} It may be enough to cite only a few of these here: "The expression 'n is a number' is to be the same in meaning as the expression 'There is a concept such that n is the number accruing to it'." (*Op. cit.*, p. 85) "0 is the number that accrues to the concept 'non-identical with itself'." (*Op. cit.*, p. 87)

This latter definition certainly becomes possible only in virtue of the fact that that of number-equality is given a forced interpretation, in virtue of which it also covers the case where no object at all falls under the concepts F and G. — The number 1 is defined as "the number that accrues to the concept 'equal to zero'." (*Op. cit.*, p. 90) Finally, *Frege* lays special stress upon the definition of the expression, "n immediately succeeds m in the sequence of natural numbers." This he does by the sentence: "There is a concept F, and an object x falling under it, such that the number that accrues to the concept F is n, and the number that accrues to the concept 'falling under F but not equal to x' is m." (*Op. cit.*, p. 89) To this are added proofs that upon each number n in the sequence of natural numbers one and only one number immediately follows, etc. These samples surely suffice to depict the spirit of this theory.

[14] {*Op. cit.*, pp. 79-80. LE} *Frege* himself seems to have sensed the questionable status of this definition, since he says in a note to it: "I think that simply 'concept' could be said in place of 'extension of the concept'." (*Op. cit.*, p. 80n.) Let us reflect on what that is sup-

DEFINITIONS OF NUMBERS IN TERMS OF EQUIVALENCE   129

Kerry's Attempt: Finally we want to discuss one more attempt, essentially distinct from those dealt with thus far, to illumine the concept of number by means of one-to-one correlation. I find the following passage in an article by *Kerry*: "The statement of the respect in which two contents are to be called equal to one another will in general be no easy matter. Rather, it is often attainable only through the creation of a novel concept. Imagine, for example, comparing two multiplicities consisting of two different kinds of objects, perhaps apples and strokes of a clock, with one another. Before one arrives at a judgment of equality in this case, e.g., holding the number of the apples to be equal to the number of the clock-strokes, one must have formed the concept of the respect in which equality obtains, the concept of the *numerable* [*Anzahlenmässigen*] – a concept which certainly can only be defined through the following stipulation: Suppose the objects of a (finite) multiplicity V, the 'numerable' aspect of which one wants to grasp, to be capable of reciprocal one-to-one correlation to the objects of another unvarying multiplicity V_1, and further that the objects of V are thought to undergo arbitrary variations. Then the 'numerable' aspect in V is that which must remain constant throughout all those variations, if the possibility of correlating the objects of V and V_1 is to be there *after* the variations have taken place as well as *before*."[15]

So: I promised to present a novel attempt at the definition of the concept of number, and I then here invoke the definition of a certain "numerable aspect," which in the context of my exposition is distinguished from "number" and is designated as the newly created concept. More closely considered, however, it is merely a

---

posed to mean. The phrase "number of the moons of Jupiter" would, according to that, mean the same as "equinumerous with the concept *moon of Jupiter*," or, more clearly expressed, "equinumerous with the totality of the moons of Jupiter." Clearly we obtain once again concepts of equal extension, but not of equal content. The latter concept is identical with the concept, "any group from the equivalence class determined by means of the totality of the moons of Jupiter." All of those groups in that equivalence class also fall under the number four. But that different concepts are present here needs no proof. One also recognizes that, through the modification mentioned, *Frege* turns into the path of the equivalence theory refuted above, which on the whole is a more natural theory.

[15] "Über Anschauung und ihre psychische Verarbeitung, Dritter Artikel," *Vierteljahrsschrift für wissenschaftliche Philosophie*, 11, 1887, pp. 78 & 79n.

matter of the novel creation of a name, whereas what is meant by it, and alone can be meant, is identical with that which *we* have understood, and which is generally understood, by "number" in the *abstract* sense of the word.[16] <124>

5   We immediately agree thus far: What this definition asserts is correct. The only thing which remains unchanged in a group V when we subject it to such variations, totally unrestricted except for the fact that its equivalence with the unvarying group V_1 remains undisturbed, really is its number. But unfortunately we
10 cannot concede that what this definition asserts is also of any use, that it accomplishes the least thing for us, or that it is instructive in any respect. What, we ask now more than ever, is number? *What is that in the given group V which remains unchanged through all those variations?* We wish to learn something about the content of
15 the concept of number, and we are told about its extension. The *Kerry* definition is nothing but a paraphrase of the proposition: The number of the group V is that which it has in common with all groups equinumerous to it. Indeed, we even hold this latter explication, plain and direct as it is, to be the more preferable one by
20 far, since *Kerry*'s comes very close to the erroneous idea that number is a partial content, an inner property. That which is everywhere the same in kind is, as our inquiry has shown, nothing that belongs to that content, but rather the form of psychical synthesis.
25 The merit of the preceding definition could perhaps be sought

---

[16] In *Kerry* the distinction between number and the numerable aspect [*Anzahlmässigem*] rests on the fact that he supposes the number concepts can be defined by means of the well-known series of propositions $1+1=2, 2+1=3, \ldots$, which concepts stand only in an indirect relationship to numbers in our sense. Now the idea that the carrying out of the comparison of two groups, with respect to equality, more and less, is possible without (successive) counting of each one and comparison of the numbers that result, and that consequently what is therein directly compared cannot be the number (in the sense of those definitions), is what has, it seems, led *Kerry* to the introduction of this concept of the "numerable aspect." This latter is supposed to Represent precisely the immediate respect in terms of which the comparison occurs. This line of thought is mistaken. Reciprocal correlation can only be called a "comparison" in a wholly inauthentic sense. It only *serves* the comparison. The "equivalence" of two groups is in no sense an equality of them. It is merely an indicator of their equality with respect to number in the authentic sense of the word. The question of what that immediate respect is in terms of which this quasi-comparison ensues, is thus without an object.

in the fact that it rigorously delimits the extension of the concept of number by means of a property (namely that of equivalence) which does not presuppose that concept. It is correct that it does do this. However, I am not aware of any very significant misgivings that have attached themselves to the extension of the concept of number, that would all be taken care of by means of its delimitation. By contrast, the content and origin of the concept have been the source of all the great difficulties that have made it into a cross philosophers and mathematicians have had to bear. <125> On these two points we learn nothing through that definition, and therefore it is of no value.

Concluding Remark: As I conclude these reflections I hope to have shown – partly by means of a more precise dissection of the essence of one-to-one correlation and its true function in the comparison of number, and partly by means of refutation of the erroneous theories founded on it – that the concept of equivalence does nothing and can do nothing for the definition or analysis of the concept of number. Above all I have thought a more thorough inquiry into the relationship between the two concepts to be required. This is mainly because at the present time there is a widespread general tendency to assign to the concept of equivalence an exaggerated importance in the respect indicated, and because at every blink of an eye there appear new attempts to clarify or even define the content of the concept of number by means of that of equivalence.[17]

---

[17] Since this chapter was written there has again appeared a series of new attempts that follow the tendency characterized above. Above all G. Heymans, *Die Gesetze und Elemente des wissenschaftlichen Denkens*, first edition, Leipzig 1890, §36, pp.146ff., is worthy of mention here. See in particular top of page 150: ". . . The concept of number-equality is historically and logically prior to the concept of number." Heymans' view, as a more detailed analysis would show, falls precisely under our "equivalence theory." — I may also cite here the work *Was sind und was sollen die Zahlen?* (1888), by the illustrious arithmetician *Dedekind*. It advances similar ideas, if not on every point, at least on the essentials. "If one carefully observes what we do in the enumerating of a group or number of things, one is led to consideration of the mind's capacity to relate things to things, to make one thing correspond to another thing, or to model one thing upon another . . . . On this sole . . . foundation . . . the entire science of numbers must be erected." (*Op. cit.*, p. VIII) The primitive number concept for *Dedekind* is, however, the ordinal number or "natural number." (p. 21) The *cardinal number* of a group is, for him, that ordinal number

which has the characteristic that the whole set of the numbers that are lower than it is equivalent to the group in question. — As much as I admire the inner formal coherence of the developments in the theory of this remarkable mathematician, it still seems to me to deviate far from the truth in its bizarre artificiality.

Chapter VIII

DISCUSSIONS CONCERNING UNITY AND MULTIPLICITY

In the last chapters we have settled all essential questions bearing upon the understanding of the psychological origination and content of the concepts multiplicity and unity, of the determinate number concepts, as well as of the concepts equal, more and less. Our next task will be to confirm the insight won by resolving the difficulties that have been discovered in these concepts, and that seem to involve us in inextricable contradictions and subtleties.

The Definition of Number as a Multiplicity of Units.
One as an Abstract, Positive Partial Content.
One as Mere Sign.

We take as our starting point the ancient definition: The number is a multiplicity of units. Back of it lies hidden, in very many authors, the gross misunderstanding which holds number to be a *specific* type of group of objects that are similar to one another. In the same way, namely, as there are groups of apples, of stones, and so forth, there would also be groups of units. According to this view the units are thought of either as concrete contents, by holding oneself to the mere names or written symbols, or (and this is the usual case) as abstract, positive partial contents, which can be selectively isolated and colligated to form groups.

As pre-eminent Representative of the latter view we quote *Locke*. He states: "Amongst all the ideas we have, as there is none suggested to the mind by more ways, so there is none more simple than that of unity, or one. It has no shadow of variety or composition in it; every object our senses are employed <127> about, every idea in our understanding, every thought of our minds, brings this idea along with it . . . ."¹ "By repeating this idea in our minds, and adding the repetitions together, we come by the complex ideas of the modes of it" ⟨the numbers⟩.² We can here dispense with a critique of this obviously mistaken theory. Still, the view of the *unit as a non-relational partial content* is a very crude one, and it already offered *Locke*'s not inconsiderable critics, *Leibniz* and *Berkeley*, occasion for comments in opposition to it. *Berkeley* sets forth repeatedly and in detail the relational nature of the number concepts, and seeks to use it as a point of leverage for his nominalistic views. Also, the *Locke*an view of the unit and number as primary qualities which also have subsistence in things outside our mind is vigorously contested by him. But while he argues for the most part rigorously and correctly in his critique, he himself comes close to holding a view whose error we will immediately bring to light, a view which tends to explain number, not as a general concept, but rather as a mere general sign.³ Approximately at the same time *Leibniz* also emphasized the relational character of the number concepts. Thus, for example, his Theophil in the *Nouveaux Essais* says: "It may be that *dozen* and *score* are merely relations and exist only with respect to the understanding. The units are separate and the understanding takes them together, however scattered they may be."⁴ He writes concisely and emphatically to Father *Des Bosses* in 1706: "Numeri unitates, fractiones, naturam habent relationum."⁵

Some authors, we said, have considered *the unit* or *the one* as a *mere sign*, which is conferred upon each of the enumerated objects. And thus the number appears to them to be a quite

---

[1] *Essay*, Book II, ch. 16, sect. 1.
[2] *Ibid.*, sect. 2. {LE}
[3] G. Berkeley, *A Treatise Concerning the Principles of Human Knowledge*. The works of George Berkeley, collected by A. C. Fraser, Vol. I, Oxford 1871, sect. 12, 13, 118-121 (where *Locke* is criticized without being mentioned by name).
[4] Book II, ch. 12, § 3, *Opera philosophica*, Erdmann, p. 238. (I have quoted the translation of the *New Essays* by Peter Remnant and Jonathan Bennett, New York: Cambridge University Press, 1989, p. 145. {DW})
[5] *Opera philosophica*, Erdmann, p. 435. In the year 1684 he still ranked number among the concepts that are common to several senses. Cp. *Erdmann*, p. 79.

specific concrete group of nothing but ones. Misinterpretation of the most primitive symbolic procedures of enumeration has <128> occasioned this error. Since, conforming to the goals of enumeration, attention must not be fixed upon the peculiar properties of the individual objects, the abstractive process has been facilitated by the following simple expedient: each object was replaced by a sign that is uniform and as plain as is possible, e.g., by a mere stroke 1. Thus there arose, as stand-ins for the multiplicities to be enumerated, groups of mere 1-strokes. But then these groups immediately turned out to be very well suited to stand in for the concepts of numbers themselves, as signs of them, presupposing that these same means of designation are used – an obvious move, and one which actually occurs. This modelling [*Abbildung*] of each one of the things to be enumerated by means of a uniform sign, meaningless in itself, in fact only mirrors the process that leads from the concrete multiplicity to the number. And *only in so far* as it does so has it sense and significance. All enumerating (which certainly has become a machine-like procedure in consequence of incessant usage) would be completely devoid of sense if the symbol "1" or the word "one" did not possess the signification corresponding to the *concept* 'one', i.e., if it did not point to the abstraction process which shrinks each of the individual determinate objects of the group to be enumerated down to the mere something or one. For only in this way do we, instead of getting an empty heap of strokes or words, attain to the *concept* under which falls the concretely present multiplicity as multiplicity, and which alone is what we really have in view. But in restricting oneself to enumerating merely as a machine-like exterior process, the logical content of thought which confers on it justification and value for our mental life was totally overlooked.

Even if we disregard the reprehensible errors supported by the definition of number mentioned, we find it to be of little advantage. Multiplicity and unity are correlatives. Consequently, it in no way provides a characterization of the nature of the multiplicity which we call "number" to say that it consists of units. But if perhaps attention is to be called to the fact that under "number" not a concrete multiplicity is to be understood, but rather one which is thought under mediation of the abstract concept of multiplicity, i.e., one whose individual elements <129> are to be attended to merely insofar as they are units, then it would surely be best explicitly to say just that in the definition. But even then the definition would still lack complete clarity, since it would fail to take into account the distinction between multiplicity in the broader and multiplicity in the narrower sense (the latter of which is based on a classification of the forms of multiplicity). In this regard the formulation by *Hobbes* is somewhat clearer: "Number is 1 and 1, or 1, 1 and 1, and so on,"[6] if only "one" is understood in the correct sense, the one made precise by us – from which *Hobbes*, with his extremely nominalistic point of view, certainly was very far removed. For the rest, with definitions of this type very little is accomplished. The difficulty lies in the phenomena, in their correct description, analysis and interpretation. It is only with reference to the phenomena that insight into the essence of the number concepts is to be won.

One and Zero as Numbers

We have in essence objected to the preceding definition only in terms of its misleading character, occasioned by that of the concepts of unit and multiplicity. Much more serious are the objections other authors have raised against it. Some find it simply inadequate, others erroneous from the ground up. The critique of those objections serves us, in turn, as the external occasion for clearing up more important misunderstandings that are documented in them.

The objection of inadequacy is supported as follows: The definition is obviously applicable only to the numbers of the number sequence from two onward. *Zero* and *one* seem to be excluded from the concept of number according to it.[7] "Let no one object

---

[6] *De corpore*, VII, 7, cited from J. J. Baumann, *Die Lehren von Raum, Zeit und Mathematik*, Vol. I, p. 274.

[7] Cp. G. Frege, *Die Grundlagen der Arithmetik*, p. 38. The sentences following in the text are taken from the same work, p. 57, where they stand, however, in a different context of thought. But since our concern is precisely with the Ideas [*Ideen*] expressed there, we insert them here.

that 0 and 1 are not numbers in the same sense as 2 and 3! Number answers the question 'How many?' And when one for example asks 'How many moons does this planet have?' then one is prepared to get the answer 0 and 1 just as well as 2 or 3, <130>
without thus transforming the sense of the question. To be sure, there is something peculiar about the number 0, and likewise about 1. But in the last analysis that is true of each whole number. It is just that with the larger ones it is less and less evident. It is totally arbitrary to draw here a generic distinction. What is not proper for 0 and 1 cannot be essential to the concept of number."

This objection does not merely concern the definition by *Euclid*, but also our own analyses of the concept of number. We have never spoken here of zero and one as numbers, and it is also clear that all of our analyses are absolutely inapplicable to these concepts. The only remaining way out is flatly to deny that zero and one belong among the number concepts, and that seems to be blocked by the above line of argument as well as by the customary usage of language in general.

But let us consider the matter somewhat more closely. Without any further explanation, and almost as a nominal definition of number, the following proposition is advanced: "Number answers the question 'How many?'." Or, to select a mode of expression which more exactly indicates the sense of the proposition: "Number is *any possible* answer to the question 'How many?'."[8] Let us reflect on the sense of this proposition. "How many?" is the question about the closer determination of a *many*. The "many" in this case obviously does not signify the contrast with a "few," but rather it simply expresses the (authentic or symbolic) representation of a collection (a totality, a multiplicity) of objects. The question is then directed upon the closer determination of the collection as a "two" or a "three," and so forth. According to our earlier investigations the numbers are to be understood as the entirety of conceivable determinations of the indeterminate multiplicity concept. Each possible way of determining this concept by delimiting it yields a novel number concept. The definition, "Number answers the question 'How many?'" appears thus to

---

[8] Cp. the remark by J. F. Herbart, *Psychologie als Wissenschaft*, Part II, top page 162.

harmonize completely with the results of our investigations.

In reality, however, this is only the case if it is correctly understood. Not every possible answer to the question "How many?", but rather every possible *positive* answer to that question, is what leads to numbers. The situation is the same here as with many other analogous definitions. For example, each answer to the question "Where?" <131> is called a determination of place. Each answer to the question "When?" is a determination of time. In these cases, too, negative answers are precluded through the sense of the definition. Nowhere at all and never are not particular cases of place or of time. Certainly it is true that these negative answers function grammatically just like the positive ones do. Therefore the grammarian also has no occasion to distinguish adverbs which determine place and time in the literal sense from those which do it non-literally. But conceptually there is an essential difference. A similar point obviously holds for the negative answers to the question "How many?" "No-many," or no multiplicity, is not a special case of *many*. *One* object is not a collectivity of objects. Therefore the assertion that there is one thing here is no assertion of number. And likewise *no* object is not a collectivity, and therefore the assertion that there is no thing here is no assertion of number. One and none – these are the two possible *negative* answers to the "How many?" Linguistically, on the other hand, they function just like numbers, and therefore the grammarian is at liberty to regard them as numerical determinations. But logically they are not that.

Accordingly one must pay close attention to the fact that the designation of zero and one as "numbers" presents a *transference* of that term to concepts of a different kind, even though they stand in close relationship with the literal numbers. For the rest, we absolutely do not mean that merely linguistic, and not also scientific motives of the greatest importance, speak in favor of this transference. It would lead to the most burdensome of intricacies in number theory – indeed, to breakdowns – if one wanted constantly to hold the literal numbers and the one and zero separate from each other, and rejected a common designation of them all (which is why the term "number" automatically presented itself in the so very obvious transference). Zero and one are

possible, and often enough actual, results of arithmetical problems. An algebraic calculus that aims to accommodate all conceivable cases of calculation can therefore make no distinction between zero and one, on the one hand, and the remaining numbers on the other. The a̲, b̲, c̲, . . . x̲, y̲, z̲ of general arithmetic must be signs which can be specified by 0 and 1 just as well as by 2, 3, 4, ...
<132>

The introduction of the 1, and more especially of the 0 (still less obvious), as numbers on a par with the 2, 3, 4, . . . cannot be prized highly enough with respect to their mathematical significance. It first made possible an arithmetical algorithm, i.e., a system of formal rules by means of which problems concerning numbers can be solved in purely mechanical operations – i.e., from known numbers and number relations unknown ones can be discovered. Certainly the decimal number system, which is the basis of the common technique of calculation (*arithmetica numerosa*), concerning itself with determinate, given numbers, would be unthinkable without this momentous expansion of the concept of number.

Now as to the intrinsic grounds of this arrangement, it resides in the homogeneity of the relations that link up all of the numbers of the broadened domain with one another. The relations of more and less obtain not merely between the literal numbers, but also between them and the one and the zero. a̲ is by a − 1 more than 1, this latter by a − 1 less than a̲. If the positing of a number a̲ is taken as a collective addition to nothing, then one can even say: a̲ is by a̲ more than 0, 0 by a̲ less than a̲. This, then, is what the integration of the one and the zero into the number sequence is based upon. If we arrange the numbers in the "natural" sequence, i.e., in such a way that each subsequent one arises from the preceding one by the collective addition of one unit, then 1 + 1 is the first number, inasmuch as it has no predecessor. But since 1 + 1 arises out of 1 in the same manner as 1 + 1 + 1 does out of 1 + 1, the 1 naturally fits in as a member of this sequence. Whether or not we call 1 a number, it belongs to this sequence of concepts. And the zero too can be adjoined to it, by taking the positing of a 1 as collective addition of the 1 to nothing. The same thing is accomplished in a still easier manner by the inverse process of subtraction. By means of the same step which leads from 3 to 2 and from 2 to 1 we get from 1 to 0.

The way in which the zero and the one belong to the series of numbers, in terms of the elemental relations and operations, makes it understandable why in the solving of problems by calculation (as soon as rules of calculation for the literal numbers had been found), not merely any number whatever, but also the zero and the one could turn up as a term in the operation or as a result. <133> But this circumstance had to lead to the broadening of the domain of numbers and to the modification of the concept of number bound up with it – an advance in arithmetic which of course did not have to come about in the form of purely logical reflections and definitions, but which declared itself instead through the introduction of the symbols "0" and "1" and their consistent utilization in calculations. If, now, one considers the fact that a uniform, rule-governed operation is only possible if every conceivable result of an operation can be formally dealt with in the same manner, then it becomes clear how this broadening of the domain of calculation really had to constitute a significant step forward in the direction toward an arithmetic.[9] For the rest, even arithmetic could of course not completely obliterate the essential conceptual distinction of the newly adjoined numbers over against the original ones. Their marginal character shows itself clearly in the exceptions they imply for most types of calculation: the addition of zero does not increase, the subtraction of it does not diminish, division involving it leads to a result without sense, likewise with raising zero to the zero power, and so on. Multiplication by one does not proliferate, division by one does not split up, and so on. These are peculiarities which are evidently of a wholly different type than those belonging to the genuine number species; for they infringe against the generality of propositions that govern all the rest of the domain of numbers (precisely that of the true numbers).

So we nevertheless may henceforth speak, and with good

---

[9] As is well known, the use of 1 in calculation, which was fairly obvious, already belongs to the pre-scientific period of arithmetic, whereas for the introduction of the 0, presupposing a relatively highly developed arithmetical understanding, the wisdom of the Hindus is to be thanked. As M. Cantor, *Vorlesungen über Geschichte der Mathematik*, Vol. I, Leipzig 1880, p. 159, observes, the Pythagoreans still held the one not to be a number.

reasons, of one and zero as numbers. We may also regard every number, zero and one excluded, as the sum of identical numbers "one." But the factually present conceptual difference must not be lost sight of. The unity of the concept is for the actual numbers – that is, those which are determinations of multiplicity – an intrinsic one. They form a logical genus in the narrower sense. The unity of the concept of number *after* its expansion is, by contrast, an extrinsic one established by means of certain relations. According to this latter concept number is any possible result of a calculation, any <134> conceivable term in the sequence of numbers, any possible answer to the question "How many?" It is clear that in each of these formulations the concept of number in the narrower signification is presupposed.

So I believe that the distinction in kind between the two types of concepts is by no means an arbitrary one, and that what *Frege* formulated in the words quoted above – "What does not apply to 0 and 1 cannot be essential to the concept of number"[10] – must therefore be branded as a false principle.

The Concept of the Unit and the Concept of the Number One

To the analysis of the concept of the number one we attach an observation which, although in itself almost self-evident, is not superfluous, as the following will show. The concept of the *number one* is, namely, to be well distinguished from the concept of the *unit* or of the *one*, of which we constantly spoke in our earlier investigations. "One" as a possible answer to the question "How many?" does not coincide in concept with one as correlative to multiplicity. Unit [*Einheit*] in *contrast* to multiplicity is not the same as unit *in* the multiplicity. Along with the concept of the multiplicity (or number) the concept of the unit is inseparably given. But in no way is this true of the concept of the number one. The latter is only a later result of technical developments. From the practical point of view these distinctions certainly are inconsequential, since to each unit in the number the number one also

---

[10] *Op. cit.*, p. 57. {LE}

accrues. The proposition, "The number is a multiplicity of units," remains true whether we take the term "unit" in the one sense or in the other.

Even if such distinctions are a matter of indifference to the arithmetician – and justly so – they nevertheless must not be such for the logician. In that the former, guided by certain considerations, includes the one under the term "number," he also modifies its concept, because those considerations influence the content of that concept. Thus "one" becomes an equivocal term. By overlooking this circumstance one easily falls into error. In fact, here lies the source of a mistaken line of argumentation that is occasionally encountered. Number, it is said, cannot have originated <135> by means of adjoining one to one. For not only is the one itself a number, but rather we even need the larger numbers for the formation of its concept. "The larger numbers do not originate out of the one, but rather just the reverse, the one originates out of the plurality."[11] It is also for this reason that *Herbart* and many who follow him reject as presumably false the common view which thinks of the number *as made up of units*. We thus would at the same time have here an argument against the old definition of number, one which holds that definition to be a ὕστερον πρότερον.

The argument would be correct if "one" always and everywhere designated the *number* one. But this is absolutely not the case. If we say that number originates by means of adjoining one to one or that it is made up of units – then one and unit signify merely the correlative to multiplicity. But from this perspective neither is the multiplicity prior to the unit nor the unit to the multiplicity. The two originate simultaneously.

It is moreover to be observed that it somewhat strains the language when one, with *Herbart, Volkmann*[12] and others, takes the term "unit" in the sense of the number one. We do not speak of the number "unit." To the question of how many apples there are we do not get the answer "unit," but rather "one," or "their number is one." That is also why the expression, "multiplicity of

---

[11] J. F. Herbart, *Op. cit.*, Part II, p. 162.

[12] W. Volkmann, *Lehrbuch der Psychologie*, Vol. II, 3rd Edition., Cöthen 1885, p. 114.

units," generally does not signify the same thing as "multiplicity of numbers *one*." To identify the two is to attach to the term "unit," in addition to the many equivocations which it possesses anyway, a new one from which it is still free in common linguistic usage.

Further Distinctions Concerning One and Unit

We have already repeatedly had occasion to see how great the temptation is in our domain to be led into error through synonyms and equivocations. The following considerations will provide new examples of this. It will often be <136> necessary for us to lose ourselves in what seem to be linguistic investigations concerning the meanings of terms in order to put an end to obscurities and misinterpretations of the concepts that interest us.

We find ourselves in such a position also with the question that is to employ us now: namely, that about the relationship between the concepts or terms *one* and *unit*. Some philosophers lay weight upon the difference between them, without, however, any agreement on the sense of it. *Leibniz* thought of *unit* as the abstractum of *one*. "Abstractum ... ab uno est unitas," he says. Nonetheless he still uses, in the continuation of the same sentence, the plural "unitates," whereas an abstractum surely cannot take the plural.[13] Moreover, this plural form is common elsewhere. "The number is a multiplicity of units," "Three units plus five units make eight units," etc. On the other hand one speaks, again, of the *concept* one, and accordingly "one" is used as name for the same concept as is designated by "unit." The plural "ones" [*Einse*] is also in use – which here could seem objectionable for the same reasons – and it then signifies the same as "units."

In spite of this confusion in linguistic usage, the conceptual distinction which *Leibniz* wants to express by contrasting *one* and *unit* remains valid. A more general reflection will also offer illumination of the case at hand. Namely, this equivocation of the terms "one" and "unit" (which became obvious through the

---

[13] *De arte combinatoria, Opera philosophica,* ed. J. E. Erdmann, p. 8.

use of the plural form) is a phenomenon that occurs, in a wholly analogous manner, in all abstract terms.

Each abstract term is used with a twofold signification. In the one it serves as name for the abstract concept as such; in the other, as name for any object falling under that concept, it designates the concrete thing under mediation of the abstract element contained in it or related to it. Language frequently operates with abstract names, but utilizes them as a designation of concrete things and processes. This becomes possible in virtue of the fact that it <137> utilizes abstract names as *general* terms, and then, through the combination of several such terms reciprocally limiting each other in their generality, singles out the concrete thing.

Thus, for example, color can signify in its own right the logical part which is common to red, blue, etc. But if we speak of "colors," of "this color" and "that color," etc., then color is a general term for every particular color species as such. In order more clearly to signal this way of using the term, instead of "color" we say "a color," "a certain color," and so on, whereas, where the abstractum is intended, we emphasize: the abstract *concept* color (or for short, the concept *color*). With the plural, of course, the necessity for such supplementations falls away, since then, *eo ipso*, only the conceptual objects can be intended. "Colors" cannot mean anything other than "certain colors."

What we have illustrated with this example holds true in complete generality. Corresponding to practical needs, most terms usually function as general terms. Interest chiefly turns toward concrete things and relationships. However, life and science certainly are not lacking in occasions to consider the abstract as such, and accordingly speech also has made provision for the clearly marked designation of it, in which the general term is used and is simply modified by means of determining expressions, by the attachment of syncategorematic signs, etc. In particular, endings such as "-hood" (Father – Fatherhood) or "-ity" (human – humanity) also were employed as the designation of the abstract *characteristics* corresponding to general representations. Of course the respective abstract concept was already formed as the general term originated. But it was the object of interest only insofar as these and those objects might contain it as a property.

The interest in the abstractum in its own right therefore very surely can have arisen only later and have given occasion for a special denomination of it.

Our exposition could evoke a certain doubt: It set out from abstract terms and showed how they are used as general terms in which, for the signalling of this use, the linguistic expression of the abstract term is often modified. It then moved on to general terms and <138> showed how, out of these, by means of certain linguistic modifications, abstract terms are formed. With this it was moreover emphasized that the general terms, as standing closest to the commonplace goals of thinking and speaking, might well belong to an earlier developmental level. If, thus, the abstract terms are first to originate from the general terms, how in turn are the general terms to be formed out of the abstract ones?

That would certainly be an obvious circle if the last mentioned second level were only a mere inversion of the first. That is, of course, not our view. After certain abstract terms were formed out of the originally given general ones, then out of those abstract terms there is not perhaps to be formed, in turn, the *same* general terms, but rather precisely different ones. The adjective "red" functions as a general term for all red things. From that there originates the abstract name redness (the red), which then in turn serves as general term for the various specific differences of red: crimson, vermillion, etc.

But the above-mentioned linguistic provisions for the perspicuous differentiation of the abstract terms which are to be formed from general terms (e.g., kind – kindness, friend – friendship) proved not to be sufficient. Following the predominant tendency of our thought, these forms, expressly fashioned to be names for abstracta, were nevertheless used in turn as general terms, and so there resulted again names that are all the more equivocal.[14] One spoke, for example, of kindnesses and friendships in the plural.

---

[14] After these discussions it may seem as if the division of names into abstract and general, upon which, e.g., J. St. Mill (*Logic*, Book I, ch. II, § 4) places such great weight, is one that is useless because it cannot be implemented. If, however, we define an abstract name as a name of an abstract concept, but a general name as a name of the respective conceptual objects, then in each case it is immediately decidable from the sense of the context whether a name is being used as an abstract or general one, in spite of its equivocal character.

This is also the situation, then, with the names "one" and "unit." Originally "one" was in any case a concrete-general term, grounded on the concept named "unit." But when the name "unit" was more and more misused as a general term, it might have happened that some, feeling the need for a designation of the abstractum, would precisely reach for the expression "one" running parallel with it, so that it too became equivocal. <139>

It is with this in mind that we also want to explicate the distinctions between the terms "multiplicity," "plurality," "group," "totality," "aggregate," "assemblage," and so on. One part of these names (such as the three latter ones) is almost exclusively used in the distributive manner. When one utters them concrete phenomena are always in view. The abstract concept they are based on is not the object of special interest. It serves merely as mediating thought, as general sign for the particulars falling under it. It is otherwise with the names "number," "multiplicity" and "plurality," grounded in precisely the same concept. Indeed, they too are more commonly used as general names, but besides that as names for the concept itself, to which they in part are predestined by their grammatical form. With the name "number" matters are complicated still further by the fact that it serves not only as a general name for any concrete group whatever, but also as a general name for each of the specific numbers, two, three, four, ... falling under the concept of number.[15]

All of these different ways of using the same names must here, as elsewhere, be kept clearly separate from one another. Then there disappear on their own many of the pseudo-difficulties that have been discovered in our concepts.

Sameness and Distinctness of the Units

Much more serious are the difficulties associated with the question about the manner in which representations of sameness and

---

[15] This was one consideration that led us, in cases where we remained within the domain of concrete phenomena, to prefer the names "totality" or "group," and, where we passed over to general concepts, to choose the names "multiplicity," etc. (Compare also the "Note" on p. 100 above.)

difference are involved in the formation of the concepts of unit and number, and about the extent to which they explicitly enter into the content of those concepts.

The *sameness* of the units was already strongly emphasized in earlier times. Thus *Hobbes* says: "Number, in the unqualified sense, presupposes in mathematics units which are the same among themselves, out of which it is produced."[16] We see that here <140> the sameness of the units is presented as a *special presupposition* of mathematics. And the same viewpoint is affirmed by many others, e.g., by *J. St. Mill, Jevons, Delboeuf, Kroman*, etc. Those investigators who have emphasized the sameness of the units form, in general, two groups. The one group (like those just mentioned) advances the sameness of the units as a prerequisite or presupposition. The others affirm it without making a requirement of it. *Locke* belongs among the latter. Since, according to him, *every* content, whatever it may be, always carries with it the simple Idea [*Idee*] of the unit, and number originates through repetition of that simple Idea, he of course does not need first to require the sameness of the units. As to the investigators of the former group, they either suppose, with *Hobbes, Mill* and the others, that the sameness of the units is a necessary hypothesis merely for the purposes of calculating and for the application of arithmetic, or else they suppose that the objects enumerated must, as such, be the same as one another in some respect in order to be countable at all.

The latter is also *Herbart*'s view. "First of all one recalls that in enumerating there is always something that is enumerated, and that the representation of this something must always remain of the same kind, in that, as is well known, things not homogeneous – e.g., pens, sheets of paper, sticks of sealing wax – cannot be enumerated together unless they are conceived of as being of the same kind (under the general concept of writing materials). Now every number refers in some such way to a general concept of the enumerated. But this concept can remain wholly indeterminate, in that for determination of number it is wholly irrelevant *what* is

---

[16] See J. J. Baumann, *Die Lehren von Raum, Zeit und Mathematik*, Vol. I, p. 242.

counted."[17]

However, other philosophers found this to be so little "well known" and self evident that they, rather, maintained the exact opposite. So *Leibniz*, when he emphasizes that number "is, as it were, a non-corporeal figure originated through the unification of things of any kind whatsoever, e.g., God, an angel, <141> a man, and a motion, which together are four." (Cp. p. 17 above.) And also *Jevons*, who even defends the thesis that number is only another name for distinctness.[18]

Frege devotes lengthy discussions, in his *Grundlagen*, to the question about the circumstances under which sameness and distinctness contribute to the concept of number. "Accordingly we stand," as he sums up his result, "before the following difficulty: If we wish to allow number to originate through the grasping together of different objects, then what we get is a mere heap, in which the objects are contained with precisely those characteristics that distinguish them from one another. And that is no number. On the other hand, if we wish to form the number through a grasping together of that which is the same, then all constantly blends together into one, and we never arrive at a plurality. If we designate each of the objects to be enumerated with a 1, that is a mistake, because distinct things receive the same sign. But if we endow the 1 with distinguishing strokes, then it becomes unusable for the purposes of arithmetic."[19]

How are these difficulties resolved according to our theory?

As concerns the role played by representations of difference in the abstraction of the concept of number, we have exhaustively characterized it in our earlier investigations (pp. 49-64). Obviously it is only what is distinct that can be combined into a totality. But in the representation of the totality nothing of difference relations is present. The elements of the totality are there in our representation simply as what they are, and do not become

---

[17] *Op. cit.*, p. 161. In the same sense, Fr. Ueberweg, *System der Logik*, 5th Edition., Bonn 1882, p. 129, explains: "The numeralia can be understood only on the basis of concept formation, for they presuppose the subsumption of homogeneous objects under the relevant concept." (Cp. also *Ibid.*, p. 141)

[18] Cp. the quotation on p. 51 above.

[19] *Op. cit.*, p. 50.

such elements by our first distinguishing them. There is not first required a special activity of distinguishing in order that they not blend into one. Only if we are dealing with highly similar contents do acts of distinguishing come into play in order to ward off the danger of confusion. That is, we then do attend to the properties in which distinctiveness consists. As to the concept of *number*, it originates from totalities in such a manner that it too, fundamentally considered, requires no special acts of differentiation. The enumerating, i.e., the sequential process through which we ascertain the number of the group, in general requires only that the objects to be enumerated are distinct <142>, but not acts differentiating them. Exceptions arise only with the enumeration of contents that are easily confused. We must in such cases watch out for omissions and for counting over again.

It is more difficult to provide a psychologically correct characterization of the role allocated to *sameness relations* in the representation of number.

Whatever view concerning the origination of the number concept may be held, the sameness of the units is a fact not to be denied. The apparent self-contradiction of many investigators will not mislead us concerning what, in the last analysis, would have to be their true opinion. Where the sameness of the units is denied, there the sameness of the objects to be enumerated is intended; and then the question becomes how a thoroughgoing sameness of the units in the number is to be compatible with the distinctness – indeed, possibly distinctiveness so great as to make comparison impossible – of the enumerated objects which those investigators had in mind. Certainly many contend, in opposition to this – such as *Herbart*, for example (compare the quotations above) – that things not homogeneous do not admit of being enumerated together. Different things, if they are to be enumerable, must always first be brought under a shared generic concept.

An easily drawn distinction will show that in a certain sense both parties are right. If we accept as content of a representation only partial representations in the authentic and rigorous sense (the "intrinsic properties," as some logicians called them), and accordingly consider representations as comparable only when they possess in common partial contents of this kind, then there are infinitely many disparate and non-comparable representations; and it is clear that then enumeration does not require comparability (in this sense), since, to the contrary, totally disparate things can be enumerated together. My soul and a triangle are two, although they have not a single intrinsic property in common. If, however, we accept as content of a representation also all of the negative and relative determinations (the "extrinsic" properties) that accrue to it, then there are in general no non-comparable representations, for there are none which are not of the same kind, at least as falling under the concept *something*. And it is precisely this subsumption under the concept *something* that we must (according <143> to our theory) carry out, with reference to each of the objects to be enumerated, in order to grasp the number of them. Insofar, it is therefore correct to say that the things to be enumerated have to be brought under a shared "generic concept" (this phrase taken, to be sure, in the most 'extrinsic' of senses).

However, the view of those philosophers who ground number in sameness surely goes much further than we can allow. According to our perspective, the representation of the number of a determinate group does not arise through our comparing the objects of that group with one another and subsuming them under the generic concept emerging from that comparison (horse, apple, tone, pencil), but rather through our bringing them – whatever it is we may be counting – *always under the same* concept, that of the "something," and simultaneously grasping those objects collectively together, objects thought under mediation of that concept and designated as same in terms of it. Thus there originates the general form of multiplicity – one-and-one-and . . . one – under which the multiplicity concretely before us falls, i.e., under the number belonging to it. According to the widely held viewpoint under discussion, each enumeration would require preliminary or simultaneous comparisons, and relations of sameness would essentially enter into the concept of number. According to our viewpoint neither one of these is the case. That abstraction (or better: reflexion) which we must bring to bear upon the members of a group in order to arrive at the number *eo ipso brings about* as consequence the sameness of the units. But neither does it itself have anything to do with comparisons, nor do the relations of sameness between the units necessarily enter into the representation of the number as explicit constituents. They are so far removed from forming *essential* psychological factors in the representation of the number that in many cases they do not come to our attention at all.

Since on this point I must stand in contradiction to the majority of investigators, and since it concerns a rather subtle issue, I will pursue the critique in greater detail.

In what way, I first inquire, is the sameness of the objects to be enumerated, with respect to some generic concept or other, to contribute to the emergence of the abstraction of number? Now we are counting apples. In that case *apple* is the generic concept. After that <144> we count horses. Then *horse* is the generic concept. Therefore the particular generic concept, which is always present, cannot be at issue. We can, consequently, grasp the situation simply in this way: Numbers originate through abstraction from groups, the members of which are represented as in some respect the same among themselves. "Groups of things the same among themselves" – thought *in abstracto* – these are "numbers."

Our earlier investigations make available to us everything we need to lay bare the psychological foundations of the abstraction process that comes into consideration here. One immediately sees that the founding relationship which makes the relational complexes here present into unitary representations, into wholes, cannot be that of sameness, but rather only the *collective combination*. Two apples: that does not mean an apple *the same as* another apple, but rather an apple *and* an apple, and so also in general. Not $1=1$ is 2, but rather 1 *and* 1 is 2, etc.

But if this is conceded, then it is obvious that the abstraction of the general concept which covers only groups of objects that are *the same among themselves* presents us with exactly the same process, except for one modification, which also supplied us with the general concept under which groups of totally *arbitrary* objects fall. All of our expositions would still be valid, provided only that at all appropriate points a certain addendum were inserted. In fact, if we set out from concrete groups of contents that are the same among themselves, from what would we have to abstract and what would we retain in order for the general concept sought to result? One sees again that nothing can be retained from the contents but the extrinsic property *that* they are contents. So the empty concept "something" or "a thing" must also mediate here.

But besides this concept and the concept of the collective combination, which *by themselves alone*, on our view, make up the concept of number, there would still have to be added here the restricting concept of sameness in kind. Not "something and something" or "a thing and a thing" would be the linguistic explication of the concept *two*, but rather "a thing and a thing of *the same kind*." Likewise, "three" would mean the same as "a thing and a thing and a thing – which are of *the same kind*," etc. Clearly it would be easy for our theory to adapt itself to the requirements here posed, <145> whereas, on the other hand, the viewpoint which we contest, if it is to be at all possible and consistently put into play, would have to entirely follow the route which we have sketched out in the foregoing investigations. This opposing theory, modified in such a way and consistently developed, would come to stand in the most intimate of kinships with our own theory. It would differ from our theory only through those restrictive addenda, by virtue of which sameness of the units in kind would be added on to our formation of the concepts as a special requirement.

All well considered, I nevertheless cannot concede the validity of those addenda. The basic thought to which they give expression – namely that only such contents are supposed to be enumerable together as are represented to be the same in some respect – is certainly incorrect. To the question, "How many are Jupiter, a contradiction, and an angel?" we immediately answer: "Three." But does it occur to us first to mull over whether or not these contents stand in some relation of sameness? Do we first call to mind that Jupiter, an angel, and a contradiction are only the same as one another insofar as each one is a "something"? Or in enumerating do we make comparisons at each step? I can observe nothing of all this. We simply count: Jupiter is one, and a contradiction is one, which yields one and one; and the angel is one, which gives one and one and one – thus three.

Of course the units are "the same" as each other. But these samenesses of theirs are a *consequence* of number abstraction, not its basis and presupposition. They arise, not through a preliminary comparison, but rather through that absolute depletion of content which number abstraction requires under all circumstances, even where contents that are compared and represented as the same are enumerated.

Certainly an apple and an apple are two apples because each is an apple. But why are they two in general? Not because the one representation content is the same with reference to the other as an apple, or in any other determinate respect, but rather because each of them is a *one* or a *something*. Two apples, two men, an apple and a man, etc., are in every case two because they Represent concrete totalities which, through the process of enumeration, fall directly under the abstract group form one-and-one, and under no other. <146>

I therefore believe that representations of sameness contribute just as little to number abstraction as do representations of distinctness. *Comparing* and *distinguishing*, *colligating* (the unification of concrete contents into totalities) as well as *enumerating* (the abstraction of the general forms of totalities) are well distinguished mental operations that must be held apart from each other. Of course there are plenty of occasions where all these operations come into play together. Thus, for example, when we count out the gold pieces from a pile of various kinds of coins.

As to those complex forms of sameness which, according to the viewpoint here contested, would be the number concepts themselves, I certainly do not deny that they Represent quite legitimate conceptual constructions, of which we also make use often enough. I only deny that they are identical with the number concepts in their *full generality*. It is because of this that they also do not have special names – names that would be superfluous after the formation and naming of the true concepts of number. To abstractly express that sequence of concepts it is sufficient simply to add on the attribute of sameness in kind. And we do indeed proceed in just that way in saying: two of the same things, three of the same, four of the same, and so forth. In this way the numbers, as the most general forms of totalities, really do constitute an important instrument for the type of thinking that combines things and brings these its combinations to linguistic expression. Thus, for example, we speak of a whole with three parts, one of which possess the characteristic X, a second the characteristic Y, and so on. First there arises the empty totality form "something and something and something." And then individualizing occurs, and the empty form is filled out with concrete content.

An entirely different question is whether or not those restricted conceptual constructions were the historically and psychologically earlier ones; whether or not, therefore, numbers first originated by means of abstraction from groups of objects the same among themselves. As a matter of fact, children do learn to count with homogeneous – and specifically even with physical – things, such as balls, cards, and so on. And it is certain that the most obvious and by far most common occasions for the forming of the number concepts involve groups of things the same among themselves. For our practice of judging, as also for our emotional life, the most essential of interests attach to them. <147> The same held no less, then, for that lower level in the development of humanity, where the formation of the first and most elemental abstractions and the origination of language occurred. Since things of the same kind were designated by the same names, one was forced – in order to name totalities of the same kind of things, and thus to establish them for linguistic interchange as well as for one's own thinking – to form, depending on the circumstances, larger or smaller totalities of the repeated names themselves, thus: of A and A, A and A and A, A and A and A and A, and so on – an all too cumbersome mode of expression. In order to avoid it, the obvious course was to fix the forms of multiplicity with the aid of an indeterminate concept and name suited to every content (something, a thing, one) and to introduce special names for those forms ("two" for: one thing and one thing, and so on). The immense abbreviation of expression is obvious. Instead of A and A and A and A one could now say "Four A's." That is, one combined the sign for one and one and one and one with the common name of what here formed the content of the one (or something). In this way the governing interest was satisfied, for the number and the shared generic concept express everything that we as a rule are concerned about in a group.[20]

By conceding all of this we in no way fall into contradiction with our principles. The limitation with which the number concepts originally emerge within the life of a people, as well as in that of individuals, demonstrates just as little here as with other concepts the scientific or even the merely practical justification of that limitation.

We would still like to devote a few words, now, to the difficulties which *Frege*, in the passage cited above, has found in the fact that not only sameness but also distinctness is ascribed to the units. "If we wish to form the number . . . through a grasping together of that which is the same, then all constantly blends together into one, and we never arrive at plurality."[21] <148> Here sameness is confused with identity. Every intuitive representation of a group of objects the same in kind demonstrates *ad oculos* that sameness and distinctness stand in no contradiction at all and very well can be given within one unifying act of thought. In a certain respect it is precisely sameness that occurs, in another it is distinctness. And depending on the circumstances attention can be focused now predominantly on determinations that are the same, and now on ones that are different. Only if the expression "grasping together of the same," whereby one intends to describe the origination of the number, required an absolute sameness – as *Frege* wrongly assumes – would a difficulty be involved here, or better, an impossibility.

"If we designate each of the objects to be enumerated with a 1, that is a mistake, because distinct things receive the same sign." However, we commit this "mistake" in every application of general names. When we call James, Charles, etc., each "a man," this is the same case as the "mistaken inscription" in virtue of which we, in enumerating, write down a "1" for each of the objects to be

---

[20] That is also why the general terms which are based on the concepts of multiplicity and number (e.g., "group") for the most part incidentally connote the sameness of the objects grasped together. Still, this is least true in the case of the name "totality," which in this respect also possesses a certain advantage.

[21] *Op. cit.*, p. 50. {LE}

enumerated. "1" is precisely the general written sign which has its foundation in the concept of the unit.

In yet another respect (as we mentioned on pp. 147-148) the sameness of the units is emphasized by many: namely, as a *presupposition of arithmetic. J. St. Mill* and *Delboeuf* express this thought with a peculiar decisiveness: "In all propositions concerning numbers," the former explains, "a condition is implied, without which none of them would be true; and that condition is an assumption which may be false. The condition is that 1 = 1; that all the numbers are numbers of the same or of equal units . . . . How can we know that one pound and one pound make two pounds, if one of the pounds may be troy and the other avoirdupois?"[22] And *Delboeuf* decrees: <149> "Equality of the units, such is the fundamental assumption of arithmetic."[23]

Little will be required to refute this erroneous view. To call the proposition 1 = 1 a presupposition of arithmetic is to totally misconceive the sense of arithmetic. Arithmetic as theory of numbers has nothing to do with concrete objects, but rather with numbers in general. It is quite correct that the usual *applications* of arithmetic have reference to the numerical relationships between multiplicities of the same kind, a case to which the relationships of commensurable quantities are reduced by way of measurement. Given such an application there certainly obtains the presupposition that, in the direct or indirect enumeration involved, only those objects are counted together, and consequently are taken as units in calculation, which actually possess the respective Moment of sameness, and precisely in the manner required. But this certainly is not a presupposition of arithmetic, but rather of the concrete problem toward the solution of which arithmetic is to serve us. Arithmetic first comes into action once all is expressed in numbers. Where numbers come from, to what problems they find application and

---

[22] J. St. Mill, *A System of Logic*, Book II, ch. VI, § 3. Compare also W. St. Jevons, *The Principles of Science*, p. 159; and, *further*, the citations from *Kroman* following below, from *Unsere Naturerkenntnis*, Copenhagen 1883. — This view is, moreover, also widely held among mathematicians.

[23] *Logique Algorithmique*, Liège/Brussels 1877, p. 33. (Also appearing in Vol. I of the *Revue Philosophique*.) *Delboeuf* also stands close to *Mill* in other respects. A few lines further we read: "Number is the scientific expression of the *sensible* idea of plurality."

Further Misunderstandings

under what presuppositions – all that has nothing to do with arithmetic.

How far misunderstandings concerning the essence of the units could go is reflected most strikingly in the peculiar requirements which frequently have been imposed on those concepts and which nearly always are closely tied to the misguided requirement of sameness of kind. As an example we here quote a passage from *Kroman*'s work, *Unsere Naturerkenntnis*. "It is," says the author, "in no way our intention to deny that number is an abstraction from reality. To the contrary, we regard it as sufficiently certain that <150> through consideration of diverse groups of natural objects of the same kind man has gradually formed his number representations and built up the sequence of natural numbers. But precisely the abstraction from all else in these groups except the number of their parts, the transformation of the parts into units that are *completely the same in kind and magnitude*, invariant, absolutely independent of time and space, warmth and coldness, etc. – just this makes the number into a self-created object, an object of the imagination. Just as there certainly is no perfectly straight line, no perfect circle, etc., so also there certainly are no units completely the same in kind and magnitude, and in any case we would never be able to experience them. However, the arithmetical units have these characteristics by definition, by the decree of mathematicians."[24]

This view, erroneous from the ground up, has its source partly in an inadequate psychology, and partly in the fact that *Kroman*, in writing the sentences quoted, was thinking only of geometric and physical *applications* of numbers. It is the same error that we had to censure in *Mill* above, and which in general corrupts the logic of arithmetic in the extreme Empiricist school. Therefore we also must totally reject the far-reaching epistemological consequences that are bound up with that view. Since number is supposedly a "self-created object," "an object of the imagination," which

---

[24] K. Kroman, *Unsere Naturerkenntnis*, pp. 104-105. *Kroman* is substantially influenced by the views of *A. Lange*. (Cp. Chapter II of the present work.)

presents only a "crude approximation" to reality, the entire certainty of arithmetic also resides in its crude approximation – a circumstance which, according to *Kroman*, is at the same time to guarantee the general validity of its theorems.

The sameness of the units, as results from our psychological theory, is obviously an absolute sameness. In fact, the mere thought of an approximation is already absurd. For it is a matter of the sameness of contents with respect to the fact *that* they are contents. To deny this sameness therefore is to deny the Evidence [*Evidenz*] of inner perception.

Having just now heard stressed the invariability of the units, their independence <151> from space and time, warmth and coldness, etc., we will no longer be much surprised by additional requirements which have been laid down by others. Thus in ancient and modern times a special weight was often laid upon the *strict self-containedness*, the undividedness or *indivisibility* of the units.[25] Of historical figures we mention here only *Locke*[26], *Hume*[27], and *Herbart*. A few sentences from *Baumann* may serve as a convenient basis for the critical illumination of misunderstandings that reign here. "What we intend to posit as a point or as no longer divided, that we regard as one. But each 'one' of external intuition, of the pure and of the empirical, we can also regard as a 'many'. Each representation is one when delimited over against another representation, but in itself it can in turn be differentiated into a many." — "*Counting* and *numbers* are thus no concepts drawn from external things, because of the fact that external things present us with no strict units. They present only bounded-off groups or sensible points, but we have the freedom of considering these themselves in turn as a "many." Sometimes we even find in the inherent nature of the given units grounds for not allowing them to stand as units, and sometimes those external units compel us not to actually differentiate them further into a "many," even though it would be mathematically possible. This

---

[25] Cp. Frege, *op. cit.*, pp. 41ff.

[26] Cp. the quotations on pp. 133-134 of the present work.

[27] *A Treatise on Human Nature*, Part II, sect. II. In the Green and Grose edition, Vol. I, pp. 337-338.

independence from our representations, this force of things opposing our representations, is at the same time a proof of the reality of those given units . . . ."[28] The errors upon which the foregoing assertions rest are grounded partly in the confusion of the concepts *unit, simplicity* and what it is to be a *point* in the strict sense – a confusion that has shown up in philosophy with some frequency – and partly in certain equivocations that infect the name "unit." That this name is excessively burdened with equivocations is something we have already had frequently to observe. It will be useful here to gather them all together for once, although we need only a few of them for the purposes of the present discussion. <152> One easily sees that it is not a question of wholly accidental equivocations, but rather of equivocations through transference.

Equivocations of the Name "Unit"

1. The name "unit" [*Einheit*] refers in the first instance to the *abstract concept of the unit*. The concept of the unit stands in the relation of correlative to the concept of the multiplicity. But this latter is nothing other than the concept of the collective whole. Thus the concept of the unit is nothing other than the concept *collective part*.

2. The name "unit" also signifies any *object insofar as it falls under the concept of the unit*. This type of equivocation is nothing peculiar to the name "unit." It shares it with all abstract names, insofar as they also are used as general names. So each member of a concrete multiplicity, in case number abstraction is exercised upon it, is accordingly a unit or one. We can also say that "unit" in this sense signifies: enumerated (or, by transference: to be enumerated) object as such, whereby is primarily to be understood actual [*wirkliches*] and not symbolic enumerating.

3. Each unit in the multiplicity is also a *one in the sense of the number one*: to each unit the number one accrues. Since, then, in place of the name "unit" (in the signification #2 here considered) the name "one" also can be used, there arose an equivocation of

---

[28] J. J. Baumann, *Die Lehren von Raum, Zeit und Mathematik*, Vol. II, pp. 669-670.

the latter name, which led to the confusing of the two concepts, and then even to using the name "unit" in the other admissible signification of "one," for "one" in the sense of the number. So for example *Herbart*, as we saw on p. 142, used the name "unit" for the number *one*, and thus confused two well distinguished concepts.

4. Since as a rule only objects of the same kind are counted, the generic concept common to the objects enumerated is also, to be sure, called "unit." Where we are dealing with a continuum, relations between stretches of it (physical parts, intervals, etc.) could be reduced to number relations by carrying out a division into equal parts. Then the common measure taken as a basis (i.e., the generic concept covering parts of the same kind) <153> is said to be the *unit of measure*, or simply *the unit*. Here too one easily perceives the basis of the transference. If, for example, pounds are counted, then a pound is the unit. In fact, if the interest is directed exclusively upon pounds and their enumeration, then the concepts "object to be counted as such" and "pound" have the same extension, and so we carry the name of the first concept ⟨"unit"⟩ over to the second. In this case "unit" indicates what is supposed to be counted as one. But it only becomes an actual unit (in the sense of #2) provided that one actually counts it.[29] The equivocation that originates through this transference has always contributed to most of the errors.

5. In *higher analysis* one speaks of units in several senses which have nothing directly to do with number and only stand in a very remote relationship to it. There one speaks, for example, of numerous types of "imaginary units." These are certain irreducible elements of calculation by means of which other expressions, that are thought as built up from them, are to be formed similarly to the way the number signs are formed from the sign 1, i.e., through formulas such as $1 + 1 = 2$, $2 + 1 = 1 + 1 + 1 = 3$, and so forth. The function of these 'units' is multiple. Often they merely serve as a technical expedient for the perfecting of the symbolic technique of

---

[29] For easily understandable reasons mathematicians tend to explain the unit precisely in this sense. Cp. P. du Bois-Reymond, *Die allgemeine Functionentheorie*, Vol. I, pp. 48 and 49, where the confusion with the unit in the sense of #2 clearly stands out. Cp. also Frege, *Op. cit.*, p. 66.

arithmetic, in which case there corresponds to them no conceptual content [*Gehalt*] whatsoever, apart from the technical one. But in the arithmetic of determinate material domains there also frequently corresponds to them such a "real signification." A similar point holds for the remaining types of units utilized by the arithmeticians, such as the negative, fractional, etc. These are to be treated in detail in Volume II.[30]

6. Arithmeticians use the symbol "1" as sign for a unit (in sense #2). The method of mechanical enumerating and calculating, which does not hold the underlying concepts before the mind, has led to the overlooking of those concepts and to considering, as with the numbers generally, so also the *unit, as a mere sign*. "Unit" is therefore defined as the sign (the <154> mark) "1," by which in enumerating each thing is "modelled." The oldest representative of this error is *Berkeley*. We find it again in more recent mathematicians such as *Stolz*,[31] v. *Helmholtz*,[32] and others.

7. "Unit" furthermore signifies much the same thing as *whole*. The occasion for this transference is obvious. Certainly nothing stands in the way of counting together unconnected and disparate contents, or isolated attributes, relative determinations, and so forth; but we have occasion for that only in very exceptional cases. As a rule we count *things* in the narrow sense – generally, composite wholes, which, through an especially intimate bonding of their parts, stand out with exceptional ease from the surroundings and draw our interest to themselves as wholes. Whatever, then, stands out as a whole in virtue of inner coherence and sharp delimitation, whatever imposes itself upon our interest and thereby becomes the main object of enumeration, that gets called a "one." But through a further transference "unit" finally came to designate the same thing as "whole."[33] For example we say that the state forms a unit.

---

[30] {This projected volume of *The Philosophy of Arithmetic* was never completed as such. But see Essay III. DW}

[31] Cp. the quotation on p. 120.

[32] Cp. the Appendix to Part I, below.

[33] Also the name "unification" and the verb "unify" (to combine into a whole) have arisen in this way.

8. Finally, "Unity" [*Einheit*] is also used in the sense of "wholeness" or "unifiedness" (one must forgive the forced word formations). For this concept we have, in general, no other current name than "unity." Here, obviously, it is a matter of a reiterated transference. In this sense we speak, for example, of the unity of the soul as one of its characteristic properties.

With the aid of the seventh and eighth significations of the name "*Einheit*" ⟨"unit"/"unity"⟩ we now can indicate the pathway by which one may come to assert the strict punctuality of the unit. Whatever is unified is one. But unification admits of degrees of completeness. The more intimate it is, the more complete it is. But the ideal of unification is indivisibility, and the ideal of indivisibility is the mathematical point. — Consequently, punctuality accrues to "strict" unity.

Against *Baumann* I further insist that even if the external world consisted exclusively of discrete mathematical <155> points, the numbers would be no more and no less abstracted from it than now, where that world is constituted in a totally different way. The mental process of number abstraction would remain exactly the same.

The Arbitrary Character of the Distinction between Unit and Multiplicity. The Multiplicity Regarded as One Multiplicity, as One Enumerated Unit, as One Whole.

According to *Baumann* the fact that the same content appears to us now as 'one' and now as 'many' is grounded precisely in that imperfection of external experience which presents us with no "strict unities." But would we not, even in the fictive ideal case, have to lay claim to the capability of detaching groups of points as new units and of enumerating them again in turn? Certainly such a group would then also appear, at will, now as a unit and now as a multiplicity.

But how is this remarkable fact itself to be explained? Already *Berkeley* emphasized it, and drew from it a trenchant argument against the realist conception of the concepts *unit* and *number* as primary qualities. ("We say one book, one page, one line; all these are equally units, though some contain several of the others. And in each instance it is plain that the unit relates to some particular combination of ideas *arbitrarily* put together by the mind."[34]) Does not a serious difficulty reside in this arbitrariness of conceptualization, which seems to obliterate any distinction between one and many?[35]

In order to get clear on this situation, let us consider it just a bit more closely. It is a fact that we are often in a position to conceive of one and the same object as one and as many, as we may wish. It is further a fact that we can enumerate multiplicities, and by doing so obtain numbers whose units are themselves in turn numbers; and on this all of arithmetic is based. It is also immediately clear that the second fact is a consequence of the first: a consequence precisely of that arbitrariness of conceptualization noted. Hence, whereas one might, on the one hand, expect that the arbitrariness would render any <156> arithmetic impossible, it turns out, to the contrary, that all arithmetic is based upon it.

However, this difficulty is mere appearance. In actual enumeration it is never doubtful or arbitrary what is to be counted as one. This is settled by the guiding interest, which also settles *how many* things we count together. Without arbitrariness in this sense there would be no number at all. If it is multiplicities that are enumerated together, then these multiplicities are the units. The contradiction lies only in the words. One must clearly distinguish the enumerations which yield those multiplicities – and only with respect to them are they multiplicities – from the enumeration which then in turn combines the multiplicities into one multiplicity. And in relationship to this new enumeration they are not multiplicities, but rather units.

We add one further comment. There is an essential distinction between multiplicity pure and simple, i.e., multiplicity thought as multiplicity, and multiplicity thought as unit. This distinction can be easily traced back to its psychological origin.

---

[34] *Principles of Human Knowledge*, section 12, emphasis added. Cp. also *New Theory of Vision*, section 109, Fraser ed., Vol. I, p. 25.
[35] Cp. also Frege, *Op. cit.*, p. 58.

What it is to conceive of a multiplicity as such we need not discuss all over again. Let us therefore immediately pay attention to that second way of taking a multiplicity, in virtue of which it is thought as unit.

A multiplicity (whether *in concreto* or *in abstracto*) is a content of representation just as well as any other content. It can therefore be colligated and enumerated with arbitrary other contents. Every (authentic) formation of number requires, with reference to each particular object, a reflexion on the psychical act representing that object, through which reflexion the object is thought as unit. Therefore the multiplicity, insofar as it functions as an object to be enumerated, is also thought as unit, since it is considered with reflexion upon the fact that it is a content, a "something."

We can speak of "multiplicity as unit" in a yet wholly different sense, namely in the sense of "multiplicity as whole." Here the elements of the multiplicity (or the units in the number) are thought as partial representations in the psychical act which has the multiplicity for its intentional object. The interest rests upon the unifiedness [*Geeinigtsein*] of the elements or units in the representation of the multiplicity or number. But the unification <157> comes about, as we have ascertained, only in the psychical act of interest and perception which picks out and combines the particular contents, and also can only be perceived in reflexion upon that act. Objectively considered, every multiplicity possesses unity in this sense. It is a whole. But not always is a special interest directed upon it. It is not always thought *as* a whole. It therefore would be misleading to say: "Every multiplicity is not merely a multiplicity, but rather a multiplicity thought as a unity [*Einheit*] . . . ."[36] The two things must, instead, be kept clearly distinct.

Herbartian Arguments

Misunderstandings of the relationship between the concepts *unit* and *multiplicity* have been the source of innumerable false lines of

---

[36] Chr. Sigwart, *Logik*, Vol. II, p. 42.

argument in philosophy. Unit and multiplicity are set in opposition to one another. Thus, one concludes, *the same thing* cannot be simultaneously one and many. This inference is correct only if one says, instead of "the same thing," "the same thing in the same respect." If we count soldiers, then a regiment is a multiplicity. But if we count regiments, then a regiment is a unit. This is no contradiction because this multiplicity and this unit are not in opposition, for they belong, as terms, in *different* relations. Only whoever would assert that a regiment is a multiplicity of regiments, or a soldier a multiplicity of soldiers, would affirm a contradiction. Things stand here as with correlative concepts generally. Only terms in the same relation mutually exclude each other.

There is sometimes associated with this misunderstanding yet another, which comes from understanding the unit also as a whole, and, nevertheless, conceiving of multiplicity as the corresponding *correlativum*.

It is upon these errors, as well as upon the confusion of the concepts *unity* and *simplicity*, that the celebrated Herbartian argument which seeks to detect a contradiction in the concept of *one thing* with many characteristics runs aground.

To the question of what the thing is we respond with a *list* <158> of its characteristics, thus with a *collectivum*. If we hold strictly to the letter of the language our response would be, according to Herbart, nonsensical, because the issue was not about a many which could merely come together in a sum, but which could merge into no unity.[37]

Already at this point we are unable to agree. We cannot find anything nonsensical in the wording, and the nonsense *Herbart* interprets into it merely marks a conceptual confusion on his own part. "*Collectivum*" in the broader sense signifies the same thing as "multiplicity," and in the narrower sense the same as multiplicity of discrete things, of individuals. By listing the characteristics we responded with a *collectivum* in the first sense, for which *Herbart* substitutes without justification the second. The linguistic expression renders the correct conceptualization without any am-

---

[37] *Lehrbuch zur Einleitung in die Philosophie*, Hamburg and Leipzig 1883, § 118.

biguity. By saying that the orange is red and round the adjectival form clearly expresses the distinction of the non-independent [*unselbständigen*] properties over against the thing. Such a sentence has a wholly different linguistic character than, for example, the following: "The commission consists of Williams, Jones and Smith." But here, where a *collectivum* in the narrower sense is meant, the wording indicates precisely that.

The conceptual confusion in virtue of which *Herbart* was forced to a false interpretation of the language also is the basis of his further argument. The answer to the question "What is a thing?" loses its apparently nonsensical character, he supposes, when we more accurately take it to be: The thing is the *possessor* of the characteristics a, b, c, etc. But then the contradiction emerges again when we consider that the possessing surely is likewise something just as manifold as the characteristics would be. The *simple* question, "What is the thing?" requires a *simple* response: one which tells us *of what*, then, it is genuinely said that it has the characteristics and *unifies* them. "If, now, we cannot reduce the manifold possessing of the many characteristics back to a simple concept which could be thought without any differentiation of the several properties, then the concept of the thing to which we, after all, must attribute this manifold 'possessing' as <159> its true quality – because we came to know it through many properties – is a contradictory concept . . . ."[38]

But to what extent is the question "What is the thing?" a simple one that requires a simple response? From the outset *Herbart* attributes simplicity to unity, and then it is no wonder when the existing multiplicity of the characteristics ⟨in a thing⟩ contradicts its unity. In order to learn what that is which makes the parts of the thing *one*, what *has* them, we do not need a simple answer, but simply the correct answer: the unification of the parts, the whole. Of each whole it is valid to assert that its parts accrue to it and that it unifies the entirety of the parts, and this is therefore also valid of that whole which we call a thing. And so the "manifold possessing" is then to be assigned to the thing as its true quality in no other sense than it is to any other type of whole. The point of

---

[38] *Op. cit.*, p. 186.

difference lies merely in the manner of the possessing, and this is also what brings us to speak in a special sense of the *unity* of the thing. What we understand thereby is nothing other than precisely this: that the characteristics in the concept of the thing are not raked together in the manner of a mere *collectivum*, but rather constitute a whole of parts bound up with one another in their content (and reciprocally interpenetrating one another). The assertion that a thing in this sense Represents [*repräsentiere*] a *unity* (namely, a "metaphysical whole"), and the assertion that it possesses a *multiplicity* of characteristics (of "metaphysical parts"), stand so little in contradiction with each other that they, rather, reciprocally require one another. Whole and part are precisely correlatives. One must not allow oneself to be deceived by the equivocation of the word "unity" which easily leads to regarding unity and multiplicity simply as mutually exclusive opposites. Unity in the sense of number (which always accompanies simplicity) and unity in the above sense of intrinsic unification [*Verbundenheit*] are well distinguished concepts.

If, now, one seeks for the source of all of these confusions, one finds that they arise from that confounding of the two concepts of collection spoken of above. If the concept of multiplicity is identified with that of a *collectivum* of individuals (which "could merely come together in a sum, but <160> could not merge into a unity"), then certainly the multiplicity of the characteristics in the unity of the thing appears to be an impossibility, for this unity is precisely more than such a collective unity. Forced in this way into confounding the concept of the unity of the thing with the concept of the simplicity of the thing, one will necessarily feel it to be contradictory that the concept of the thing does not permit itself to be grasped as a simple concept, to be thought "without any differentiation of the several properties."

Chapter IX

THE SENSE OF THE STATEMENT OF NUMBER

Contradictory Views

We now move on to the debate concerning the true subject of the statement of number. It is indicative of conceptual confusion when discord can prevail on *this* point, and one can hardly believe how far the opinions of philosophers deviate from one another here. *J. St. Mill* explains: "The numbers are, in the strictest of senses, names of objects. 'Two' is certainly a name of the things which *are* two: two balls, two fingers, and so on."[1]

Herbart thinks quite otherwise: "Each number then," he says, "refers . . . to a general concept of the enumerated, ... If one in thought joins to the number 12 the general concept of a chair or of a dollar, one will find that the number determination attaches itself undividedly and immediately to the concept."[2] And most recently this viewpoint has found a zealous advocate in *Frege*. "The statement of number," he concisely and clearly says, contains "an assertion about a concept."[3] Akin to this understanding is that of

---

[1] James Mill, *Analysis*, Vol. II, p. 92n. In the wording of the text (p. 91), James Mill says exactly the opposite of what his son says in the note: "Numbers . . . are *not* names of objects. They are names of a certain process; the process of addition; . . ." The opposition lies, however, only in the words, and is based upon the contrasting signification with which the two Mills use the terms "note" and "connote." In explaining a few lines later that the names co-designate ("connote") the things enumerated, *James Mill* expresses – in a very misleading way, to be sure – the same view as his son: namely, that the numbers are asserted of things. (Moreover, the two also agree in a crudely externalistic conceptualization of the numbers. The process of addition is placed by James Mill (*Ibid.*) on a level with walking and writing.)

[2] *Psychologie als Wissenschaft*, Part 2, p. 161. Cp. also the passages cited on p. 148 above from the same work as well as from Ueberweg's *Logik*.

[3] Frege, *op. cit.*, p. 59.

Schuppe. According to him <162> number is based, as earlier discussed, upon comparisons and differentiations of the contents enumerated. But the subjects of number predication are, as he supposes, "not the things compared and found to be distinct," but rather "that which is the same in all the things distinguished." "We have to conceive of . . . the substantive under the attributive number determination, and of the subject under a predicative one, as something identical that is appearing or being thought of with respect to several places or several times – whether it concerns an element in what is appearing or the species of such an element, or even if it is the concept of a thing, e.g., of the general traits characteristic of a dog or a cat."[4]

Other investigators, again, interpret the number assertion in another way. They consider the numbers as predicates of groups. *J. St. Mill,* not true to himself, also repeatedly expresses himself in this sense. Only a few lines after the previously cited passage it is said: "Numerals . . . denote the *actual collections* of things."[5] Finally, it also occurs that two, three, four . . . are regarded as predicates of number in general. If, for example, we speak of three apples, then three would not be a determination of the apples, but rather of the *number* of the apples. The expression, "Of apples there are three," would be *identical* in signification with the more accurate one, "The number of the apples is three." *Sigwart* can be mentioned as a representative of this point of view. "When ⟨three⟩ is the predicate," he asks, "is it truly a predicate of the things of which it is asserted, and not rather a *predicate of their number* . . . ?"[6]

Refutation, and the Position Taken

Which of these views are we to favor?

---

[4] Schuppe, *Erkenntnistheoretische Logik,* p. 410.

[5] [*Analysis,* Vol. II, p. 93n {LE}] Cp. e.g., also *Logik,* Book III, chap. xxiv, § 5 [*Gesammelte Werke,* Vol. III, p. 342]: "When we call a collection of objects two, three or four, . . ."

[6] Chr. Sigwart, *Logik,* Vol. I, p. 168n.

The view that the number is a predicate of the enumerated things requires no lengthy examination. Two certainly is *not* a "name of the things which are two," for otherwise each of them would, precisely, be two. That is why we also do not say: "*the* things (balls, fingers, etc.) are two," as we do say: "They are colored, heavy, <163> etc.," but say instead, "*Of the* things there are two." That the numbers are not attributes of things is also shown in other aspects of linguistic expression: There are no numerical adjectives, as there are adjectives for all types of attributes.

Herbart's view requires and is worthy of closer examination. According to it the *statement of number refers to a concept*. *Herbart* himself has not carried through with the detailed justification of this view. So we will turn to arguments by *Frege*, who has made every effort to develop *Herbart*'s view more fully and to provide a positive as well as a critical demonstration of it.

The following is his main argument: If we consider objects as bearers of number, then it seems as if different numbers accrued to the same object. This changes immediately when we, rightly, posit the concept as the true bearer. "When, considering the same external appearance, I can say with equal truth: 'This is one group of trees' and 'These are five trees,' or 'Here are four squadrons' and 'Here are 500 men,' what is thereby changed is neither the individual nor the whole, the aggregate, but rather my nomenclature. But that is only a sign of the replacement of one concept by another. That makes it . . . clear that the statement of number involves an assertion about a concept."[7]

The argument is based upon observations that are correct. But they do not prove what is here to be proven. It is correct that numbers do not attach to any objects as their properties, and insofar the objects are not their bearers. But they nevertheless are their bearers in another and more justified sense. The number owes its origination to a certain psychical process which attaches itself to the objects enumerated, and in this sense is "borne" by them. If one focuses upon these bearers and attends to the types of abstraction processes which are here present, then the

---

[7] Cp. G. Frege, *op. cit.*, p. 59. {LE}

difficulty brought out above also disappears. The number is univocally determined once the totality upon which we exercise that abstraction process is determined. But the objects do not, by themselves alone, determine the totality. The same objects <164> can be represented in different forms of totality. Instead of thinking them all collectively combined without any weighting of preference, we can, depending on the direction of our interest, pick out now this group and now that by itself, and thus form totalities of totalities in more or less diversified ways. We now can, further, enumerate these partial totalities, and then the final numerical result is a sum of numbers. But we also can regard each of these partial totalities as a unit, and then we obtain a different numerical result again, and so on. Hence, depending on the direction of the interest which guides the formation of the totalities, there result different numbers or combinations of numbers, each one of which is univocally determined, however, by the totality form presupposed in its enumeration. Now with this shift of interest is also connected the *shift of the concepts* under which we segregate the objects as to groupings and enumerate them. In general, however, we only enumerate things which are of interest to us as bearers of this or that characteristic. Shared generic concepts thus as a rule guide our interest, determining the concrete case of totality formation. With the shift of interest there will also occur as a rule a shift in the concept under which we think the objects enumerated. But this is not always so. Indeed, from a group of homogeneous objects, e.g., apples, we can assemble, entirely at random and arbitrarily, clusters of two, three, etc., without in so doing having had in mind any concepts marking out just these two, three, . . . objects.

It is indeed obvious that there must be present a motive for the interest which picks out and unites precisely *this* group; and that motive will consist in certain conceptual Moments marking out exactly the contents of just this group. But does that mean that thereby, and in the course of the associated act of enumeration, we must undertake a logical subsumption and explicitly think of the individual contents as being the objects of these concepts? A spatial configuration, for example, directs our interest in such a way that within a pile of apples we pick out a group of four. Must we therefore logically subsume each apple under the concept of a spatial content belonging to such a configuration? One therefore must not confuse: the incidental noticing of certain conceptual Moments (abstract partial contents) in <165> external intuition as the psychological motive for the formation of clusters, and the logical subsumption of the members of the cluster under the respective concept.[8]

But even if we assume it were always true, as asserted, that we constantly colligate and enumerate objects only to the extent that they fall under a shared concept, it clearly follows from our deliberations that number can in no way be regarded as a determination of that concept. The circumstance that the objects a, b, c, among those momentarily before consciousness, precisely mark themselves out through the shared characteristic α constitutes the psychical stimulus that leads to our picking them out in unison. But in order to count them, i.e., to ascertain the number form belonging to this totality, *that* involves new motives, and indeed such as have nothing to do with the determination of α. If we have an interest in a, b, c only insofar as they are of the genus α, then their number with the adjoined index α includes everything we need to retain for the further uses of the totality a, b, and c in thought. We thus obtain a representation simplified in content, from which are eliminated the individuating differences of those

---

[8] When *Kerry* ("Über Anschauung und ihre psychische Verarbeitung," Sixth Article, *Viertelj. f. wiss. Phil.*, Vol. 13, 1889, p. 392) says: "Wherever I am to enumerate there must be present a . . . concept the objects of which I enumerate," he obviously commits the confusion characterized above. And when he further, as proof, points out that without such a "guiding concept" we "would run the risk of also counting things which we ought not count, and of omitting something which we should have counted in" – thereby is conceded as much as we could ever wish, namely, that one can also count without such a guiding concept, whether it then may be a 'risky' counting or not.

Already on this basis we cannot concede the soundness of the following definition added on: "I call that psychical effort whereby an object is subsumed under one of the guiding concepts for a proposed enumerative task a positing of the unit." (p. 394) If I have before me a pile of apples, I do not need to carry out separate subsumptions under the concept *apple* in order to grasp each one as one apple while enumerating step by step. And what, at most, could that accomplish for me? The confirmation of the fact that each of the things before me is an *apple*. But that each is *one* apple, which alone matters, I do not cognize in this way, unless I once again turn away from the generic concept *apple* and rise purely and simply to the "one." And as here with the concept of unit, so also in the case of the other elemental arithmetical concepts, I cannot agree with *Kerry*'s analyses.

objects that are, at the time, of no significance to us. But these are interests which in no way have in view a closer *determination* of the concept (taking this expression however broadly one may wish.)

Disregarding the psychological motives which carry us into number abstraction, the direct examination <166> of this abstraction also shows that the number does not relate to the concept. The number is the general form of multiplicity under which the totality of objects a, b, c falls. Therefore it is clear that this totality (this multiplicity, group, or whatever else one may call it) forms the subject of the numerical assertion. Considered formally, number and concrete group are related as are concept and conceptual object. *Thus the number relates, not to the concept (Begriff) of the enumerated objects, but rather to their totality (Inbegriff).* Its relationship to the generic concept of the enumerated is simply the following: If we count a group of homogeneous objects, e.g., A, A and A, we at the outset abstract from the intrinsic nature of their contents, thus also from the fact that they are of the genus A. We form the totality form one, one and one, and subsequently note that "one" in this case is to have the signification "one A." Thus, it is only after the enumeration, which as such is totally indifferent to the circumstance that the objects are A's, that the generic concept links up with the number as a defining factor. It determines the unit, i.e., the representation of the "something" enumerated, which is at first void of content, as a something falling under the concept A.

The relationship between number and the generic concept of the enumerated is thus in a certain manner the *opposite* of what *Herbart* and *Frege* maintained. The number does not say something about the concept of the enumerated, but rather the concept says something about the number. To be sure, the expression "assertion of a number" is commonly referred only to relations of the number to other numbers. But that is not the only type of assertion which is possible. Here too is an assertion of number: one to the effect that the units which it enumerates have certain intrinsic characteristics. For thereby the number itself (we only speak of the general representation, not of the number as abstractum.) is determined to the point where it is a concrete representation.

Frege finds a further confirmation of the view that the number is attributed to concepts in the usage of the German language whereby one speaks of "ten man," "four mark," or "three cask." The singular here is thought to indicate that the concept is meant, not the thing.[9] <167>

This is a rather rash interpretation. *Frege* overlooks the fact that with the names "mark," "cask," "man" not the abstract, but rather the general concepts are intended. That use of language will, accordingly, suit our viewpoint much more readily, according to which, for example, "ten man" [*zehn Mann*] would say just the same thing as: "Ten, of which each is one – a man," or "Ten things, each of which is a man." The more usual way of speaking is: "ten men," "ten casks," etc. The underlying conceptualization here is something quite different. The plural "men" indicates the indeterminate multiplicity, the attached numerical attribute determines or classifies it, determines the *how many*. One now also understands why the number attribute never appears in the plural, whereas there is no lack of plurals for numbers in the substantive forms (tens, hundreds, etc.) If the number word stands in the attributive position, then it always refers to the *collectivum* as a whole, which is indicated by the attached plural (men, casks, etc.). Things are very different with other attributives. We do not say "good ⟨group of⟩ men" ["*gut Männer*"], but rather "good men" ["*gute Männer*"], because the goodness is ascribed not to the *collectivum*, but rather to each individual man. It is therefore just as manifold as are the men, and this the plural formation correctly expresses. It is consequently incorrect when *Frege* calls the common way of speaking misleading because, for example, in the expression "four noble steeds," "four" and "noble" were treated as properties of the same type. "Four" certainly is not the plural form of an adjective which, thereby, linguistically indicates a distribution to each of the steeds.

Certainly there are linguistic expressions which must be called misleading as to their form. One says "many people" ["*viele*

---

[9] *Op. cit.*, p. 64.

Menschen"], instead of the more correct expression "much people" [*"viel Menschen"*]. We speak of the multiplicity and number of people as, for example, we speak of colors and sizes of people. It is true that here a distinction is to be made. But it consists in nothing other than taking the former expressions collectively and the latter distributively.

What has led to interpreting the statement of number as an assertion about a concept is the following notion. Enumerated objects can always be so precisely determined by shared properties (whether intrinsic or relational) that the genus concept thus generated applies only to the objects enumerated *hic et nunc* <168> and to no other. If we for example judge, "The king's coach is drawn by four horses," then the concept *horse* would not satisfy this requirement, for there are still other horses as well. But it is otherwise when we choose the concept *horse drawing the king's coach*, and especially when we do not forget to take note of the place, date and time of day. This concept applies only to the four horses here concretely given and enumerated. If one proceeds in such a way in all cases, then there obviously arises a relationship of univocal dependence between the concept so determined and the number of the enumerated objects. Along with the concept the number is also determined, and indeed univocally. If we take the concept *horse*, there is yet nothing decided concerning the number of horses. But if we select the concept *horse drawing the king's coach*, then with it there is already *eo ipso* given that the number of the (so determined) horses is four and can be only four. If now the further observation is made that each change in the concept also conditions a change of the number, then the dependence of the number upon the generic concept of the enumerated is evidently a total one. So it becomes natural to think that number could signify absolutely nothing other than a certain property accruing to that concept. Thus the number statement would assert something of a concept (namely, the concept of the enumerated, made precise in the manner indicated), and to explicate what it asserts would then be the task of logical analysis.[10]

---

[10] This, no doubt, was *Frege*'s line of thought.

But on a more exact examination of the matter it is clear that the sole relation the number has to that concept is that it numbers its *extension*. With the concept, its extension is also determined; and since this latter, in the cases here considered, presents us with a delimited, finite multiplicity, then of course a number also accrues to it. It is not to the concept *horse drawing the king's coach* that the number four accrues, but rather to the extension of that concept, i.e., to the totality of those horses. Only speaking indirectly can one possibly say that the concept has the characteristic that the number four accrues to its extension. That the respective number assertion does not in the least intend the expression of this complicated conceptualization surely requires no proof. <169>

Now that we have clearly determined, in the foregoing discussions, the sense of the number assertion and its relationship to the concept of the enumerated, and have sided with *Mill*'s view cited on page 170, *Schuppe*'s theory, which regards as the subject of the assertion that which "is the same in all the things distinguished" (which indeed can only be a shared conceptual Moment) requires no separate refutation.

On the other hand, a few words might be addressed to the view last mentioned above, which takes two, three, etc., to be predicates of number in general. There is no doubt that we very often take the general concept of number as an intermediary. We say for example that the number of the apples is three. In this case we first subsume the concretely present totality of apples under the general concept of number. However, this is certainly a detour, the occasion for which is frequently lacking. If, for example, we say that of apples there are three, then the number is directly attributed to their totality.

APPENDIX TO THE FIRST PART

The Nominalist Attempts of Helmholtz and Kronecker

Already in Chapter VIII we had occasion to become familiar with a nominalistic interpretation of the number concepts. It was based upon the confusion of the concept *one* with the sign "1," which is assigned to each object to be counted in the most elemental type of symbolic enumeration. The numbers are then mere symbols for the groups of ones thus originating. Hence *Berkeley* concludes that "those things which pass for *abstract* truths and theorems concerning numbers are in reality conversant about no object distinct from particular numerable things, except only *names* and *characters* which originally came to be considered on no other account but their being *signs*, or capable to represent aptly whatever particular things men had need to compute."[1] Whereas *Berkeley* had set out from the primitive and more original forms of symbolic enumeration, and had regarded the higher forms as grounded on them, two new nominalist attempts, remaining at an even more superficial level, start out from those higher mechanisms of enumeration – to us now commonplace – whose symbolic character they totally misunderstood by seeking in them the original source of the number concepts. These attempts have an especially strong claim on our interest because they originate from eminent scientists and serve as the foundation for carefully thought out theories of arithmetic. Both *v. Helmholtz* and *Kronecker* aim in their analyses to attain to <171> the true foundations of general arithmetic. This aspect of their efforts we must leave

---

[1] *Principles of Human Knowledge*, section 122, cp. also section 121. The word "originally" in the quotation is directed against the Philosophers. They misunderstood the natural and original sense of the number names and number symbols by referring them after the fact to (imagined) abstract Ideas. (Here *Husserl* quotes from Fr. Ueberweg, *System der Logik*, 5th Edition, Bonn 1882, p. 88. The German has not been translated, but simply replaced by *Berkeley*'s English. The emphasis is *Husserl*'s. {DW})

aside in the following critique, since to this point we have not yet taken in hand the question about the foundational concepts proper to general arithmetic.

"Enumerating," says *Helmholtz*, "is a procedure which rests upon our finding ourselves in a position to retain in memory the sequence in which conscious acts have occurred one after the other in time. We can, to begin with, consider the numbers as a sequence of arbitrarily chosen signs for which only a determinate manner of succession is established by us as the lawlike, or, in the common way of speaking, the "natural" one. The designation "the 'natural' number sequence" has indeed linked itself only to a specific application of counting, namely to the determination of the *cardinal number* [*Anzahl*] of given real things. In that we cast these things one after the other into the pile of the counted, the numbers, by a natural process, follow one after the other in their lawlike sequence. This has nothing to do with the ordered sequence of the number symbols. As the symbols are different in the various languages, so also their ordered sequence could be arbitrarily specified, if only some one definite ordered sequence were irrevocably established as the normal or lawlike one. This ordered sequence is in fact a norm or law given by man – by our ancestors, who developed the language. I stress this distinction because the presumed 'naturalness' is bound up with a defective analysis of the concept of number."[2]

"According to the preceding discussions, each number is defined only by its position in the law-governed sequence. The symbol *one* is assigned to that member of the ordered sequence with which we begin. *Two* is the number which directly follows upon one in the law-governed sequence, i.e., without another number coming between. *Three* is the number that follows in the same manner immediately upon two, etc. There is no basis for discontinuing this sequence at any point <172> or for turning back in it to a symbol already used earlier."[3]

---

[2] <"Zählen und Messen," in *Philosophische Aufsätze, Eduard Zeller zu seinem fünfzigjährigen Dr.-Jubiläum gewidmet*, Leipzig 1887, p. 21. {LE}> The psychological discussions which *Helmholtz* adds on, regarding the "univocality of the sequence" of the signs and their connection to time and memory, we can pass over as irrelevant to our purposes.

[3] *Ibid.*, p. 23.

Helmholtz then continues on in this manner. The signs of the fixed sequence beginning with "one" can serve to mark arbitrary sequences of objects, among others also any segment of "the number series" itself. In this way he secures the possibility of defining *addition* as a purely symbolic operation, from which there result theorems – or, more accurately stated, equivalences of symbols – of the type:

$$(a + b) + 1 = a + (b + 1).$$

In general, he succeeds in proving all of the basic formulae of the algorithm for calculation with positive whole numbers, from which all of its rules can be deductively derived, only keeping in mind that they all appear as mere equivalences between certain complexes of symbols (valid in the sense of initial definitions of symbols).

But what is this empty play of symbols to us, one will ask in astonishment? *Helmholtz* is well prepared for this question, which *Paul du Bois-Reymond* had already raised against earlier efforts of the same tendency.[4] "Apart from the test thus made of the inner consistency of our thinking, such a procedure would of course primarily be a pure play of mental acuity upon imaginary objects, . . . if it did not admit of such extraordinarily useful applications. For by means of this symbol system of numbers we give descriptions of relationships between real objects, which, where they are applicable, can achieve any degree of precision required. And by means of this system, in a vast number of cases where natural bodies come together or interact under the governance of known laws of nature, the numerical values measuring the effect are found in advance by calculation."[5] Now, although *Helmholtz* goes into these applications in detail in the latter parts of his treatise, and <173> considers as his main task to find the answer to the question, "What is the objective sense of the fact that we express

---

[4] *Die allgemeine Functionentheorie*, pp. 50-51. Cp. in another connection also *J. St. Mill*'s objections against nominalist theories of arithmetic, *Logic*, Book II, Ch. VI, § 2 (*Gesammelte Werke*, Vol. II, pp. 274ff).

[5] *Philosophische Aufsätze, Ed. Zeller . . . gewidmet*, p. 20 {LE}

relationships of real objects by means of concrete numbers [*benannte Zahlen*] as magnitudes, and under what conditions are we able to do this?"[6] we nevertheless find absent anything that would lift his formulations above nominalism.

The numbers are defined by *Helmholtz* first and foremost as arbitrary symbols. But vainly do we seek in the further course of his expositions for what it then is that these symbols do genuinely *signify*. In the different cases they can designate the most heterogeneous of objects, and yet the designation of those objects is no arbitrary one. Wherever we use the term "five," it occurs in *the same sense*. In what is it therefore grounded that the most dissimilar of representational contents are designated in the same sense by these signs? In short, what is the *concept* which mediates each use of the signs and constitutes the unity of their *signification*?

From the way in which the *Helmholtz*ian numbers/signs are introduced and applied, it is not difficult to reconstruct the overlooked concepts which found them, if we only keep constantly in mind the question of what signs of such a type are in fact capable of standing for. In enumerative symbolization we use some segments of the sign sequence beginning with "one." A multiplicity of objects must therefore be present, of which each, presupposing univocality of designation, receives one and only one of the signs. By means of the store of designations, assumed to be unlimited, we then can symbolize the elements of any conceivable group. But if this is not to be a totally senseless act, then in the things or the group itself there must be found something that is specifically touched upon by these signs. It cannot be the concrete contents. They change from group to group. It also cannot be the being-one [*Eins-sein*] of the individual content. For the symbols which we assign to the different individual objects are different, and must therefore possess a different conceptual basis.

Only one characteristic of the signs have we left yet unexamined: their rigorously determinate succession, in which they indeed always <174> come into play (according to rule) in enumerative symbolization. If we bring this succession into

---

[6] *Op. cit.*, p. 20.

consideration, then the signification of the signs immediately comes to light. If their order is the basis of the designation, then, in virtue of its univocally determinate position in the symbol sequence, each sign must indicate the corresponding position of the group member it stands for in the serial order of the whole group. In a word, each sign is a sign of an order, it is the sign of an *ordinal number* in the usual sense of the phrase. The *signification* of each sign lies accordingly in its *place value*. "One" would be the symbol for the *first* member of a sequence as such, i.e., for the beginning member. "Two" is the symbol for the *second* member of a sequence as such, i.e., for the member immediately following the first. "Three" is the symbol for the *third* sequence member, i.e., for the one immediately following the second. It is clear that these designations are not in and for themselves applicable to unordered groups. If, however, an objective order is lacking, those designations can still find application with regard to the extrinsic temporal succession in which the members of the group are traversed.

From the explanations through which *Helmholtz* introduced his designations, and the applications which he prescribed for them in the enumerative process, we have, by these reflections, determined the concepts which must underlie them and constitute their logical signification. It becomes obvious that he confuses the concepts *one, two, three*, etc., i.e., the number concepts in the *common* sense of the word, with the ordinal number concepts (*first, second, third*, etc.), not to mention the fact that he explains these latter nominalistically, as mere symbols. We can therefore, after what has been shown, also not agree when he says: "The number sequence is imprinted upon our memory very much more firmly than any other sequence .... We therefore also chiefly use it in order to fix other ordered sequences in our memory by attachment to it: i.e., we use the symbols as ordinal numbers."[7] The function of the *Helmholtz*ian symbols is not to fix the memory of other ordered sequences, but rather to designate the positions of the members in any such ordered sequence as such, and indeed in virtue of the abstract concept <175> of position. And this is not

---

[7] *Op. cit.*, p. 22.

merely its "chief" function, but rather its *only* function. *Each utilization of these symbols signifies a utilization of the ordinal numbers in the true sense of the word*, whether directly or indirectly.

What has just been said will, of course, also be valid for that special way of utilizing the "number sequence" through which *Helmholtz* believes he can define the "concept of the *number* of elements in a group," and so we have then for good reason reckoned him[8] among those investigators who explain the concept of cardinal number as a mere derivative of the concept of the ordinal number. Let us now place the *Helmholtz*ian definition itself before us: "When I utilize the complete sequence of numbers from 1 up to n in order to coordinate a number to each element of the group, I then call n the *number* of the members of the group." This coordination produces a determinate ordering of the group members, and it is then a *theorem* "that the number of the members remains unchanged throughout variations of the ordered sequence of the members, if omissions and repetitions of them are avoided."[9]

Does this definition really correspond to the number concept with which we are so familiar? The number sequence was set up as a series of arbitrary symbols. Could the expression "number n of a group M" really signify nothing other than the property of this group that – for some arbitrary ordering and term-by-term symbolization of it by means of the sequence of those signs – n is the sign of the last and highest member?

As our preceding analyses have shown, to the *Helmholtz*ian signs there can correspond, in virtue of their peculiar nature, only the ordinal number *concepts*. Should recourse to these concepts – which are not adequately appreciated by the great physicist – perhaps lead to an improved sense for the above definition? That too I am unable to see. If I say, e.g., "the number of these apples is four," I certainly do not then have in mind the circumstance that, given some ordering of the apples, the last element is the fourth, but rather precisely that one and one and one and one apple

---

[8] See p. 12 of this work.
[9] *Op. cit.*, p. 32.

is present. <176> For the true concept of number, which contains nothing of an ordering of the units counted, the above mentioned theorem is of course just as irrelevant as it is indispensable from the standpoint of the *Helmholtz*ian definition.

After all these developments, no lengthy demonstration is required that the startling polemic *Helmholtz* brings against the designation of the ordering of the numbers as a *natural* one[10] cannot be sound. At no point do the nominalistic misconceptions into which this brilliant researcher has fallen manifest themselves more clearly than here. If it is conceded that the numbers are nothing other than arbitrary symbols in an arbitrarily established order, then certainly the designation of this ordering as a "natural" one is misleading. Any other ordering of the symbols could just as adequately have been taken to be the number sequence. However, those who speak of a natural ordering in the domain of numbers surely do not mean the ordering of arbitrary symbols, but rather of certain concepts designated by means of them. Whichever we consider, whether the "ordinal" numbers or the "cardinals" (both terms taken in the true sense), we always come to the result that the sequential order is one grounded through the nature of these concepts themselves. With the cardinals, for example, the principle of order consists of each number next in the sequence being greater by one than the preceding number.

The nominalistic attempt by *Kronecker* is very little different in essence from the one just discussed, as is clear from the following quotation:

"I find the natural starting point for the development of the number concept in the *ordinal* numbers. In these we have a supply of certain designations arranged according to a fixed serial order, which we can assign to a group of objects that differ and, simultaneously, are distinguishable by us. The entirety of the designations thus utilized we grasp together in the concept of 'the *number* of objects' of which the group consists, and we unambiguously affix the expression for this concept to the last of the designations used, since their succession is rigorously determined. <177> Thus, for example, in the group of letters

---

[10] *Op. cit.*, p. 32.

(a, b, c, d, e) the letter a can be assigned the designation of 'first', the letter b the designation of 'second', and so on, and finally the letter e the designation of 'fifth'. The entirety of the ordering numbers thereby utilized, or the 'number' of the letters a, b, c, d, e, can accordingly – with reference to the last of the ordering numbers used – be designated by the number 'five'."[11]

Is, therefore, "five" really to be nothing other than a sign for the totality of the signs "first," "second," "third," "fourth," "fifth"?[12]

The source of the noteworthy misconceptions into which these two illustrious investigators (as *Berkeley* before them) have lapsed resides, then, as was already mentioned at the outset of this critique, in the misinterpretation of the process of symbolic enumeration, which we carry out as a blind routine. Therein we proceed in such a way as to correlate mechanically the number names with the members of the group to be counted, and then take the last name required as that of the number sought. In actuality the names serve us in the first place as a mnemonically fixed sequence of symbols devoid of content; for during the enumeration their conceptual content [*Gehalt*] is totally absent from our consciousness. Only after completion of the process, and in the light of its true purpose, does the (authentic or symbolic) number concept enter into consciousness as the signification of the resultant number word. Now these great mathematicians have confined themselves to the external and blind process, have misunderstood its symbolic function, and thus have confounded symbol and thing. That they associate the number names with the same concepts as all other men do is beyond doubt, and in *Helmholtz* we find plenty of passages that can be given their fully intelligible interpretation only by reference to the true concepts of numbers. It had to be very powerful scientific interests that could lead to the overlooking of the concepts in such an astonishing way. And these interests are, at least in *Helmholtz*, transparent enough: The opinion that only the interpretation of general

---

[11] "Über den Zahlbegriff," in *Philosophische Aufsätze, Ed. Zeller . . . gewidmet*, pp. 265-266.

[12] An analogous objection also affects *Dedekind*'s definition of number. Compare above, p. 131n, as well as pp. 21, 27 and 54 of the writing by *Dedekind* cited there.

arithmetic as a coherent symbol system <178> could eliminate all of the immense difficulties which infect this remarkable science, would of necessity produce the tendency to reinterpret the number concepts, commonly thought to be its true roots, in the nominalistic sense. Here is not yet the place to go more deeply into the questions just touched upon.

SECOND PART

THE SYMBOLIC NUMBER CONCEPTS AND THE
LOGICAL SOURCES OF CARDINAL ARITHMETIC

Chapter X

OPERATIONS ON NUMBERS AND
THE AUTHENTIC NUMBER CONCEPTS

After the discussion and resolution of the subtle questions connected with the analysis of the concepts *unity*, *multiplicity*, and
number, our philosophical investigation proceeds to the task of
making psychologically and logically intelligible the origination of
a calculation-technique based upon those concepts, and of investigating the relationship of that technique to arithmetical science.

The Numbers in Arithmetic Are Not Abstracta

It will be useful first to settle a logical difficulty that comes up
with reference to all calculating. One says that 2 and 3 is 5. But
the concept 2 and the concept 3 always remain, however, the
concept 2 and the concept 3; and the concept 5 never emerges
from *them*. And does it make any sense at all to speak of calculating with the same numbers: of their addition, multiplication, etc.?
After all, gold and gold always remain, again, gold. Why does
not 5 and 5 remain, in turn, 5? How, therefore, could one conjoin
number *concepts* in operations, since each remains identically
what it is; and since each concept, in and for itself, is only a single
one, how are we ever to conjoin *identical* concepts?

The answer is obvious. The arithmetician absolutely does not
operate with the number concepts as such, but rather with the
generally represented objects of those concepts. The signs which
he combines in calculating have the character of general signs,
signs formed with the number concepts as their basis. Thus, 5
does not signify the concept (the abstractum) *five*; but rather 5 is a
general name (or else a calculational sign) for any arbitrary <182>
group as one falling under the concept *five*. 5 + 5 = 10 means the

same as: a group – any one, whichever it may be – falling under the concept *five*, and any *other* group falling under that same concept, when united yield a group falling under the concept *ten*.

The Fundamental Activities on Numbers

If we understand by "enumeration" the process of number formation, then we can say: numbers originate through enumeration of multiplicities. Sometimes one also says that they originate through enumeration of things (which, of course, are thought as members of a multiplicity), or, by a further extension, through enumeration of units.

However, numbers do not originate merely in this direct way, through simple enumeration, but rather in an indirect way as well, through calculational operations that include multiple enumerations. The *Fundamental Activities*, which we bring to bear upon all numbers, and through which alone we can form new numbers from given numbers, are *Addition* and *Partition* [*Teilung*].

In order to explain the former it is usually said: numbers can be formed, not only through enumerating units together, but rather through enumerating numbers together as well. This is a misleading mode of expression. Were one to enumerate numbers together in the customary sense, as one does apples, then the enumeration of 2, 3, and 5 would yield, not 10, but 3 instead. The enumerating intended here obviously does not bear upon the numbers, but rather upon the units in the numbers; and, indeed, in such a way that the peculiar associations which the units form in the different numbers to be "enumerated together" are dissolved and all units are joined together to form one number.

Numbers can, further, originate through partition. Any number is by nature a sectioned whole with units as its natural parts. But any number (0, 1, and 2 excepted) also admits of yet other partitions, partitions in the arithmetical sense, i.e., into numbers.

In Chapter V we set forth the psychological basis upon which the addition and partition of numbers rests. <183> It is a fact that we can *simultaneously* retain and collectively unite several totality representations, and thus can form totalities *of* totalities;

and likewise that we can retain a given totality, and yet simultaneously can unify groups of its elements through representations of particular totalities, thus enabling us to represent a plurality of totalities *within* the given one. And the same is also the case with the general representations of totalities, the numbers.

Addition

Let us now consider the fundamental operations somewhat more closely. *To add* is to form a new number by collective combination of the units from two or more numbers. For reasons which we discussed earlier[1], mathematicians consider *one* to be a special case of number. Accordingly each number appears to them, in its inner constitution, to be an additive conjunction of units or of numbers 1. Considered from the viewpoint of formal arithmetic, this is correct. One in fact can allow the collective combination of units to be regarded as that special case of *addition*, where each addendum is equal to one. Nevertheless, the addition of *units* is no *logical* specification of addition in general. Without a concept of number already given there is no concept of addition, and, also, no concept of a number 1. It would accordingly be quite wrong to explain number, as is often done, as an *additive* conjunction of units, instead of as a *collective* conjunction. The combinatory sign "+" signifies the former when set between number signs. When set between "ones" it signifies the latter. That this deceptive ambiguity of the sign is disregarded within arithmetic is deeply grounded in the nature of that science. Here we encounter the first example of an essential distinction upon which we yet will very frequently find reason to insist: the distinction between *logical* and *mathematical* generality. Considered from the mathematical point of view, a collection of units functions as one special case of a summation of arbitrary numbers, while <184> from the logical point of view the concept of the sum nevertheless *presupposes* that of the collection of units.

There is still another – moreover a related – equivocation that

---

[1] Cp. Chapter VIII, pp. 138-141.

has given occasion for confusion of these two types of combination: namely, that lying in the little word "and." In ordinary discourse it serves only for the designation of the *collective* combination. In the linguistic rendition of arithmetical calculations, on the other hand, "+" is commonly read as "and." We write "7 + 5" and read: "seven and five." "And" in this sense indicates *addition* as in the rigorous concept defined above. If one does not keep the two concepts distinct, one easily comes to misinterpret "+" and "and" in the second ⟨additive⟩ sense as "and" in the first ⟨collective⟩ sense, and thus to misunderstand the meaning of the arithmetical sign. Thus, *F. A. Lange*[2] supposes that he can justify the *Kant*ian view of the judgment "7 + 5 = 12," against the well-known attacks of *R. Zimmermann*,[3] by interpreting "7 + 5" as a mere combination of 7 and 5. In this he rightly appeals to *Kant*, who found expressed in this composite sign the "juxtaposition" ["*Zusammenstellung*"] of the two numbers instead of an addition.[4] But it is quite beyond doubt that "7 + 5" *cannot* have the signification of a collectivity [*Kollektivums*] whose terms are 7 and 5. Otherwise 7 + 5 would always remains only 7 + 5; and then propositions such as "7 + 5 = 8 + 4 = 9 + 3, etc.," and likewise even the proposition "7 + 5 = 12," would be Evidently false. Thus, it is not the mere collocation of 7 and 5 that is designated by the complex sign "7 + 5," but rather their additive unification. It signifies: a number that simultaneously embraces the units in 7, and those in 5, and these alone. Given this, the proposition "7 + 5 = 12" actually holds true, and, to be sure, as a proposition which one can prove to be necessarily true from the concepts 7, 5, 12 and the concept of addition.

A special case of addition, to be given a general characterization, is supposed to found what, according to the common explanations, amounts to a new fundamental operation of arithmetic. The *multiplication* <185> of a number a by a number b is

---

[2] *Geschichte des Materialismus*, Book Two, 3rd Edition, p. 119.

[3] "Über *Kant*'s mathematisches Vorurtheil und dessen Folgen," *Sitzungsbericht der phil.-hist. Classe der kais. Akademie der Wissenschaften*, Vienna 1871, Vol. 67, pp. 16ff.

[4] *Critique of Pure Reason*, Transcendental Doctrine of Elements: Part II, First Division, Book II, ch. 2, sect. 3 (*Hartenstein* edition, Vol. 3, p. 157).

customarily defined as the additive conjunction of so many of the same number a as there are units in the number b. If we name the result of this operation the product of a and b, then we can concisely designate the product as a sum of b numbers a. The multiplication of a by b is, thus, not a "conjunction" in the ordinary signification of the word. One speaks of a "conjunction" only in cases of separated contents, and, accordingly, does not designate a whole as conjoined with its parts. One number a is not conjoined with another number b, but rather there are linked several – and indeed b – numbers a. The a in "a × b" is a plural: "4 × 3": i.e., four threes. But not four threes pure and simple; rather, four threes to be additively conjoined.[5]

Is *multiplication* really a *mere special case of addition*, namely, of equal addenda? It is very tempting to say yes. However, one still does not designate sums such as

a + a, a + a + a, a + a + a + a ...

as multiplications or as products. Only by the enumeration of the terms added do we obtain the multiplier, and thereby the possibility of forming products

2a, 3a, 4a, ...

What, then, is the purpose of such a distinction between sum and product? For what reason do we have, besides the enumerations within the individual numbers, also the enumeration of these numbers themselves? The answer seems clear: As the numbers in general function as abbreviating general signs for the facilitation of our thinking and speaking, so they function also in the role of multipliers. As in abbreviation of complicated names of the form "A and A and A and A" we say "four A's," so in the

---

[5] The relation between the number and its accompanying name is commonly regarded as a multiplicative one. "4 apples" = "four times one apple," as "4 × 3" = "four times one three." In fact, with multiplication we have an enumeration of equal numbers, instead of, as in the other case, an enumeration of equal *things*. Nevertheless, the two cases are to be well-distinguished. And surely it is the numbers that are added, but not the things, especially since for things addition has no sense.

abbreviation of "3 + 3 + 3 + 3" we say "four times three," whereby the addition is tacitly understood.

Accordingly the whole distinction between multiplication <186> and addition seems to reside in a new mode of designation which is possible for special forms of addition. The "product" offers a conveniently abbreviated symbolic representation and denomination for special forms of sums in which the terms to be added are equal numbers: forms made possible and mediated by the enumeration of those equal summands. — But if this is so, why do we speak of a special *operation* of multiplication? The abbreviated mode of designation may, as such, be very convenient and useful, but it is, after all, no operation. It symbolizes briefly and precisely the manner in which the number is to be formed. But in doing this it only states the problem, and does not provide its solution. To actually [*wirklich*] obtain the intended number there is no other way than actually carrying out the additions upon which the symbolization is based. But these additions are in no way distinct from any other additions. Equal numbers are not added in a manner different from unequal numbers. And so we remain in the dark as to why arithmeticians speak of multiplying as a novel fundamental operation upon numbers.

A product can be multiplied by a number again, the resulting product once again, and so on. Expressing this symbolically we can form:

$(a \cdot b) \cdot c; \quad ((a \cdot b) \cdot c) \cdot d; \ldots,$

for which one simply writes:

$a \cdot b \cdot c, \quad a \cdot b \cdot c \cdot d, \ldots.$

Here too we are concerned with methodologically abbreviated symbolic representations and denominations of complicated sum formations, mediated by the successive enumerations of equal summands in levels founded one upon another. And also here we have, once again, mere designations of tasks, and not the operations which carry them out.

But the special case where the factors are equal to one another once again gives occasion to speak of a novel calculating operation: *Exponentialization*. But we again find nothing other here than symbolic representations and symbolizations of a higher level, mediated through a *new type of enumeration*, namely, enumeration of equal factors. The products   <187>

$$a \cdot a, \quad a \cdot a \cdot a, \quad a \cdot a \cdot a \cdot a, \ldots$$

obtain, for thinking and speaking, an extraordinarily simplified symbolic expression through enumeration of the factors. The uniform repetitions of $\underline{a}$ times $\underline{a}$ times $\underline{a}$ times $\underline{a}$ ... – which in case of larger numbers of factors are ultimately overwhelming (besides being difficult to distinguish) – we gladly replace in ordinary language by "$\underline{a}$ multiplied by itself, two, three, ... $\underline{n}$ times." And, using symbols, these expressions may be simplified even further. We need only to attach a numerical index to the symbol $\underline{a}$, which then symbolically expresses how many times $\underline{a}$ is to be utilized as a factor, and we have a complex sign which in all rigor signals the mode of formation of the intended number. For this purpose the "powers" of arithmetic

$$a^2, a^3, a^4, \ldots,$$

serve as maximally abbreviated inscriptions for the products written out above. Already with quite simple examples the usefulness of this succession of symbolizations is striking:

$$4^3 = 4 \cdot 4 \cdot 4 = (4 \cdot 4) \cdot 4 =$$
$$(4+4+4+4) + (4+4+4+4) +$$
$$(4+4+4+4) + (4+4+4+4).$$

And then we would still have to replace each 4 with $1 + 1 + 1 + 1$, and actually carry out the additions. The complication grows, as one easily recognizes, at an exceedingly fast rate with ⟨the growth of⟩ base and exponents. And thus it becomes clear that, by this simple expedient, we are in a position to designate with precision, by one short symbol, additive combinations of such complexity that their fully actual inscription or denomination (which certainly could come about only through a successional process) would not be possible in days or years, if at all.

However, does not precisely this circumstance limit the usefulness of the mode of designation within narrow confines? Ultimately we surely have to do with actual [*wirklichen*] numbers, and thus with the actual addition which underlies the sign structure and which alone yields the number designated. If we no longer can write and state the addition, then we also can no longer authentically think it, much less then carry it out. And nevertheless arithmetic does not sense itself <188> to be at all hindered by such constraints, and it claims to be able to calculate where one can no longer speak of an actual [*wirklichen*] representing, but rather only of an indirect symbolization.

Partition

The second fundamental activity which we can bring to bear upon numbers is that of *partitioning*. Addition links a given plurality of numbers into one new number. Partition segregates a given number into a plurality of partial numbers, and hence is the inverse operation to addition.

Here one must be struck by the fact that arithmeticians cite only two special cases of partition as particular calculating operations: *subtraction* and *division*. To subtract a number b from a number a means that, after segregation of b units out of a, we bring together the units still remaining into a new number c. This number c is the result of the subtraction, and is called the "difference" of the two numbers. Subtraction therefore Represents the solution of the following partition problem: If a given number a can be split up into two partial numbers in such a way that b is one of the two, what is the other one?

How is it possible, one will now ask, that the arithmeticians who totally disregard the general operation of partition can, rather, in its stead regard subtraction (which is indeed a mere special case of that operation) as the operation inverse to addition?

At first glance one might perhaps be inclined to place the above subtraction problem alongside the following addition problem: If we add to a given number a another number b, what is the resulting number? Obviously there is to be expressed through this formulation a certain precedence of the a over against the b, analogous to that of the minuend over against the subtrahend. But what, then, is the ground for this favoring? Is there a difference when this problem is contrasted with the reverse problem, where a and b are interchanged? Certainly, it is replied. I can still produce the sum, in that at one time I begin with a and then adjoin the units of b, and the other time begin with b and adjoin the units of a. However, precisely <189> this we cannot concede. Like the concept of number, the concept of addition of numbers also contains nothing of temporal succession. In the relationship between minuend and subtrahend we are dealing with a "precedence" that is quite different from such an extrinsic (logically accidental, although perhaps psychologically necessary) one. Here there really does reside a conceptual distinction, which is attested to, for example, by the fact that the operations (a − b) and (b − a) are mutually exclusive as to their logical possibility.

Once again, therefore, analysis of the fundamental arithmetical concepts seems to force us to reject as erroneous the view taken by arithmetical science. It does not seem tolerable to consider addition and subtraction as inverse operations.

Out of the partition of a number into equal parts there arises the fundamental arithmetical operation of *division*. The problem to be solved here consists in this: to partition a number into a given number of equal parts, and to determine the common numerical value of the parts obtained. This value is the "quotient," the number to be partitioned is the "dividend," and the number of the parts is the "divisor." For example, let 20 be divided by 4. According to the explanation this means: to determine the common numerical value of the four equal partial numbers resulting from the partition. The solution reads: 5 is the quotient.

One wholly inexperienced in arithmetic could represent the carrying out of the division of a by n in this way: one removes a group of n units from a, and then another group of n units, and from the two groups forms a group of n twos. Then again, one picks out a series of n units from a and forms, with the preceding, a group of n threes. And so on. If a is divisible by n, we finally obtain the product a = (n · q), which distributes all of the units of a amongst the n partial numbers q, and the problem is solved. That arithmetic is far removed from proceeding in this way – just as little as it would occur to it to carry out multiplications and work out powers through actual computation of the underlying additions – is well known.

We forego raising analogous misgivings also for finding roots and logarithms: the operations inverse to exponentialization. <190> In all areas there emerges the same characteristic conflict with arithmetic, whose sense and essence seems to become more and more obscure the more we wish to illumine it.

Arithmetic Does Not Operate with "Authentic" Number Concepts

If we understand by *operations* actual [*wirkliche*] activities with and upon the numbers themselves, then there are no operations other than combination [*Verbindung*] and partition. But what arithmetic *calls* "operations" corresponds not at all to this concept. They are indirect symbolizations of numbers, which characterize the numbers merely by means of relations, instead of constructing them through operations. But if in arithmetic we were dealing with the actual numbers, then the evaluation of those symbolizations would always require recourse to the actual activities underlying them, and thus to the carrying out of actual additions and partitions. But in all of arithmetic not even a trace of this can be found.

Obviously we are not on the right path. The presupposition from which we set out at the first, as from something self-evident – namely, that each arithmetical operation is an activity with and upon actual numbers – cannot correspond to the truth. All too hastily we allowed ourselves to be guided by the common and naïve view which does not take into account the distinction between *symbolic* and *authentic* representations of number, and which does not do justice to the fundamental fact that all number representations that we possess, beyond the first few in the number series, are *symbolic*, and *can* only be symbolic. This is a fact which totally determines the character, sense, and purpose of arithmetic. In that those who study the logic of arithmetic also overlook this important circumstance, or have not appreciated its significance, a deeper understanding of that discipline would necessarily have to remain closed off to them. And so, then, we find expounded almost everywhere the false theory that, out of adding and subtracting conceived of as actual [*wirkliche*] activities, the higher operations proceed through mere specialization. <191> Multiplication is supposed to be nothing but a special way of adding, and exponentialization nothing but a special way of multiplying. "All of the calculation symbols in arithmetic," says *Dühring*[6] for example, "are accordingly only more precise determinations of the peculiar combinations in which the *activities* corresponding to the simplest signs, plus and minus, find an application of a specialized nature. The rich diversity that the lower and higher operations yield does not concern the original materials or, in other words, the basic activity in itself, which pervades all else; but rather the diversity only presents novel turns within that medium." And upon the possibility of a multi-faceted variation in these "turns," arithmetic is supposed to rest: ". . .there would be no distinct arithmetic whatsoever if the various forms that are possible in the grouping of units were not an issue." The preliminary reflections above already make clear the unfeasibility of such views; and we shall, in the further course of our investigations, supply yet more direct positive proofs that these "new turns," these diverse "forms in the composition of units," are nothing more than turns and forms of the *symbolism*, grounded upon the fact that all operating which reaches beyond the very first numbers is only a symbolic operating with symbolic representations.

If we had authentic [*eigentliche*] representations of all numbers, as we do of the first ones in the series, then there would be no arithmetic, for it would then be completely superfluous. The most complicated of relations between numbers, which now are discovered only laboriously, through intricate calculations, would then along with the number representations be simultaneously present

---

[6] E. Dühring, *Logik und Wissenschaftstheorie*, Leipzig 1878, p. 249.

to us with that same intuitive Evidence as we have, say, with propositions such as 2 + 3 = 5. To those who know what "2," "3," "5," and the signs "+" and "=" signify, this proposition is immediately clear and Evident. But in fact we are extremely limited in our representational capacities. That some sort of limits are imposed upon us here lies in the finitude of human nature. Only from an infinite understanding can we expect the authentic representation of *all* numbers; for, surely, therein would ultimately lie the capability of uniting a true infinitude <192> of elements into an explicit representation. Nevertheless, finite beings would be conceivable who could achieve actual representation of the millions and trillions – indeed, even the light years of the astronomers. Such a case would suffice to deprive the development of arithmetic of any practical occasion. Indeed, the whole of arithmetic is, as we shall see, nothing other than a sum of artificial devices for overcoming the essential imperfections of our intellect here touched upon.[7]

In truth, we are removed from this hypothetical ideal case just mentioned by almost the whole of its accomplishments. Only under especially favorable conditions can we represent authentically concrete multiplicities of approximately a dozen elements. Authentically: that is, factually (and *as* intended) grasping each of their terms as something separately and specifically noticed, and together with all of the others, in one act.[8] Accordingly, twelve (or a lower number close to it) is also the ultimate limit for the conceptualization of *authentic* number concepts. Nonetheless, no one feels restricted by these constraints, which were, in fact, first discovered by psychological analysis. Inside and outside of science we speak as though we could continue the sequence of numbers to infinity, i.e., beyond any limit attained. In addition, these concepts are even regarded as the logically most perfect ones in the domain of human knowledge.

---

[7] In the light of this, the famous line by *Gauss*, to the effect that "God arithmetizes," is not compatible with the concept of an infinitely perfect being. *Dedekind* (see the motto for his book, *Was sind und was sollen die Zahlen?*) rephrases it: "Man continually arithmetizes," which is unacceptable for other reasons. I would merely say: "Man arithmetizes."

[8] W. Wundt, *Grundzüge der physiologischen Psychologie*, Vol. II, 2nd Edition, Leipzig 1880, p. 214.

But *how* can one speak of concepts which one does not genuinely [*eigentlich*] have? And *how* is it not absurd that upon such concepts the most secure of all sciences, arithmetic, should be grounded? The answer is: Even if we do not have the concept given in the *authentic* [*eigentlicher*] manner, we still do have it given – in the *symbolic* manner. The discussion of this essential distinction, and the psychological analysis of the symbolic number representations, are to form the task of the following chapters.

Chapter XI

SYMBOLIC REPRESENTATIONS OF MULTIPLICITIES

Authentic and Symbolic Representations

5   We will first briefly explain the distinction between *symbolic* and *authentic* [*eigentlichen*] representations, which is fundamental to all further discussions.

A *symbolic* or *inauthentic* representation is, as the name already indicates, a representation by means of signs. If a content is not
10 directly given to us as that which it is, but rather only indirectly *through signs which univocally characterize it*, then we have a symbolic representation of it instead of an authentic one.[1]

We have, for example, an authentic representation of the outer appearance of a house when we actually look at the house; and
15 we have a symbolic representation when someone gives us the indirect characterization: the corner house on such and such side of such and such street. <194> Any description of a perceptual

---

[1] In his university lectures *Franz Brentano* always placed the greatest of emphasis upon the distinction between "authentic" and "inauthentic" or "symbolic" representations. To him I owe the deeper understanding of the vast significance of inauthentic representing for our whole psychical life, which before him, so far as I can tell, no one had fully grasped. — The above definition is not identical with that given by *Brentano*. I have thought it especially necessary to emphasize the univocality of the characterization, in order to keep *inauthentic* representations clearly distinct from *general* representations. In fact no one would designate the general representation "a man" as a representation (even as a symbolic one) of a determinate man Peter. It contains only a part of the defining properties appropriate to the characterization of the latter, and only through the addition of further properties is it to be filled out in such a way that we may designate it as an (inauthentic) representation of the individual, which then is capable of surrogating for the authentic representation of the same. — I hope to communicate more detailed investigations concerning symbolic representations and the methods of cognition grounded thereon in an Appendix to the second volume of this work. — Cp. to this present chapter also A. Meinong, *Hume-Studien*, II, Vienna 1882, pp. 86-88, (*Sitzungsber. d. phil.-hist. Classe der kais. Akademie der Wissenschaften*, 1882, Vol. CI, pp. 656-658).

[*anschaulichen*] object has the tendency to replace the actual [*wirkliche*]² representation of it by a surrogate sign-representation. Characteristic properties mark the object in such a manner that it can be recognized again as occasion demands, and thus all judgements adhering to the symbolic representation can subsequently be carried over to the object itself. Accordingly, the symbolic representation serves us as a provisional surrogate for the actual representation, and, in cases where the authentic object is inaccessible, even as a permanent one.

However, it is not merely objects of perception that can be symbolized, but abstract and general concepts as well. A determinate species of red is authentically represented when we find it as an abstract Moment in a perception. It is inauthentically represented through the symbolic determination: that color which corresponds to so-and-so many billion vibrations of aether per second. If we associate the name "triangle" with the concept of a closed figure bounded by three straight lines, then any other determination which belongs in univocal exclusiveness to triangles can stand in as an adequate sign for the authentic concept – e.g., that figure the sum of whose angles equals two right angles.

External signs can also serve in symbolization. Thus, by "C^3" the non-musician will merely represent to himself the indirect characterization: that tone which musicians indicate by means of the sign "C^3." Psychologically considered, external signs mediate every time language comes into play. But so far as *logic* is concerned, such signs come into consideration only in cases where the concept of that which is to be designated as such by an external sign belongs to the essential content of the symbolic representation.

Additionally I note that the authentic representation and a symbolic representation correlative to it stand in the relationship of logical equivalence. Two concepts are logically equivalent when each object of the one is also an object of the other, and conversely. That – for the purposes of our interests in forming judgments – symbolic representations can surrogate, to the furthest extent, for

---

² "*Wirkliche Vorstellung*" is used by Stumpf for *sensation*. In this and following passages "actual" and "actually" bear an associated sense of strong connection with reality. {DW}

the corresponding authentic representations rests upon this circumstance. <195>

Sense Perceptible Groups

After these preliminaries we turn to an in-depth study of the origination and signification of symbolic representations in the domain of number. For this purpose we must first consider more closely the function of inauthentic representing in the formation of *multiplicity representations*, in the course of which we may restrict ourselves to multiplicities of sense perceptible contents.

A sensible group [*sinnliche Menge*] – from now on we allow this convenient, though not wholly correct expression – first presents itself to a discriminating interest as a unified intuition, as a whole. But this does not distinguish the sensible group from the individual sense perceptible thing. Such a thing is also a whole, with regard to the fact that subsequent analysis discovers in it a multiplicity of parts: namely, of properties. However, in the sensible group the parts are, precisely, not contained in the manner of properties, but rather in the manner of *discrete partial intuitions*. And, indeed, they are of such a type that under given circumstances they draw a dominant and unitary interest to themselves.[3] Precisely because of this our primary intention is directed toward the formation of a totality representation that grasps each of these partial intuitions for itself and draws it into unison with the others. Our intention is directed to such union, but we lack the corresponding mental capacities to completely satisfy it with larger groups. Indeed, the successive grasping of the terms in the group one by one is still possible, but no longer their comprehen-

---

[3] Certainly representations of physical parts can also enter into a thing representation. But in this case the attention rests upon the combination of those parts with the whole – upon their belonging to it – which makes them into *properties* of it. In an entirely different manner do representations of physical parts belong to a group representation. Each one stands on its own, and not as property of the whole. Here the parts are, precisely, intuitively separated, so that their combination within the intuition of the whole recedes into the background. Moreover, where sharp separation of the partial intuitions is absent, it depends upon the direction of interest whether we speak of a group or of an intuitive thing (a tree/a group of branches).

sive collection. And, insofar as we in such cases still speak of a group or multiplicity, this obviously can only be in the symbolic sense. But we also commonly stay with symbolic representations even where an authentic <196> group representation could still be formed. And where that is not possible, we still do less in the way of authentic activity than we would be capable of. We omit the actual [*wirkliche*] apprehension of all the individual terms, carrying out only a very few steps, if any at all, and are thus satisfied with a still much more inauthentic subsumption under the multiplicity concept than in other cases. This mostly happens when the objects are similar to each other, and are of interest to us only as belonging to their immediately recognizable kind. The fact, verifiable at a glance, that this *hic et nunc* given intuition is a multiplicity of things of kind A is by itself enough to satisfy our interest, provided that it does not also require an exact determination of the "how many."

Attempts at an Explanation of How We Grasp Groups in an Instant

Consideration of the most obvious examples already shows that, in every case, serious and striking difficulties bar the way to the understanding of the Moments that mediate the symbolization.

We step into a large room full of people. One glance suffices and we judge: A group of people. We look up into the starry heavens, and in one glance judge: Many stars. The same holds true for groups of wholly unrecognizable objects. How are such judgments possible? For the actual [*wirklichen*] representation of groups we need, according to the foregoing analyses, a psychical act which represents each individual term of the group for itself and together with all of the others: thus, just as many psychical acts as there are contents, unified by a psychical act of second order. And only with respect to this form of psychical combination of individually grasped contents do the names "group," "multiplicity," "totality," etc., acquire their signification. Are we perhaps actually exercising in one surveying glance this complicated psychical activity, and also, above all, enacting a reflexion specifically upon it? For in the above examples we have not merely an apprehension of the group, but also its subsumption under the concept of group. This would indeed place strong demands upon our psychical capabilities. In the purposeful carrying out of a collective combination under the most favorable <197> of conditions – i.e., with exertion of all our mental power, and presupposing contents which are especially easy to perceive and which present themselves for apprehension in a succession that does not move too quickly – we take in not more than a dozen elements. And are we here to be mastering perhaps a hundred, quite without effort, almost instantaneously, and in any case unconsciously? I said "unconsciously" because we notice nothing of a lightning fast sequence of individual apprehensions and linkages; and we surely cannot assume that such a comprehensive and vigorous activity would be *forgotten* again in an instant.

The hypothesis suggested is, therefore, too improbable for us to base anything on it. There can be no doubt that the concrete multiplicity representation in these cases is no authentic one, and that the subsumption under the general concept of multiplicity, which is given with the use of the name "group," could take place only in the symbolic manner. But the question now becomes: What grounds and sustains this symbolization?

Here first off the following attempt at an explanation presents itself: That "one glance" of which we spoke above is not to be taken in an entirely rigorous sense. The apprehension is not exactly instantaneous, and we sometimes note how the moving eye on its own picks out this or that individual object or a small cluster here and there. Instead of carrying through the entire process of collection, we thus content ourselves with the mere rudiments of one. We seize upon whichever of the individual objects directly impose themselves, conjoin them, but break off again immediately by forming the surrogate representation: a total collection of objects which would be produced by fully carrying out the process just begun.

But now it is, in turn, the possibility of this symbolic representation that poses the difficulty. How can the apprehension and bringing together of some few terms serve as a symbol for the total collection intended? What enables us to know that the

process of collection can be continued by only so much as a single step, that beyond what has in fact been colligated there still remains something more to be colligated? What enables us to know that a "total collection" is to be intended? For this nothing
5 less would be necessary than the already realized subsumption of the sensible intuition of the group under the concept of group. If it already is known that the intuition considered <198> appertains to a group – that upon it, therefore, those psychical activities could be brought to bear which are peculiar to the concept of group –
10 then we might very well actually [*wirklich*] carry out (on the most available partial intuitions) some few steps of the symbolically represented collection process, and end with the familiar "etc.", sparing ourselves the complete execution of the actual formation of the group. In the opposite case, all of this is deprived of an
15 intelligible basis.

Symbolizations Mediated by the Full Process of Apprehending the Individual Elements

The resolution of these difficulties will come more easily if we first also examine more closely *those* symbolic representations of
20 groups in which the inauthentic subsumption under the concept of group does not come about in instantaneous immediacy (or else under the mere mediation of the individual apprehensions of a few members of the group), but in which, rather, there is accomplished by the authentically requisite psychical activities that which after
25 all must be accomplished: namely, *the successive apprehension* (even though not the unitary grasping together) *of all members of the group*. It is to be expected that *these* symbolic representations, as closer to the corresponding authentic ones, will form, as it were, the bridge between these latter and those other, more distant
30 symbolizations.

Certainly we can now no longer hold together in one act the successive apprehensions of the members of the group. Only a small number of those members remain at any one time sharply distinguished within the scope of the colligating activity. While
35 ever new members are grasped and conjoined, others of those

earlier picked out fall away again. The acts which represented them by themselves more and more blend into the background of consciousness, and finally disappear completely.

Nonetheless we do possess a determinate concept of the unity of the *whole* process. Even if only the last and very limited segment is actually present to us, we still have an awareness that this segment is not the whole process. <199> The course of idea association leads us back again along the chain to the earlier steps, or at least to the recollection that earlier steps had been taken. Executed anew, the process may well take up the members of the group in a different order of succession. But it is nevertheless the same members to which it refers, and the same unity of intuition within which it unfolds. We are capable of recognizing both of these ⟨as the same⟩. It may also happen that terms grasped one time are overlooked the second time. Yet nothing bars us from adding to the concept of the process the requirement that it take in all conceivable members. In numerous cases there also is no lack of means (as we shall yet discuss) for safeguarding the process in such a way that no member of the group can be passed over. And thus, from all this, we may form the symbolic representation of a completed process which, in *one succession or another* (indeed, it makes no difference), brings all conceivable members of the intuitive whole into our grasp.

The represented unity of this process which – even though not as one act, still as one succession – conjoins the separate apprehensions of all the individual members (or else is symbolically represented *as* conjoining them all) can then in further consequence serve as the symbolic substitute representation for the unity of the actual [*wirklichen*] collection, truly aspired to but unrealizable.

New Attempts at an Explanation of Instantaneous Apprehensions of Groups

Let us now see whether the understanding of these symbolic group representations, standing closest to the authentic ones, can really help us with those more removed symbolic representations whose illumination engaged us above, but without success. In one glance, we said, their character as groups is acknowledged. Is the authentic or the symbolic representation of the process just now described, which decomposes the unitary intuition into a succession of separate apprehensions, supposed to have somehow contributed to this symbolic subsumption under the concept of group?

One might suppose that the "one glance" signifies the lightning-fast execution of such a *process of separate apprehension* <200>, which then, in the manner described above, serves as symbolic replacement for the authentically intended collection. But one must reject this view for reasons similar to those earlier applied to the analogous supposition that during the instantaneous apprehension the actual *collection* comes about unconsciously. The deliberate carrying out of the process requires a considerable amount of time under the most favorable of circumstances – and the more so, the more elements there are to be traversed. But here an involuntary traversal is supposed to transpire in an instant, regardless of whether the members of the group are many or few. That is unacceptable.

The second attempt at an explanation, stated above, can also be transferred here in an analogous manner: instead of carrying through the whole process, we satisfy ourselves with a mere fragment of it. We select the first individual objects that impose themselves, and conjoin them, but break off immediately by forming the surrogate representation: a totality of objects which the process just begun would, in its full execution, bring to successive individual apprehension.

But here too the attempt fails, and for similar reasons. Once again we must ask: How can the first two or three steps of the process serve as a sign for the full process presumably intended? What enables us to know that the process of separate apprehension can be continued by even so much as a single step? What enables us to know that a "complete process" is to be intended? It is clear that nothing less would be required than the already realized subsumption of the intuition before us under the concept of group. Indeed the steps carried out suffice in order to speak, with reference to them, of a group. But that is not the group which is here in question. The totality of the few members grasped does not exhaust the group intuition before us; and of this fact we have a sure knowledge from the outset. We are aware that, besides the members picked out, many others also exist. And it is precisely this awareness that first gives our representation of a process rudiment, and of a full process to be intended, its sustaining substance.

Thus the essential difficulty remains unchanged. We must *already* have knowledge that the unitary intuition at hand is a group in order that that concept, which is supposed first to supply the symbolizing Moment and to explain the <201> indirect subsumption of the intuition under the concept of group, should receive any intelligible sense at all. It almost appears as if we are forced, after all, to return to the 'unconscious processes' rejected above.

Hypotheses

Only one way out is conceivable here: In the intuition of the sensible group there must be present *immediately graspable tokens* through which its character as *group* can be recognized, in that they indirectly guarantee that the process described above can be realized. With these tokens, then, the name and concept of the group could also be immediately associated.

That such characteristics could not appertain to the individual members of the group is clear. Each member could also exist for itself, and exactly as what it is in the group. It does not receive any new positive property by being together with the other members. — Should we then perhaps look to the sensible *relations* that combine the members of the group two at a time into pairs? But these too cannot offer the token markings sought. The particular relations cannot provide them for the same reason as the particular group members cannot. And all of the relations together can much less do so, since their diversity is much greater than that of the group members founding them.

Only if we might assume the following would matters be otherwise: that of the relational complexes encompassing the total group all or some particular ones should fuse into fixed unities which would then impart to the whole appearance of the group an *immediately perceptible specific character* – a sense perceptible quality of second order, so to speak. These quasi-qualitative characters, which in contrast to the element relations conditioning them would be the πρότερον πρὸς ἡμᾶς, could then provide the support for the association in each case. They would indirectly guarantee the existence of a relational complex, and therewith that of a multiplicity of relational terms founding it. Even if these quasi-qualities differed from case to case, if, depending upon the various kinds and the specific <202> differences in the element relations grounding them, they themselves appeared in various kinds and differences, there nevertheless could exist from group to group (e.g., corresponding to the classes of relations involved) elementary resemblances between the quasi-qualities which would then mediate the association. But we must also hold open the possibility that perhaps in each case of immediate knowledge of a group just one single quasi-quality, always showing up in the same way, would serve the desired end. However that might work, there then actually could result in *one* glance an instantaneous, even though wholly inauthentic, subsumption under the concept of group.

In place of a fusion of the (primary) *relations* contained in the unitary intuition of the group, one also could have recourse to the possible fusing of the intuitively separated group members themselves (perhaps also, further, the fusion of these with the "background"); and, indeed, either of the group members as wholes, or of some abstract positive Moments in them. Since any such fusion would be a relation, this hypothesis would fall within the domain of the foregoing one, if we may assume that the total fusion of the partial contents would itself constitute a fusion outcome of the elemental fusions conceivable within it.

Finally, yet a third hypothesis would be possible, which would base itself upon the quasi-qualitative character of the one as well as of the other type, by allowing that now the first, now the second, and then again complex characters proceeding from the interweaving of the two types, present us with the reproductive Moments in question.

Also with these latter hypotheses we must, in order to make things quite right, leave open the possibility that perhaps *one* wholly determinate (whether simple or complex) quasi-quality achieves the desired end, in that by showing up in each case of immediate apprehension of a group it would provide the support for the association with the concept *group*.

Given the close affinity of these hypotheses we can cohesively express the resulting explanation of the psychological origin of the symbolic group representation in question as follows:

In all cases where within one unitary appearance we <203> find before us intuitively separated parts, the entirety of which, picked out in one process of successive individual apprehensions, ultimately exhausts the whole, we obtain (as has been explained) a well-grounded symbolic representation of the collection corresponding to that unitary appearance. But precisely in such cases we also constantly find certain pronounced characteristics which, arising out of the fusion of the partial contents or of their relations, are immediately noticeable in the manner of sense perceptible qualities, and which perhaps (in case, namely, they admit of various species and differentiations) group themselves into "quality spheres," as it were, through striking similarities. Since from early life on we have run through apprehensions of individuals with the most heterogeneous types of sensible groups, those characteristics (or else their various generic types) had to become associated with the concept of such processes – and, in further consequence, with the concept *group* – and to thus erect in each case the bridge to the immediate recognition as a group of what is at first a unitary sensible intuition of the type here considered.

The Figural Moments

Thus, by careful consideration of the peculiar difficulties presented by the understanding of the immediate apprehension of larger groups *as* groups, we have been forced to hypotheses which appeared to be the only means of salvation from the here – if anywhere – unacceptable hypothesis of unconscious psychical activities.

Now everything will depend upon the *testimony of experience*.

That, above all, it abundantly confirms the *existence of quasi-qualitative Moments* of the type presupposed by our hypotheses is shown by many examples, which can be multiplied at will. One only needs to have once noticed them in order to find them again everywhere. In numerous cases they are also clearly expressed by the language of common life. One speaks, for example, of a *file* of soldiers, of a *heap* of apples, of a *row* of trees, of a *covey* of hens, of a *flight* of birds, of a *gaggle* of geese, and so on. In each of these examples we speak of a sensible group of objects like each other, <204> which are also named in terms of their kind. But not this alone is expressed – for the plural of the name of the kind would by itself suffice for that. Rather there is expressed a certain *characteristic property* of the unitary total intuition of the group, which can be grasped at one glance and which in its well-distinguished forms constitutes the most essential part of the signification of those expressions introducing the plural: "file," "heap," "row," "covey," "flight," "gaggle," etc.[4]

In every case, the differences of these quasi-qualitative Moments stand in functional dependence, now upon the inner characteristics of the respective partial intuitions, now upon certain relations and relational complexes that connect the partial intuitions to each other, and now upon both of these at once. Indeed, we will even attempt to justify the view that these Moments are to be plainly considered as units in which the peculiarities of the contents or of their primary relations fuse with one another. I say "fuse" ["*verschmelzen*"], and wish thereby to stress that the unitary Moments are precisely something other than mere sums. We grasp the quasi-qualitative character of the whole intuition as something simple, and not as a *collectivum* of contents and relations. But what is simple to our first apprehension turns out upon subsequent analysis to be something complex. We discover the intrinsic and the relational peculiarities appertaining to the respective quasi-quality; we clearly see (at least in the cases easier to analyze) that those peculiarities form its *parts*. And we always can bring to Evidence the fact that they actually and exclusively condition the specific character of the quasi-quality. — If we afterwards find that which originally seemed simple to be something that is in truth complex, we do not thereby apprehend it as a

---

[4] That, moreover, sense perceptible similarity also belongs among these characteristic properties we will soon prove.

mere multiplicity. Complexity is not multiplicity pure and simple, but rather is a multiplicity of parts united into a whole in the narrowest sense of the word. There is therefore no disadvantage in the fact that we pick out the quasi-qualitative Moment in the manner of something simple, and that it nevertheless is subsequently to be analyzed into a multiplicity of parts noticeable in their own right. <205>

Let us now begin the more detailed exposition by considering certain distributions of objects within the visual field.

In reference to such a distribution there is first to be confirmed the fact that we grasp its *configuration* just like we do a quality: in *one* glance, without there occurring, in and with it, an analysis into the particular relations which condition the figure, and without such an analysis even being possible. The assertion that the representation of the figure consisted in the representation of the *sum* of those conditioning relations, would in fact involve the requirement – in general quite unfulfillable – that in an actual representation of a totality we inclusively take in all of the individual elemental objects with their reciprocal relations. It is obvious that only subsequent analysis shows us that the Moment of the figure is necessarily conditioned upon these and those relations. Each variation in relations of position conditions a variation in the figure, and conversely. But we observe the variation of figure *before* it comes to our consciousness that these or those positions have been changed. This is illustrated with especial clarity by arbitrary cases of simple or complex sequential orders. The intuition of the whole is modified, depending on whether the particular terms and the particular sequences are closer together or further apart, are equally spaced out, or whether they run parallel to each other or at an angle. The figural Moment leaps out at us immediately, and only upon subsequent reflection do we notice the conditioning relations, changing as they do from case to case.[5]

---

[5] One sees that what is here meant by the "figural Moment" of the intuition (even under the merely provisional restriction to spatial characteristics) is something more inclusive than "figure" in the ordinary concept of the *Gestalt* or shape of a spatial intuition, in contrast with its size or position: a concept from which, also, the idealizing conceptualization of the geometrical concept of figure takes its start. Any displacement or rotation of a spatial shape in our field of vision already conditions a modification of that Moment in the

With this example we also can make clear what relationship the relations, just shown to *condition* the figural Moment, have to one another and to that Moment itself. As just emphasized, the sequence character is what first presses itself upon us as a unitary abstract characteristic of the intuition. But with subsequent analyses we observe <206> quite well that this quality is nothing *simple*. We grasp the simple relations which join any pair of neighboring members in the sequence, and we also grasp the relations of second order that combine *those* pairs of simple *relations* which are after a fashion contiguous to one another, in that they have in common a term that is identically the same. And it is not first through the activity (coming later) of relating the terms that we set up these [second order] relations. They are there and undoubtedly *come along with* the unity of the figure. That this unity is more than the mere sum of the relations we gladly concede. But that is after all true for any unity that is more than a mere collective unity.

One also sees that our way of saying that the relations *fuse* into the unity of the quasi-quality is wholly correct. The fusing here is the exact analogue to that which *Stumpf* discovered in the qualities of simultaneous sensations.[6] In fact we rediscover here again the essential determinations of *Stumpf*'s concept: above all, that the elements fusing can also show up, as what they are, outside of that fusion, and that in the fusion they are not "in the least modified . . . but a new relationship emerges between them, which institutes a closer unity than occurs between the terms of a mere sum . . . ."[7]

Also, it is to be confirmed here that fusion shows up in *varying degrees*. And likewise we can also here designate as a consequence of fusion that in its higher degrees the total impression, other circumstances being equal, approximates one of a truly

---

intuition which we would subsume under the concept of the figural Moment. Likewise, of course, for each change of size while retaining the shape. But certainly what is meant nowhere so clearly leaps to the eye as with the figure in the ordinary sense of the word.

[6] Moreover it does not escape *Stumpf* himself that the concept of fusion has a wider validity, in that he explains: "Fusion . . . [is] that relationship of two contents – *especially*, contents of sensation – in terms of which they form not a mere sum but rather a whole." *Tonpsychologie*, Vol. II, p. 126.

[7] *Ibid.*, p. 64.

simple quality and becomes more and more difficult to analyze. In the present case, the point continuum gives the highest degree of fusion.

We have indicated just now the manifold *variations* which the figural Moment undergoes with variation of the relations determining it. Collectively, those variations stand in the same <207> relationship to one another as do the manifold specific differences of a genus of sensible qualities. We find between them elemental similarities of many levels, from which there develops a "genus" concept in the rigorous *Aristotelian* signification of the term. The general concept of the configuration is the exact analogue to the concept of a genus of sense perceptible qualities. Equality signifies in the configurations, too, nothing other than extreme similarity.[8] The specific differences in our genus concept depend upon the combination of the differences in the elemental relations of distance and direction ⟨between the elements⟩ which fuse into the specific differences, and they therefore Represent a much more diversified continuum than the species of these elemental relations themselves.

Up to this point we have considered only the, so to speak, "geometrical" relations of distance and direction. But other relations as well have a noticeable influence upon the quasi-qualitative character of certain distributions of objects in the visual field. The individual objects stand out more or less sharply from one another and from the intuitive background; and this "standing-out" is also a sense perceptible, relational Moment which admits of many graduated degrees and, corresponding to them, can influence in various ways the quasi-qualitative character of the total group intuition – as can be tested on easily accessible cases. Upon a certain degree of separateness there depends the very possibility of apprehension of any other figural Moments: indeed, the possibility of an apprehension of a group itself. If the members of the group immediately border on one another, if they completely fill up one part of the visual field, then their color qualities must differ by more than a certain interval in order that they not fuse into an unanalyzable unity. If, on the other hand, the members of the

---

[8] Cp. C. Stumpf, *Tonpsychologie*, Vol. I, p. 111.

group are strewn across the field of vision, then this same condition on their color quality holds in comparison with the color of their "background," while among themselves they can be of the same quality.

The qualitative relationships themselves, in turn, condition in various ways the character of the total group appearance, and in so doing they often fuse into very noticeable quasi-qualitative Moments, <208> that usually are, to be sure, essentially conditioned by the Moment of configuration. A characteristic case is, for example, that of the qualitative sequence in fusion with the sequence in the sense of loci. If the same objects form a different configuration, then in general the character of the color constellation will also be quite different. Only if the objects are of the same color does the Moment of configuration not influence the Moment of the quality.

Qualitative sameness, and hence in general the sense perceptible sameness of the entire membership of the group, is one of the most conspicuous of quasi-qualitative Moments. That sensible sameness actually grounds such a Moment, one recognizes in each example. It gives to the intuitive group as a whole a specific character that is recognized without each particular thing having to be compared with each other one. We judge at one glance: a group of apples, nuts, people, and so on, without having to undertake the corresponding $n(n-1)/2$ comparisons, and in most cases without being able to do it in actuality. The memory of earlier comparisons cannot guide us in this judgment; for (to mention only one thing) even with groups of totally unfamiliar objects such an exhaustive comparison is absolutely not required.

To this point we have considered in the visual field only distributions of immobile objects. But any kind of movement or qualitative change in the individual objects also imparts to the whole an immediately perceptible quasi-qualitative character. In certain of our examples this character is also expressed in the language used, as when we speak of a flight (of birds). For the rest, one recognizes everywhere the peculiar character and immediate perceptibility of this Moment, as soon as it has only once come to our attention. — And precisely such Moments arise, it seems, from all other primary relations that fall within the realm of the visual sense.

Besides the relations, the *intrinsic characteristics* of the partial intuitions are also efficacious as elements of fusion. In examples resembling the chess board this is present beyond all doubt. The configuration of the black squares is exactly the same as that of the white. Within each configuration <209> the squares are identical to each other in shape, size and color, and thus ground in the one case as in the other a figural Moment of sameness. Nevertheless, the unitary total character of the two appearances is a sharply distinguished one, and this precisely in virtue of the distinct colors of the squares in the two cases.

An analogous point holds also for figural Moments of separated partial intuitions. Any change in them, e.g., in the size or form of the individual members of the group, also modifies the immediate character of the total appearance.

Everything we have stated here for groups within the field of vision can obviously be carried over to sensible groups of every type; likewise to groups in general, whether groups of sensible objects represented in phantasy, or groups of psychical acts. In the latter case, for example, temporal succession, and, in general, temporal configuration (the exact analogue of the spatial), forms a Moment of this kind.

In association with their most prominent special case, it would perhaps not be inappropriate to select the phrase *"figural* Moment" for these peculiar characteristics of unitary intuitions analogous to sense qualities.

What was stated above of the spatio-figural Moment of visible groups holds true for each one of these figural Moments of group intuitions: to each one there belongs, besides its specific character, a certain generic character.

The various types of figural Moments occur, corresponding to their specific differences, in the most diverse of mixtures – or, more exactly, fusions. Cohesively fused in the relevant intuition, they are separated out only by abstraction. The Moment of temporal configuration, for example, fuses with the Moments of quality and intensity; and the compound has at the outset a unitary figural character which only through analysis decomposes into its components. A melody involves an intricate complexity of such a type. A quite simple example is offered by a sequence of similar objects of vision. Here one very easily separates the sequential Moment from the Moment of [qualitative] sameness.

While the unification of the figural genus-characters plays a considerable role in the formation of general group representations <210>, or of group-types so to speak (consider the general signification of such names as "file", "gaggle", "covey", "heap", etc.), it is the unification of their specific differences that gives the intuitive group its individually unifying character.

The various figural Moments offer different degrees of resistance to abstractive isolation. The one is noticed more easily, the other with more difficulty. Many cannot be independently apprehended at all, and their existence can only be indirectly inferred by the way the total quasi-quality varies with the variation of the elements conditioning it. Here we can speak of a differing degree of fusion which the various figural Moments enter into. An especially powerful stimulus is exerted upon our discriminating attention by all types of series, arrays, systems, and by all configurations constructed out of relations of distance and direction – of which the spatial ones make up only one special case. Complexes of temporal points, intensities and sense qualities provide further examples.

Where a multiplicity of separate objects come together in one intuition, there the figural Moments which appertain to all conceivable partial mulitiplicities in it stand in competition. In that we detach one *determinate group* in its intuitive unity, precisely that figural Moment is victorious which exerted the strongest stimulus upon our apprehension. Sometimes this victory is only momentary: we grasp now this, now that group within the total intuition to which they all belong, depending on whether the figural Moment of this one or that one prevailed.

The Position Taken

Our purposes do not require a comprehensive investigation of these remarkable – and until now almost totally neglected[9] – features of unitary phenomena, and <211> especially of sense perceptible groups. What has been discussed suffices to assure us of the existence of the figural Moments of which our hypotheses made use in the explanation of the symbolic apprehension of groups.

Which hypothesis, then, are we to prefer? The first reflected merely on figural Moments that proceed from the fusion of relations, the second merely on those which proceed from the fusion of the absolute contents. However, these two sorts of Moments show up in such an intimate connection that from the start we are inclined toward the third hypothesis, which allows for Moments of both sorts – perhaps in a unitary fusion.

The assumption (which all hypotheses held open) that certain *constant* Moments that show up in every case of direct cognition of groups mediate the association with the concept of group, is one we shall certainly have to reject: even though there may be no shortage of such Moments. Of this sort, for example, is the Moment of separateness; for we can only speak of a sense perceptible group where, within one unitary appearance, the elements stand out clearly from each other or from the background. But such Moments always show up in fusion with other Moments that are sometimes more powerful; and it is not required for the explanation of the group apprehensions in question that analysis should first set in and that the association of the group concept

---

[9] The foregoing investigations had been worked out for almost a year before the penetrating work of *Chr. Ehrenfels*, "Über Gestaltqualitäten," (*Vierteljahrsschrift für wissenschaftliche Philosophie*, 14, 1890, pp. 249-292) appeared, in which the "figural Moments" investigated above only incidentally, in the interest of an explanation of the indirect apprehension of groups, are subjected to a comprehensive investigation. Unfortunately, the essay mentioned is not accessible to me as I prepare these pages for the press, so that I must forego a more full acknowledgment of it. As he states at the very opening of his presentation, *Ehrenfels'* research was instigated by E. Mach's *Beiträge zur Analyse der Empfindungen*, Jena 1886. Since I had read this work by the gifted physicist right after its appearance, it is quite possible that I too was partly influenced in the progress of my thought by reminiscences from that reading.

should always come about directly through the meditation of those former, constant Moments. In that especially striking figural Moments of both types (as, for example, configurations in the narrower sense in fusion with Moments of sameness or quality) have provided impetus for the detachment of unitary intuitions of groups – and, in further consequence, the process of term-by-term apprehension has joined in – the symbolic concept of group had necessarily to associate itself with them, or rather with their characteristics typical of the genus. And this without there first having to occur an analysis of the <212> partial Moments and an isolation of certain ones that are everywhere the same.

If we have in fact carried out the traversive process of exhaustive (or supposedly exhaustive) apprehension of the individuals in many discrete distributions of objects within the visual field, eventually we come to recognize immediately each new distribution – and even without that process – as a group. The analogy of all the configurations to each other – or the analogy with respect to complex Moments that easily ground recognizable group-types – mediates the association. In the case of groups within other sense fields, other Moments may serve us. But it is always the similarity within a certain range of qualities that offers us, so to speak, the most available support for the association. This is also recognizable in the examples from which we set out, where the figural character led to special names such as "file", "herd", "flight", etc. Two files, herds, or flights are never exactly the same; but their similarity grounds the genus concepts, which, directly cognized, then also mediate the direct knowledge of the character *group*.[10]

---

[10] Our theory of the inauthentic apprehension of groups also explains the fact, noticed by *Stumpf*, that "... the accurate differentiation of one plurality from another already ... <is> a higher function than the perception of a plurality in general." (*Tonpsychologie*, Vol. II, p. 371) That perception is certainly, as a rule, a symbolic one, mediated through the figural character of the whole intuitive unity of the group; and there then occurs no individual apprehension of the members of the group, as will in general be required for the purposes of an exact comparison of the group with other groups, or for its enumeration (which *Stumpf* especially has in view in the above passage). And it is indeed very likely that a figural Moment would still be reproductively efficacious, even when the group members in the intuition are so intimately fused that a clear separate apprehension of them becomes completely impossible.

SYMBOLIC REPRESENTATIONS OF MULTIPLICITIES   225

The Psychological Function of the Focus upon Individual Members of the Group

The conceptualization of those inauthentic representations of groups here considered is usually accompanied by a few steps of
5 individual apprehension of some of the members of the group. It is not without psychological interest to analyze more closely the particular function of this process.

It supplies, on the one hand, an approximation to the actual [*wirkliche*] formation of the group, <213> and to the actual sub-
10 sumption under the concept of group, by actually carrying out the requisite psychical activities upon at least a few select members. As earlier explained, the rudimentary process then serves as a symbol for the full process intended, whereby the unitary figural quality of the group intuition guarantees us that the process begun
15 can be continued – especially since the intuitive group-like unity of the members picked out is recognized as part of the total group intuition.

On the other hand, the individual apprehension also gives us the genus concept of the members. In fact, it frequently first deter-
20 mines the segregation of the group as a whole. As soon as interest is directed upon one thing merely in virtue of a certain characteristic, then with one stroke there stands forth the entire totality of the objects of this genus which still remained in the intuitive background, provided only that they protrude sharply enough in
25 order to be at all capable of forming an easily noticeable group unity. And it all depends upon the interest, which now turns to this, now to that genus concept, and extracts with it now this, now that group unity from the unanalyzed background. The fusion of the homogeneous contents in the intuition forms, quite before any
30 analysis, a certain unity, and thereby mediates this peculiar type of association. The single content hangs, as it were, upon a chain, which it pulls after itself as soon as it comes into conscious focus. If, for example, we attend to a white square of the chess board, then the whole configuration of the white squares stands out; if we
35 attend to a black square, then the same holds of the black squares. And conversely: if we wish to retain the one or the other configuration, then we must focus upon at least one of its squares. If we

attend to one line on this page, then the entire sequence of lines already emerges. If by chance our interest falls upon the white space between two lines (as in estimating their distance), then it is the sequence of these bands which stands forth in a unitary manner. Parallel bands of alternating colors, or series with members of alternating shapes, in general furnish clear examples.

I mention, finally, that in the course of a thinking which proceeds more rapidly <214>, external intuition also frequently can become the symbolic replacement for the authentic group representation merely in virtue of its figural character – and without any rudimentary processes. If for example we speak, with reference to a certain intuition, of a regiment of soldiers, a row of trees, or the like, then usually there is lacking any truly plural representation, and only the unarticulated intuition is there. But as soon as it is necessitated by the course of further thinking, there come into play the psychic activities which the plural expresses.

What Is It that Guarantees the Completeness of the Traversive Apprehension of the Individuals in a Group?

In the cases considered up to now the interest that drives the apprehension finds its satisfaction in the fact that a multiplicity of objects of a determinate genus is given, a multiplicity delimited by the exterior frame of the intuition (i.e., by an imposing figural Moment of it). But in other cases the interest is turned toward each of the individual objects enframed, and it therefore aims at *exhaustively traversing* all of those objects. How that comes about would, again, be an unsolvable problem without recourse to the figural Moments of the intuition. Since authentic collocation is an impossibility for larger groups, where do we get the certainty that we actually have grasped *all* members of those groups? The correct answer to this question is of special importance for the psychology of the enumeration process, for it simultaneously answers the question about the possibility of a *complete enumeration*, one by one.

Now here once again there are certain figural characteristics within the unitary group-intuition which confer upon our interest in the distinct apprehensions an ordered and secure procedure. For example, if the group possesses the character of a simple sequence, then it is the linkages of the elemental series relations (and these are themselves figural Moments) which guide us. If the sequence is bounded, we begin with one of the boundary terms, for these exercise, other circumstances being equal, the first and strongest stimulus on the discriminating interest. <215> The boundary term belongs to a first elemental linkage, which imposes itself upon our apprehension in the manner of a unitary figural Moment. Through analysis of that linkage we reach the first adjoining term; and in this latter a new elemental linkage connects up with the first, on the basis of which we attain to the second adjoining term, and so on. In that two contiguous linkages are always simultaneously present to us – and, indeed, as clearly distinct – the new one can be recognized as such; and the progression becomes univocally determined. And this univocality guarantees the completeness of the traversive apprehension of the individuals.

The matter is not much more complicated in the case of a double sequence. The total group here immediately breaks down into a group of groups, of which each is characterized as such by a special figural Moment (the sequential Moment), while the entirety of the groups shows itself, again through such a figural Moment, to be precisely a group *of* those groups. The progression from sequence to sequence, and the progression within each sequence from term to term, is a univocally determined one, on grounds analogous to those in the previous example, just as soon as, in each case, a boundary sequence is chosen as the sequence to start from, and in it, as in each following one, a boundary term is chosen as the beginning term.

Now it is, quite generally, similar with arbitrary groups. *Either* the given group possesses from the start not merely a figural Moment singling it out as a whole, but also [interior] articulations conditioned by special figural Moments – natural articulations or ones which impose themselves given dominant habits of thought. These articulations mark off partial groups within the group, and allow the overall group to appear as a clearly ordered sum of groups, whose ordering, of course, is again made possible by a figural Moment. *Or else* the group is at first without any clear articulation or ordering. In that case we must provide artificial aids through arbitrary formation and arrangement of the groups. In the course of this we frequently employ, for the greater security of the procedure, extrinsic marks, such as framing by boundary lines, assigning numbers to the groups, and the like. For this the groups must be so chosen that the progression from member to member is sufficiently determinate to guarantee the completeness of the apprehension of the entirety of their <216> members. In virtue of the ordered relationship of the groups to one another, an exhaustive traversal of the members of the overall group is then also secured.

Apprehension of Authentically Representable Groups through Figural Moments

Before leaving this subject I would further observe that even in the apprehension of small groups, where an authentic collocation can still be spoken of, figural Moments often play a not inconsiderable role. Any, even the smallest, group – of, for example, visual objects (and hence of sensible contents in general) – is characterized as an intuitive unity by a figural Moment; and hence that Moment determines the frame for our successive apprehensions of the individuals. Further, even with groups of four or five objects there frequently, though not necessarily, occurs an articulation into sub-clusters. We then grasp them under the form $2 + 2$ or $2 + 3$ (or else $2 + 2 + 1$). With groups whose number of elements is greater than five, such articulations are a sheer necessity if an actual [*wirkliche*] group apprehension is to occur at all. And where, in virtue of the nature of the intuition, such articulations do not of themselves stand forth, we must artificially introduce them into it.

That in such forms of apprehension a certain facilitation resides is an indubitable fact, but also a remarkable one. Accordingly, the apprehension of a collection of collections seems easier than the apprehension of a simple, unarticulated collection, even though in the former case psychical acts are required that are of a higher order than in the latter (pp. 96-97). However, the psychical acts which come into play in the former case are extraordinarily facilitated by the figural Moments, which connect the contents of the clusters in a characteristic manner. In order, for example, to get a clear apprehension of six successive strokes of a clock, we break them up into two clusters, each with three strokes. Through a graduated emphasis we thus evoke, corresponding to the clusters, especially noticeable figural Moments (each compounded of an intensive and a temporal Moment) <217> which, as primary connections running parallel, give to the collection of clusters a fixed framework, so to speak, or an external support. In addition, it seems that the colligating act enters into an association with the corresponding figural quality. Through such means one still masters groups which, in an unclustered form, could no longer be sustained as actual collections – perhaps because the conjoining interest would tire from the uniformity of their composition.

But nevertheless, one must still regard group representations originating in this manner as authentic, since all members actually [*wirklich*] appear unified into an explicit representation through one single act. The supporting figural Moments in all of this belong, indeed, to the psychological, but not to the logical, content [*Gehalte*] of the representation. As from the temporal succession, so also we can abstract from the figural Moments which single out either the total group or its subgroups, and attend to the mere "togetherness" of the members in one representation. And this is something that we indeed *must* do when the genuine intention of the *group* representation is in question.

The Elemental Operations on and Relations between Multiplicities Extended to Symbolically Represented Multiplicities

As can be easily seen, one can also extend the concepts of the elemental operations and relations to symbolically represented multiplicities, in which once again the figural Moments will often serve as mediators. If, for example, several sense perceptible groups are simultaneously given to us, marked as such by the familiar symbolizing Moments, then there accrues to them also a total intuitive unity. They have it in virtue of a figural Moment encompassing them all, which in turn characterizes the whole as a group. This sense perceptible relationship of intuitions supplies a sure support for the symbolization of the additive conjunction between the corresponding actual collections. Likewise we can form the concept of partial groups with reference to an intuitively presented group. The total intuition articulates into partial intuitions, to each of which accrues a clearly noticeable figural Moment characterizing it as being a group. <218> The sense perceptible relation of partition may in turn serve as support for the symbolization of the intended actual group relations. With respect to this state of affairs we can in a good sense speak of adding and partitioning, and likewise, obviously, of increasing and diminishing. For arbitrary, symbolically represented groups we can raise the question of comparability, whereby the concepts of equal, more, and less appear in obvious symbolic applications (and therefore in no need of further discussion). With the substantiation of these relationships in given cases of comparison, symbolic devices come into play to which we have already called attention upon another occasion.[11] Thus, we have the coordination in pairs of the successively selected group members; or, instead, the reciprocal denumeration, provided that the symbolic extension of the number sequence is already realized.

Finally, I emphasize that the modifications which the multiplicity representation undergoes through all of the symbolizations described do not affect its *logical* content. *Multiplicity* remains the concept of a whole, of a determinate collection of separate contents. Only, in the cases now considered, the segregation of contents and their collocation, instead of coming to actual [*wirklichem*] realization, remains either wholly or for the most part a mere intention [*blosse Intention*].

Infinite Groups

The symbolic group representations which we have considered

---
[11] Cf. Chap. VI, pp. 111-115.

up to now still do not cover the full extension of which the concept of group or multiplicity is capable through symbolic means. There remains yet one especially noteworthy extension for us to analyze, one which extends the original concept in such a manner that it surpasses not merely the, so to speak, contingent limits, but also those necessary to the essence of all knowledge, and thereby also attains to what is fundamentally an essentially new content. An extension of the capacity for representation <219> which would put it in a position to grasp, in the collective manner, groups of a hundred, a thousand, or a million elements in genuine engagement is quite conceivable. And hence our intention [*Intention*] which underlies the symbolic representation of such large groups gives no occasion for logical scruples. That intention tends toward the actual [*wirkliche*] representation of collections which, if not within our scope, yet fall within that of an idealized capacity of human knowledge. In surrogating for the actual collocation there can then serve, in many cases, sequential processes which at least bring all members to a successive individual apprehension. We are able to bring *each single* group element to representation in its own right in temporal succession, even though not in one all-inclusive act. But all of this is impossible in the cases to which we now turn. We speak of totalities, groups, and multiplicities also where the concept of their authentic formation, or of their symbolization through sequential exhaustion of the individuals involved, already contains a logical impossibility. We speak of *infinite groups*. The extensions of most general concepts are infinite. The group of the numbers in the symbolically expanded number series is infinite, as is the group of points in a line, and, in general, that at the limits of a continuum. The thought that some conceivable expansion of our knowledge capacity could enable us to have the actual representation – or even the mere sequential exhaustion – of such groups is unimaginable. Here even our power of idealization has a limit.

But how then do these symbolic concepts come about? What constitutes their psychological and logical substance [*Gehalt*]?

In each case where we speak of an infinite group, we come upon the symbolic representation of a process of concept formation that can be continued without limit. A clear principle is given according to which we can transform (or can also *symbolically represent* as transformed) any concept already formed, of a certain given genus, into a new concept rigorously distinguished from the first, and this latter again, and so on, in such a way that there is the *apriori* certainty of never turning back to the concept with which we began or to the concepts already generated. We form, then, with reference to the results <220> of this process, successive group representations which constantly expand, and if the principle of formation actually is a determinate one, then the concept of the constantly expanding group of concepts also receives a wholly determinate content. It is *apriori* determined, by means of rigorous conceptual Moments, what this constantly expanding group includes or can include, and what it cannot. That is, for any given object of thought it can be unambiguously decided whether or not it can be a member of this process or of this formation of groups.

Let us consider, for example, the concept of the infinite group of numbers. The process of adjoining one unit to an arbitrarily given number is an operation whose concept apriori guarantees that it leads to a new and determinate number. If we start from the number one, then this principle of formation leads to two, three . . ., and to new and ever new numbers, without turning back and without limit. The conceptual determination: "a possible result of this process," is, like that of the process itself, a rigorous one; and therefore the possible results of the sequential construction of concepts indicated possess a common characteristic that binds them together, analogously as the collective unity (or an intuitive unity replacing it) binds together the members of a group. Thus, when we speak of the totality of all natural numbers, we represent first of all a group in the usual sense – namely, the numbers of an initial subsequence of the number sequence (symbolized through the intuitive series of signs, or the like). To this is joined the supplemental representation that this sequence, in virtue of its principle of formation, can be extended *in infinitum*, whereby each new member would be determined by means of this process. If we speak of the infinite group of points in a line, we first represent some distribution of points on it, and then add the supplemental idea of an unlimited process through which we can think any pair of neighboring points as mediated by new and ever new points. And similarly in other cases.

It is now easy to indicate the Moment which has provided occasion for the transposition of the concept of multiplicity over to formations that, in their logical character, are essentially distinct. Already in the symbolic representation of groups in the ordinary sense there often surrogates, as we saw earlier, the Idea of a process whose unity <221> receives its determinateness through some figural Moment of intuition. It is similar here; only now it is a more removed, a conceptual, principle that confers upon the process its determinateness, and that gives a certain support to the representation of all that is attainable through that process, of all that it "embraces." But whereas in the first case its finitude belonged to the concept of the process, so that in the succession of steps one of them must be its *last*, here, to the contrary, what belongs to the concept of the process is its being unlimited: the concept of a last step, and thus of a last-reached member of the group, becomes senseless. In the first case, it was sometimes possible actually to exhaust the process completely, and, indeed, perhaps to form the corresponding authentic collection. In the latter case, the mere thought of that is absurd. The two are essentially distinct logically. But the analogy, of which we spoke above, awakens a natural inclination to insert into the representation of the infinite group the intention toward the formation of the corresponding actual collection – in spite of the absurdity of the thought. Hence there arises a concept which is, as it were, imaginary, but whose contra-logical character causes no harm in everyday thinking precisely because the inconsistency it contains does not come into play in the usual course of events: as, for example, when we represent to ourselves the "all S" of the universal judgment as a closed set. In other cases things are different, where the imaginariness mentioned is itself an effective factor in influencing the judgment. This much is clear: For a rigorously logical understanding we may impute no more to the concept of the infinite group than is actually, logically, admissible – hence, above all, not the absurd intention toward the formation of the actual group. The representation of a determinate, unrestricted process is logically irreproachable, as is the Idea of all that which falls within its scope, which it encompasses by means of its own conceptual unity. So much and no more may, therefore, be assumed within the concept of the infinite group. But with this there also clearly stands forth the fact that in it we have an essentially new concept which is no longer a concept of a "group" in the true sense of the word, although it does contain this concept (e.g., in that of the process) as an essential constituent.

Where we speak of groups in what follows, *finite* groups shall always be meant, unless the opposite is expressly stated.

Chapter XII

THE SYMBOLIC REPRESENTATIONS OF NUMBERS

The Symbolic Number Concepts and their Infinite Multiplicity

The symbolic representations of *groups* form the foundation for the symbolic representations of *numbers*. Had we only the *authentic* representations of groups, then the number series would at best end with twelve, and we would not even have the concept of a continuation beyond that. Along with the obvious lack of restriction on the symbolic expansion of groups, the same is also given for numbers, as we will soon see.

The numbers are the distinct species of the general concept of multiplicity. To each concrete multiplicity, whether it be authentically or symbolically represented, there corresponds a determinate multiplicity of units: a number. If we think of each member of the multiplicity as subsumed under the concept *unit*, then the concept of the collection of all of these units is indeed a completely determinate one. The collection is changed with each member or sum of members that we may add to or take away from the given group. In the symbolic sense we thus can say of any arbitrary group that a determinate number accrues to it even before we have formed that number itself; indeed, even when we are not in position to undertake the actual [*wirklichen*] formation of it. Likewise we may with good reason state that two arbitrary groups must be of the same or of a different number, whether we can conceptualize the number or not.

We are even justified in judging that the domain of number encompasses *an unbounded manifold of species*. <223> In fact, if we start with any arbitrary symbolic group representation, then we possess (at least Ideally) the capability of expanding it without limits by continually adding new and ever new members. If we can do nothing else, we can at least think of the members proper to the group as mirrored in a constant reiteration, and accordingly form the concept of the progressive expansion of the group by means of the members of its reflections. Certainly this symbolic concept formation involves a strong Idealization of our powers of representation. Indeed we cannot, in fact, form the required reiterations *in infinitum* and arrange them in sequence. We lack time and strength for the ever-renewed mental activity required, as well as symbols for keeping the formations of those reiterations distinct. However, we can, by way of Idealization, disregard these limitations on our abilities and conceive the concepts, which are symbolic also in this respect. If now the given group is symbolically expanded through such means, then there belongs – again, in symbolic representation – to each level a determinate number, different for each new level. The earlier group-formation is indeed part of the one newly formed; and hence the same is also true of their numbers. The manifold of conceivable number specifications is therefore infinite, like the manifold of conceivable group levels.

In this way, along with the limits that restricted our representing of groups, the limits on the conceptualization of number concepts have also fallen away. In a symbolic but wholly determinate sense we can speak of numbers where their authentic representation is forever denied to us; and on this level we are even in position to establish the Ideal infinity of the realm of numbers. But in no wise is our investigation thereby already completed. The remote symbolizations which we have attained to up to this point certainly can be of no use, given their vague generality, for the purposes of enumerating and calculating. For those purposes we require symbolic formations richer in content, which, coordinated in their rigorous distinctiveness with the true – but to us inaccessible – number concepts "in themselves," are well suited to stand in for those concepts. <224>

The Non-Systematic Symbolizations of Numbers

Assuming that ten was the last authentically representable number, then there are various possibilities for denumerating those groups that are not exhausted by the numbers up to ten. Any arbitrary decomposition of the group – or any decomposition of it that is of itself suggested by the character of the group intuition – into authentically enumerable partial groups leads to the symbolic construction of the concept of a number which is additively composed from the authentically representable numbers of the partial groups. What we thus grasp in concrete cases we can generalize, and there result symbolic number formations such as "10 + 5," "9 + 6 + 8," "7 + 10 + 5," and the like.

In these formations an important function is fulfilled by the number names or number signs. In spite of the articulations, we can no longer hold such large groups of units clearly distinct in unitary representation. The composition of the signs is our crutch. As we reflect step by step upon their signification, the individual numbers in the sum enter our consciousness in the form of a determinate succession. Even if, as the new number turns up, the previous one blurs into obscurity – and, accordingly, the actually [*wirklich*] intended sum-representation cannot be realized – we still have the sensible composition of the names (or written symbols) as the fixed framework within which the succession of the conceptual elements in the sum, mediated through the symbolization, can always be generated in the same determinate manner.

One immediately notices that this procedure, which first offers itself for singling out determinate, symbolic number forms, is still highly incomplete. If we split up larger groups into partial groups, each of which does not exceed the number ten, we soon arrive at so many repetitions of the same partial numbers that we are hardly better off with the distinguishability of the emergent symbolic formations than with the corresponding unarticulated sums of units. This inconvenience can be remedied, at least to a certain extent. The number of the elements in the sum must remain as small as possible. If, nevertheless, any arbitrary group is to be denumerable, then not merely the numbers which can be authentically represented, but also the <225> symbolic numbers already formed must be admitted as mediators of sum formation. It is clear that to this end a special name would have to be introduced for each such number composition, since without the support of external signs the graduated structure of symbolizations grounded one upon the other would have nothing to rest upon. Proceeding in this way, one could even restrict oneself to sum formations of two terms. If, for example, one has introduced the symbolic formation "$p = 10 + 5$," one can then construct, say, "$p + 8 = p'$," and then again "$p' + 10 = p''$," and so forth, whereby each later construct has the whole sequence of earlier ones for its basis.

Given such symbolizations, nothing more seems to stand in the way of the universal expansion of the original number domain beyond any limit. But, closely considered, we do not get very far by the means indicated. If at each step new types of sum-formations were used for the construction of the symbolic number concepts, then the manifold of number forms would soon become so vast that it would be inconceivable that memory should master them. To this are added other and no less serious deficiencies. I only point out the problem of number comparison. Given that wholly systemless expansion of the domain of number there will, in general, turn up whole sequences of symbolic number formations each of which corresponds to one and the same actual number. This is easily made clear by examples. One and the same group admits of various articulations, each of which will lead to a new symbolic number form; while the identity of the actual number corresponding to them all is guaranteed through the identity of the group in question. But one certainly would never suspect this from the diverse forms (e.g., $10 + 5, 9 + 6, 8 + 2 + 5$, and so on). Accordingly, such systemless sum formations are totally useless for the purposes of number comparison. If we use the form $p + q$ for one group, and the form $p_1 + q_1$ for a second, then merely by looking at them we cannot yet decide whether or not the same number corresponds to the two groups, or to which the greater and to which the smaller number corresponds. With this the chief aim of all enumeration would be missed. <226>

The Sequence of Natural Numbers

In order to overcome these deficiencies we require above all a rigorously systematic principle for the construction of the symbolic number forms that are destined to supplement the narrow domain of authentically representable numbers. Only such a principle does not place any excessive burdens upon our memory. If the progression from the given numbers to ever new ones ensues by means of constant application of a uniform and unambiguous principle of construction, then we need to inculcate just this principle alone, and not the forms to be constructed. The procedure which supplies them is unambiguously determinate in all steps. It would further have to be ensured that by this procedure there should never result more than one symbolic number form for each of the actual numbers to be symbolized. For only then are we in position to infer – from the divergence in the number forms issuing from the comparative enumeration of two multiplicities – the difference of the corresponding actual numbers. If the ordering of the numbers procured through the systematic mode of origination agrees with the order in terms of more and less, we could, in addition, decide the question as to which of the two numbers would be the greater, and which the smaller, directly from their mere position in the system.

All of these requirements can be most simply satisfied, as is already done for the most part, by the procedure of *successive number formation through the addition, in each case, of one unit to the number already formed*. If we think of the authentically given numbers as so arranged that each arises out of the preceding one by increase of one unit, then we obtain the sequence:

$$1; \; 2 = 1 + 1; \; 3 = 2 + 1; \; 4 = 3 + 1; \; \ldots ; \; 10 = 9 + 1.$$

That this sequence (taking ten to be the final number authentically representable) can be symbolically continued is clear. Certainly we can immediately form the inauthentic representation of a new number which issues from ten as ten does from nine: *viz.*, through addition of one unit. If we call the number which is symbolized by the sum $10 + 1$ "eleven" (11), then that number is given and defined by the equation $11 = 10 + 1$. Likewise we can further define: $12 = 11 + 1$, $13 = 12 + 1$, $14 = 13 + 1$, and so on. Thus do we obtain <227> the sequence of number definitions – which can be continued indefinitely – by means of which we can count any arbitrary multiplicity, provided that the range of concept formation and terminology has been developed sufficiently far.

The enumeration of a given multiplicity would proceed in the following manner: one begins with any member, numbers it as one, goes over to a second and forms $1 + 1 = 2$; the transition to a third, and the addition of the corresponding "one" to the "two" just formed, yields $2 + 1 = 3$; and so on, until all members are exhausted. The univocality of this procedure is assured. With whichever member we begin, and whichever direction the successive enumeration through the members may take, the result must always be the same. The multiplicity to be numbered actually remaining identically the same, the number of the units belonging to it, correspondingly, does so too. Thus, various enumerations would merely yield diverse symbolic forms of composition of the same number from partial numbers. Now the symbolization of any number utilizing the number less than it by one, or utilizing the whole segment of the number sequence that precedes it, is indeed a univocal one, because the law for the formation of the sequence is completely univocal. Consequently, *in concreto* there can belong to one and the same multiplicity only one entirely determinate symbolic number form from the sequence.

In concept, the continuation of the number sequence *in infinitum* is restricted by nothing. It can therefore be regarded as actually continued beyond any given limit, and therefore any multiplicity of however many units can be regarded as denumerable. However, the method contains one flaw, which would confine its application in practice within very narrow boundaries. Each new step of symbolic number formation requires a new step in naming. If we were to choose ever new names (and that surely is unavoidable), then this would entail burdens for our memory which, already with numbers we currently are accustomed to regarding as moderate in size, would be unmanageable. And it would never be able to bear the load if all the new names necessarily had to be independent of each other: if we could not succeed in symbolizing all numbers of the sequence by means of a limited – and not too large – number of basic signs, following a uniform, easily understandable and clearly surveyable principle.
<228>

One could perhaps suppose that such a principle would be

obvious to us from the mode of formation of the sequence itself. One need only make the designations into a true mirror image of the conceptual formations; and then even one unique basic name, that of the unit, would already suffice, through its continued reiterations, for the naming of any arbitrary number. This is surely true. But such a mode of designation would be so crude that even the mode most opposite to it, which names each new number by means of a *novel* and independent sign, would remain preferable. How awkward and cumbersome would be the names arising from only five or six – not to mention twenty or thirty – repetitions of the name "one." And who could distinguish, without special precautions, even nineteen repetitions of "one" from twenty – to say nothing of the denominations for large numbers. Just so little as an actual [*wirkliche*] formation of number is possible beyond the limits imposed upon us by the *de facto* weakness of our capacities for representation, so little also would this crude symbolization – running parallel to the numbers by means of groups of ones – be practicable or useful for an exact classification and denomination of the numbers.

The System of Numbers

But in what way, then, are we to actualize the Ideal of a denomination for numbers that first makes possible a more significant degree of practical mastery of the domain of number? How are we to find a transparent, simple principle which permits us to construct, from a few basic signs, a symbol system that confers on each determinate number a convenient and easily distinguishable sign, and simultaneously distinctly marks its systematic position in the number sequence?

At first glance there seems to be here only a question about nomenclature. But closer consideration of the situation shows that the difficulties run much deeper, and that our problem is much more concerned with concept formation than with the mode of naming. How would we ever construct a system of number designation grounded upon some few basic *signs* unless a system of concept formation grounded upon certain *basic* concepts would correspond to it in rigorous parallelism? <229>

And yet one other perspective is of importance. In concept, we supposed, the simple number series could be thought of as continued *in infinitum*. Very well. But for all that, it is actually carried out and given to us only within the limits of what we name. How are we to hold the uniform steps of number formation distinct in their limitless succession – where each new step indeed presupposes the whole series of earlier ones – without the support of accompanying denominations? The concept 50 is given to us through the formation 49 + 1. But what is 49? It is 48 + 1. What is 48? It is 47 + 1, and so on. Each answer signifies a displacement of the question one new step backward; and only when we have arrived in the domain of authentic number concepts can we rest satisfied. But what support would this chain of conceptual formations have without the aid of sharply differentiated names?

Thus we cannot refer to the number sequence as to a sequence already *given* beyond any specified limit – whose members need only be symbolized in a different way than before, more ingeniously or more conveniently. Rather, it is here much more a question of another method of concept formation itself, one which is no less clear and systematic than the earlier one, and yet is at the same time more comprehensive – and which thus makes possible a more extensive and easier mastery of the domain of numbers in thought and word. The principle of simple series formation through successive addition of one unit was too crude; and this forced upon us the alternative of either providing each number with a novel symbol entirely unsystematically, or else, in exact imitation of the enumeration process, denominating all the numbers by means of one unique name (through successively expanding repetitions of it) – a systematic but totally unusable principle of naming.

So let us attempt to construct a better method of number formation and number designation – i.e., one better adapted to our logical requirements.

Every symbolic mode of formation aims to give a relational determination of the number by means of operations on numbers – and thus, in the last analysis, through addition and partition of known numbers, whether these are given in authentic representation or already by means of symbolization. <230> If the system of designation truly follows along the pathway of the formation of concepts, then a corresponding composition of the number name out of the names of the elemental number concepts will result, in which the combination is indicated by operation signs. If the concept formation is systematic, then the name formation is too, and conversely. If in this latter a restricted number of signs is to suffice for all situations, then correspondingly the concept formation also may use only a limited number of concepts to combine, through operations, into new ones. Under any circumstances there are given to us the authentically representable numbers and, at most, those standing next to them as well. If any, it is these which are to be accepted as the elemental concepts; and so the task will then be to derive from them – according to a unitary principle, constantly uniform in its applications – number after number in rule governed sequence. This is to be done in such a way that we have certainty that in the system a completely determinate position must accrue to each conceivable number (or, in other words: that any conceivable multiplicity must be denumerable by means of the system).

The unsurpassable simplicity in the formation of the sequence of natural numbers – which, as we have seen, already satisfies the greater part of these requirements – makes it desirable to preserve its principle to the extent feasible.

Therefore let us consider the numbers

$$1, 2, \ldots, X$$

in their natural sequence as the beginning segment of the system given to us; and we then, to start with, attempt novel formations following the old sequence principle:

$$X+1, \quad X+1+1, \quad X+1+1+1, \ldots$$

We must avoid the old principle of designation – according to which for $X + 1$ a new sign X', and for the next number $X' + 1$, in turn, X'', and so on, would be posited – by restricting ourselves solely to the signs $1, 2, \ldots X$. So either we retain the above

244   PHILOSOPHY OF ARITHMETIC

designations – which, because of their awkwardness, will not do – or we symbolize more simply:

$$X+1, \quad X+2, \quad \ldots, X+X,$$
$$X+X+1, \quad X+X+2, \quad \ldots, X+X+X,$$
5 $\quad X+X+X+1, X+X+X+2, \ldots, X+X+X+X,$
.... <231>

But this mode of designation also does not suffice. The further we go the more tedious becomes designation by the accumulating sums of X's. A new means of abbreviation presents itself at this
10 point: the simple denumeration of the X's leads to the multiplicative symbolization in thought and sign; that is, to:

$$2X, 3X, 4X, \ldots$$

respectively for:

$$X+X, \ X+X+X, \ X+X+X+X, \ldots$$

15 Accordingly we get the sequence:

$$1, \ldots, X, \ X+1, \ldots, 2X, \ 2X+1, \ldots, 3X, \ 3X+1, \ldots, 4X,$$
$$4X+1, \ldots, XX, \ XX+1, \ldots, XX+X, \ldots, XX+2X, \ldots,$$
$$XX+XX;$$

or again (in multiplicative formation):

20 $\quad$ 2XX; and then further:
$\quad$ $2XX+1, \ldots, 3XX, \ldots, XXX, \ldots, XXXX, \ldots, XXXXX, \ldots$

But once again the formations showing up here (in the decimal system: 10 times 10, 10 times 10 times 10, 10 times 10 times 10 times 10, and so on) become so unwieldy that new abbrevations
25 are wanted. The enumeration of the factors leads to the formation of powers:

$$X^2, X^3, X^4, \ldots$$

After the introduction of these formations the sequence from $XX = X^2$ on reads as follows:

$$X^2, X^2+1, \ldots, 2X^2, 2X^2+1, \ldots, 3X^2, \ldots, (X-1)X^2, \ldots,$$
$$X^3, X^3+1, \ldots, 2X^3, 2X^3+1, \ldots, 3X^3, \ldots, (X-1)X^3 \ldots,$$
$$\ldots\ldots\ldots\ldots\ldots\ldots\ldots\ldots\ldots\ldots\ldots\ldots\ldots\ldots\ldots\ldots\ldots\ldots\ldots$$

One sees how the sequence continues, and how in doing so the iteration of the symbolic formations introduced last would again and again lead to new formations. For practical needs it suffices if we cease with exponentiation.

The following table may serve to bring to clear intuition the mode of enumerating systematized in this fashion:

$1, \ldots,$	$2, \ldots,$	$3, \ldots,$	$X-1, \ldots$
$1X, \ldots,$	$2X, \ldots,$	$3X \ldots,$	$(X-1)X, \ldots$
$1X^2, \ldots,$	$2X^2, \ldots,$	$3X^2 \ldots,$	$(X-1)X^2, \ldots$
$1X^3, \ldots,$	$2X^3, \ldots,$	$3X^3 \ldots,$	$(X-1)X^3, \ldots$

The enumeration is articulated into a sequence of levels. In the *first* the simple enumeration through the sequence from 1 to $X-1$ occurs. In the *second* we already have a multiform enumeration. On the one hand we have the multiplicative enumeration, through which X is replicated step by step 1-fold, 2-fold, ... $(X-1)$-fold, and, on the other hand, the <232> additive increases by numbers of the first level $(1, \ldots, X-1)$, which mediate between each two members of that multiplicative sequence. It is similar with the *third* level. In the multiplicative enumeration the X^2 now functions as the unit enumerated. Between each two of the numbers so formed there intervenes, in always the same manner, an additive enumeration, and specifically: that step by step increase through all terms of the second level, etc. Finally, one more mode of enumeration cuts across the system in a vertical direction, so to speak (i.e., running longitudinally through the sequence of levels). This is the enumeration in the exponentiation of X – which, however, has no essential systematic value, since no new number-forming operation follows upon exponentiation.

The numbers within each level form a sequence ordered according to magnitude. Each number is greater by one than the preceding one, but the first [of each level] is greater by one than the last one from the preceding level. Thus all levels link together and form a unique and endlessly continuing sequence of numbers which corresponds exactly to the natural, primitive number sequence. But while in the latter the relation of "greater than" referred to was the one and only principle for forming and ordering numbers, with the former, other and more complicated principles enter in its place, through which each number is systematically brought forth through a definitional chain, not from the mere number one, but rather from the sequence of numbers $1, 2, \ldots, X - 1$. Expressed in the language of modern analysis, this systematic mode of formation consists in each number being represented and, accordingly, named as an "entire, whole-numbered function" ⟨i.e., Polynomial function⟩ of a determinate "base number" X, fixed once and for all by convention, with coefficients that belong exclusively to the number segment $1, 2, \ldots, X - 1$, or disappear if the term in question does not occur at all. Each number is, therefore, symbolically given in the form of an aggregate,

$$a_0 + a_1 X + a_2 X^2 + a_3 X^3 + \ldots,$$

in which each of the a's assumes one of the values $0, 1, 2, \ldots, X - 1$. The numbers $1, 2, \ldots, X$ are usually called "ones," and the powers of X ($X^0, X^1, X^2, \ldots$) "units of the 0, 1st, 2nd, … degree, or power units [*Stufenzahlen*]."

In such a way we have attained to a principle of formation for numbers and number signs which actually does satisfy the logical <233> requirements imposed: — It makes possible the systematically uniform continuation, beyond any limit, of the narrow domain of numbers given to us. To accomplish this it requires, through the introduction of the symbolic formation principles of multiplication and exponentiation, no other building blocks than the numbers and signs $1, 2, \ldots X$. It encompasses, in concept, the entire domain of number: that is, there is no actual number to which there would not correspond, as its symbolic correlate, a wholly determinate systematic formation equivalent to it.

Relationship of the Number System to the Sequence of Natural Numbers

We have pointed out above that our number formations run, in their systematic arrangement, step by step parallel to those in the "natural" sequence of numbers. This circumstance leads to an observation which needs to be emphasized. If, namely, the series of natural numbers is thought of as developing at all points in parallel with the sequence of our system, then it seems that nothing would stand in the way of regarding the designations of the latter as designations of the corresponding numbers in the former. To each natural number there corresponds a wholly determinate systematic number (equal to it), and to this latter, in turn, corresponds a wholly determinate designation mirroring its manner of formation. The systematic number concept would therefore be the mediator between the natural number and the systematic denomination.

However, we find the view objectionable according to which the number system is a mere tool to provide a systematic nomenclature for the natural numbers, intending economy of symbolism. We refer back to the earlier discussions (pp. 241-242). The situation is not after all such that the sequence of natural numbers was first given to us, and we subsequently sought for a symbolization adequate to its concept formations. Only a tiny opening segment of that sequence is given to us. Certainly we can conceptualize the Idea [*Idee*] of an unlimited continuation of it, but the *actual* continuation, even for only the moderate range <234> involved in the ordinary practice of calculating, already places demands upon our mental capabilities which we cannot fulfill. The impossibility of being able to solve more demanding problems of calculation in such a primitive way was the source [*Quelle*] for those logical postulates whose satisfaction led to a new and farther reaching method of concept formation. And so the number-systematic arrived at (specifically, our ordinary decimal system) is not, then, a mere method of symbolizing concepts which are given, but rather one of constructing new concepts and simultaneously designating them along with their construction. Of course we can form the Ideal of an unrestricted continuation of the simple number sequence by correspondingly Idealizing our mental capacity. We can, further, think of the sign formations of the number system also as a symbolism for the parallel members of the (Ideally expanded) number sequence. But one must consider well the fact that these all are only modes of representation and expression which are inauthentic in the highest degree and have their source [*Quelle*] in the Idealizations mentioned. To interpret them in another, more authentic sense would be to distort the entire *sense* and purpose of the systematic formation of numbers. All logical technique is directed toward the overcoming of the original limits of our natural mental abilities through the careful selection, arrangement, conjunction and persistent repetition of the activities which make that overcoming possible, and which, considered in isolation, are capable of accomplishing only a very little. So also in the case at hand. We first ran up against the limits of our capacity for collective combinations, and we overcame them through various kinds of symbolizations. Then we ran up against the limits of memory, and overcame them, again, through symbolizations – but through far more ingenious ones, which integrated into the harmonious structure of the number system.

And yet one thing more must be emphasized: namely, that the so-called "natural" number formations are not the least bit more natural than those which are systematic in the narrower sense (e.g., the decimal). In both cases it is a matter of symbolic formations for those species of the concept of number that are not accessible to us in the authentic sense; and it is purely a matter to be decided by logical adjudication – i.e., adjudication adapted to the goals of knowledge of the domain of numbers – as to which method of symbolic formation is to be preferred.

These points had to be discussed in some detail, <235> because the prejudices to be resisted here are widespread, and are even shared by scholars who, in other connections, have given special attention to the logic of this matter.[1]

---

[1] Cp., e.g., H. Hankel, *Zur Geschichte der Mathematik in Alterthum und Mittelalter*, Leipzig 1874, pp. 10 and 12.

The Choice of the "Base Number" for the System

To this point we have taken no position concerning the "base number" X of the system. Not as if the choice of it were logically indifferent, but rather only because our reflections up to now were not affected by it. We must now fill in this remaining gap.

It is easy to characterize the manner in which different choices of the base number influence the logical character of the system. We need exactly as many elements (concepts and signs) for the construction of all the numbers and number signs as the base number X possesses units. (These elements are, in fact, the numbers of the sequence 1, 2, . . ., X.) If, therefore, the requirement of a smallest possible number of elements were the governing principle, then obviously the choice of $X = 2$ would be most advantageous. But yet other factors essentially come into consideration, and, above all, the extent of the number domain to be mastered. The greater the base number is, the smaller is the number of repetitions of elements necessary for the construction of any given numbers, and the simpler and more easily discernible is their expression. It would be a grave defect in systematization if numbers which we were enjoined to form and master were constructed and expressed in a form which – in virtue of a too frequent repetition of the elemental enumerations – would threaten their differentiation or make it altogether impossible. In this respect, however, the dyadic system of *Leibniz* would not serve much better in our number domain than that of the natural number sequence.

The greater, therefore, the base number, the more inclusive is the number domain that can be actually mastered. But certainly only under one condition: namely, that we may regard the numbers 1, 2, . . . X as actually [*wirklich*] given – that they therefore do not first come to us through mediation of remote <236> and excessively complicated symbolizations. Herein lies the essential reason why we are restricted as to the number of elements; for also on this point we quickly run up against the limitations of our mental capacities.

One might at first want to lay down the requirement that the elemental numbers must still fall within the domain of numbers that are authentically representable. Such a far reaching restriction would, however, be unnecessary. We will soon have occasion, in fact, to see how it proves in general feasible and preferable to substitute certain symbolizations even for the number concepts accessible to us in the form of authentic representations – indeed, to put it plainly, to substitute external signs. With this (at least for such ends as we here pursue) the authentically representable numbers lose their essential distinction over against other numbers that can *only* be represented symbolically; and nothing hinders us any longer from taking even a part of these latter as elements in the systematic. But that in no way means that we are totally freed up to substitute for the sequence 1, 2, . . . X an arbitrarily large segment of the number sequence as the elemental one. This is simply because the elemental sequence must be structured in conformity with the principle of the natural number sequence, and must be accessible to us without remote systematic expedients (for otherwise we indeed would have complication upon complication). In this regard it is questionable whether on the average we can trust our memory, for the full certainty required of reproduction, with much more than three or four dozen elemental signs. Nevertheless, we would reach out beyond our favorite number ten, and would have enough possibilities to give us a choice regarding X.

Conforming to the logical requirements drawn into consideration up to now, the system would be the more preferable the larger the X adopted. For the larger the number of elements, the less complicated will be the expression for each new number to be formed. However, yet other logical requirements can influence the choice of X, the base number of the system; and such do factually arise out of our drive toward the greatest possible convenience in *calculations*. Each number system grounds calculation mechanisms peculiar to it, and the best system will be that one which allows for the shortest and the most convenient of such mechanisms. From this point of view, those systems whose base number <237> is divisible by the greatest number of other numbers, and whose addition and multiplication tables do not place too great demands upon our memory, prove to be especially advantageous. This is why mathematicians consider the duo decimal system as preferable to the decimal system now adopted. We shall not here enter into a more detailed investigation of this problem, not in itself a difficult one.

The Systematic of the Number Concepts and the Systematic of the Number Signs

We have seen how each coherently formulated number system satisfies a whole series of logical requirements, and accordingly possesses a corresponding series of logical perfections. There is only one which we have not yet mentioned. This one, although an incidental consequence of the others, perhaps must after all be designated as the most important, but in any case as the most noteworthy, among them all. The following reflection may serve for its characterization.

The systematic which we have developed above presents, as one immediately sees, a two-fold aspect. On the one hand, it provides for each number a systematic mode of formation (as symbolic replacement for the missing authentic number concept) utilizing certain elemental numbers, 1, 2, . . . X, that are given. And on the other hand, it provides, starting from the number names "1," "2," . . ., "X," a systematic mode of formation for the number *name* appertaining to each one of the numbers. A rigorous parallelism governs here between the method for the continuation of the sequence of number *concepts*, and the method for the continuation of the sequence of number *signs* – and this not merely in general, but rather for each individual step, one after the other. And the systematic of the signs is no less consistently self-contained than is that of the concepts. Let us abstract from the signification of the designations "1," "2," . . ., "X," as well as from the designations of the operations of addition, multiplication, and exponentiation, and take them as totally arbitrary symbols without signification (as, for example, with the counters in a game). Let us replace number definitions and operation rules, which are the regular medium of systematic procedure, with corresponding, conventionally fixed formulas expressing the equivalences of <238> sign combinations. One will then recognize that, in this way, there actually originates an independent system of symbols which permits the derivation of sign after sign in a uniform pattern without there ever turning up – nor could there ever, as such, turn up – other sign formations that appear in other circumstances, accompanying a conceptual process, as designations of the concepts here formed.

It is not difficult to bring to Evidence the inner ground of this peculiar relationship. The essence of the systematic number formation consists in this: that it constructs all other number concepts by means of some few elemental concepts and propositions (numerical formulas and rules of operation). If, now, the system of designations faithfully reflects these conceptual formulations, then each number designation will also have to be one formed, in exact correspondence, from the designations of the elements and operations, in which certain formal rules for the replacement of sign compounds by others will come into use – rules that correspond to those propositions concerning the relations of concepts. The sequential process of formation of the designations will proceed in such a way that, in typical form, sign is step by step annexed to sign (e.g., $X + 1$, $X + 1 + 1 = X + 2$, $X + 2 + 1 = X + 3$, and so on, and likewise $X^2 + 1$, $X^2 + 2$, $X^2 + 3$, . . .), whereby certain composite signs are always replaced by simpler ones (e.g., $1 + 1$ by 2; $2 + 1$ by 3, etc.; according to the formulas $1 + 1 = 2$, $2 + 1 = 3$, $3 + 1 = 4$, . . .). After attaining a certain level there occurs, conforming to determinate types, a simplifying transformation of the articulated sign attained (e.g., $X + X$ becomes $2X$, $X + X + X$ becomes $3X$; XX becomes X^2, and XXX becomes X^3, . . .), whereupon the uniform process of annexation begins again. And so on. If now, on the one hand, all number formations are systematically rigorous consequences of the elemental concepts along with their forms of combination and transformation, so, on the other hand, the parallel sign formations will also have to be systematically rigorous consequences of the elemental signs along with *their* forms of combination and transformation. Certainly in the *former* case the transformations unfold on the basis of knowledge proceeding with necessity from the relevant concepts; while in the *latter* case the transformations of signs will indeed proceed according to certain *types*, but in a wholly external and mechanical manner. If, now, we detach <239> these types from their conceptual supports, and if we fix them once and for all in the form of conventional sign equivalences (in the manner of the rules of a game), then it is clear *apriori* that we now possess all that is necessary for the independent development of the systematic of the signs, and that no result can come about which would not find its correlate on the side of the systematic of concepts.

The Process of Enumeration via Sense Perceptible Symbols

The double aspect of the systematic, which we have here explained, has as a consequence the fact – most noteworthy from the logical perspective – that both in problems of practical enumeration of given groups, as well as in those of derivation through calculation of numbers from numbers, the solution can be obtained in a purely mechanical fashion. This happens in that one substitutes the names for the concepts, and then by means of the systematic of names and a purely external process, derives names from names, in the course of which there finally issue names whose conceptual interpretation necessarily yields the result sought.

Already with the expansion of the number sequence in the primitive form first discussed it is clear that the enumeration of a group by means of successive steps through the sequence in no way requires the step by step formation of the (authentic or symbolic) number *concepts*. Even the successive subsumptions of the individual group members under the concept of *unit* become superfluous. Instead of all that, it suffices that we have a step by step symbolization of the group members by means of the sequence of number names, and the last name used in the symbolization will necessarily have to be that of the concept sought, as well as the one which in fact does result when the conceptual route is followed. How the relations – relations expressed in the so-called "number definitions" (regarded as mere symbol equivalences) – of the members following one after the other in the sequence of number signs can serve in the discovery of infinitely many number formulae and the corresponding number propositions, is not to be further considered here in greater detail. It is enough to have pointed the way.

In all of these respects matters stand precisely the same with the so much more thoroughly developed system of numbers and number signs treated in the second place above – except that it admits of correspondingly more perfect <240> mechanisms of calculation. In enumeration one quite simply follows the systematic of the designations, ultimately attaining to a composite sign whose mode of formation conceals within it precisely that of the concept sought. The same holds, as will be shown, for calculating: it is not an activity with concepts, but rather with signs.

The great importance of this perfection of the number system is obvious. It opens to us the prospect of mastering problems that, given even a perpetual action of the mind upon the concepts themselves, would remain totally unsolvable because of the abundance and complexity of the highly abstract conceptualizations to be mastered. An enormous savings of higher psychical activity becomes possible, and therewith an enormous expansion of the working capacities of the intellect in general.

Expansion of the Domain of Symbolic Numbers through Sense Perceptible Symbolization

The first and most significant result of the parallelism between the system of concepts and that of signs consists in a singular expansion undergone by the number domain itself.

Up to now we have spoken as if the type of symbolic number formation developed above actually made possible, through a complex of indirectly characterizing (but themselves authentically represented) relative determinations, an unrestricted expansion of the domain of number. This was an incorrect way of speaking, adopted for the sake of a simplified presentation. More exactly considered, the narrow confines of consciousness impose insurpassable limitations upon us, even given the implementation of those artificial devices. Ultimately we no longer survey the sequence of linkages. Indeed we can, advancing step by step, actualize ever new relationships. But to retain in our consciousness, as one cohering whole, the totality of those already actualized – that we cannot do. It was, therefore, already an Idealization of our finite mental capacities, undertaken with respect to the extent of the relational chains to be actualized, when we spoke of the unrestricted continuability of the systematically sketched <241> number sequence – an Idealization wholly similar in kind to that which we confirmed as the basis of ordinary talk about the infinitude of the *"natural"* number series.

How is it, then, that one does not become aware of these limits, even though there already show up in occasions of practical life many numbers, manipulated with confidence, which must go beyond those limits? The answer is: the symbolic number formations of the system are precisely not thought of as compositions of purely abstract determinations. The sensible signs involved here are not, in the manner of language signs, mere accompaniments of concepts. They participate in a far more striking manner in our symbolic formations than we have validated to this point. Indeed, so much so that they ultimately dominate nearly the entire field. In fact, the rigorous parallelism between the system of number concepts and that of number signs makes it possible to regard the systematic continuations of the series of signs as the Representatives [*Repräsentanten*] of the systematic continuations in the sequence of (inauthentically represented) concepts. But with the composition of sensible signs which we still can survey in one glance and retain, we reach much farther than with the composition of those extremely abstract determinations that correspond to the signs as their significations. Disregarding the smaller amount of psychical labor required for the comprehension and unification of the sensible signs, in contrast to the more remote abstractions, the figural Moments that impart a unitary character to even very large sign complexes and extraordinarily facilitate their comprehension in one grasp are of special importance. In written signs it is the linear, and with oral word signs the temporal and acoustical linkages that here come into consideration. Each such sign complex supplies, in its intuitive unity and typical form, the stable foundation for that chain of conceptual transformations which constitute the "interpretation" of the compound sign. It moreover is also necessary to point out that sequences of sensible signs more easily imprint themselves upon our memory than do such abstract concepts, and therefore the fixed associational sequences of the former can also serve in the symbolic representation of the latter.

If, now, a number is defined through such a systematic complex <242> of sense perceptible signs, then the cohesiveness of this complex forms the means of symbolization for the sequence of conceptual steps which, otherwise, does not hold together. In what way do we have, for example, the concept of a twenty-place number? We obviously first think the mere concept: a certain number which corresponds to *this* sign complex. If one inquires about the exact content [*Gehalt*] of the concept – thus about the sense of this symbol – then there begins a chain of explications which has its support in the unity and the special mode of formation in the composition of the sign. Only in such a way do we gain the possibility of a continuation of the number domain which, incomparably more far-reaching than any before, is able completely to satisfy the requirements of ordinary life and science. Of course even now we are not absolutely unlimited as we follow the route of mere signs. But we no longer feel the limits, for they do not hinder us in dealing with the problems that fall within the range of our interests.

Then the question arises as to why we do not really notice at all the difference between the numbers it is still possible to symbolize conceptually, and those which can *only* be symbolized *signitively*. The answer is clear. It is because even with small numbers too we at once find it more convenient to stick to the external symbols; thus, even where the conceptual content could still be brought before the mind. It is a fact that *in praxi* all numbering and calculating could dispense with recourse to the underlying concepts. The concepts are of course indispensable for anyone who for the first time grasps the essence and aim of the number system, or who at some later time has the need to bring to consciousness the full conceptual content of a complex number sign. Reflections on concepts are the sources [*Quellen*] out of which arise the rules of all arithmetical operating. But the mere sensible signs continually underlie practical activities. The explanation of this fact, to whose curious character we are rendered insensible only by everyday habit, will engage us in the next chapter. <243>

Differences between Sense Perceptible Means of Designation

The reflections in which we have just engaged call attention to a logical distinction of the highest significance for our scientific domain: the distinction between the sense perceptible devices of designation. It may at first glance provoke doubt when we say that it is a major task of science to busy itself with such an apparently subordinate matter as the choice of sensible signs. However, it will not be difficult to disperse those doubts. We will soon see that the distinction between oral and written signs is so essential for arithmetic that an inescapable restriction to the former would have made a development of arithmetic on a larger scale impossible. In fact, we will later find occasion to show how even such an apparently immaterial difference as that between writing on paper with ink and pen, or writing with a stylus upon a dust-covered tablet, can essentially influence the progress of arithmetical methods. And is that not supposed to be a logical difference? *Any* difference which influences the technical mastery of a domain of knowledge is quite certainly a logical difference.

That the employment of sensible signs has, in general, extreme importance for the arithmetical domain already follows, with all clarity, from our reflections to this point. Now those means of designation which more perfectly correspond to their purpose will be the logically more perfect ones. But in this regard number signs that can be enduringly fixed for numbers – in particular, written signs – are remarkably superior to the number words. A systematically written symbol can, without losing its ability to be taken in in one glance, be incomparably more comprehensive than a systematic word symbol. It is easier to manipulate, and, in the enduring manner in which it stands fixed, it places no special demands upon our memory. If the result of an enumeration leads to a very complicated number *word*, then we are immediately in danger of forgetting it again, even if we have perfectly grasped it. But a written symbol stands fast and can be newly grasped at any moment. Moreover, such differences are even much more essential for the purposes of calculation than for those of enumeration. The more complicated tasks of calculation <244> would from the start be impossible to carry out if, limited to enumerating with words, we depended upon the defectiveness of our memory. The best signs, therefore, are written signs that are easily fixed, are formed to have the greatest possible capacity for comprehension in one glance, and at the same time are the shortest and most distinct ones possible. The familiar "place system," from India, corresponds to this Ideal in the most perfect manner. It attains to its insurpassable surveyability and brevity through replacing the written number words with numerals and utilizing the perceptual linear order as the systematic means for designating the order of the power terms – thus making superfluous a special designation for the power unit.

A decadal number has the form:

$$a_n \cdot 10^n + a_{n-1} \cdot 10^{n-1} + \ldots + a_2 \cdot 10^2 + a_1 \cdot 10^1 + a_0 \cdot 10^0$$

The Indian symbol system writes it in the form:

$$a_n \, a_{n-1} \ldots a_2 \, a_1 \, a_0,$$

in which the order of the numerals from right to left corresponds to the order of power terms, and designates them. The $(m+1)^{th}$ numeral a_m in the series (counted from the first one on the right) indicates through its position alone that the units which it counts belong to the $(m+1)^{th}$ degree. Certainly such a mode of designation only became possible through the invention of the numeral zero, which has the function of marking the absence of a determinate power term, and precisely by that means retaining the completeness of the sequence of degrees upon which the appraisal of the place value rests.

The Natural Origination of the Number System

One might suppose that the discovery of this ingenious and portentous two-fold systematic of the numbers and number signs – since it serves ends both comprehensive and thoroughly pondered, and is only to be justified through intricate processes of thought – must be the product of a mind of genius, fully conscious of its goal, such as is conceivable only upon the native soil of a highly developed national culture. And still, almost all peoples who were in general sufficiently developed to sense the need for a broader expansion of the number domain – thus already at a quite low cultural level, and long <245> before any scientific reflection – were brought to number systems that on the whole (disregarding incoherencies scattered here and there) follow the fruitful principle presented above. And not less striking is the independence of discovery by different peoples, which is deducible with certainty from the existing differences (e.g., in the choice of the base number of the system) found alongside all commonalities.

The explanation of this noteworthy phenomenon will furnish us with the first example of a general fact which so often comes to light in our scientific domain. It will make intelligible how, in general, a sign system that is artificial in its type and constitution, and whose consciously intended invention and theoretical justification would require abstract reflections of the most complicated sort, can come about through a course of *natural psychological evolution* – and this already at a mental level from which the explicit formulation of the problem alone must no doubt lie at an inconceivable distance.

The periods within which the origination of number systems and number sign systems falls are unknown to any historical tradition. Therefore there can be no thought of a reproduction of the historical development. We nevertheless possess sufficient clues – through the original signification of many number names, through our knowledge of the practices of enumeration among half-civilized and savage peoples, and above all through the understanding of the general traits of human nature which here come into consideration – in order to reconstruct the psychological evolution of such formations of systems in an *aposteriori* fashion that is still correct in all essential points.

The systematic of numbers and number denominations grew out of systematic enumeration, and this latter in turn out of certain natural habits of enumeration which, in virtue of the homogeneity of human abilities, must have originated among the most diverse of peoples at a certain level of culture.

Let us transport ourselves into the early stage of the development of a people. The repeated interest in sensible groups of objects the same in kind had already led to the apprehension of a certain analogy, and therewith of a shared characteristic founding it; and thus it had led to the concept of multiplicity, which at this level, of course, being much less abstract than on our own, restricted itself to multiplicities of homogeneous and sense perceptible contents. The drive to <246> communicate concerning the events of practical life, in which determinate groups of such objects played a great role, led here (when circumstances were particularly favorable) more easily than in other areas to the thought of an imitation by sensible means of the things represented. This thought would be immediately suggested by the hands. These visibly prominent organs, which the individual chiefly employed in both serious and playful activities, and which (depending upon the position of the fingers) presented varying sensible group formations (the clusters of fingers), must accordingly have come immediately to mind for the imitation and symbolization of corresponding groups of arbitrary other objects.[2] Thus the "finger numbers" arose within sign language as the first number signs.

Indeed we can very well claim still more: it is as a rule only on this path of the sense perceptible that a sharp differentiation and classification of the determinate number forms could first come about at all. In a certain manner one of course already possessed the number concepts when the analogy of different groups equinumerous to one another and to groups of fingers was grasped. But only through a constant back-reference from groups of the most various types to the finger groups, sharply distinct in sensible appearance, did the finger numbers rise to the level of Representatives of general concepts, of general characteristics of groups

---

[2] I speak above of an "imitation" of groups with respect to their number by means of clusters of fingers. This mode of expression, although incorrect, is nevertheless appropriate here, because it is suited to the mental level concerned. The psychical activities brought to bear upon the sensible groups supply concepts which the more naïve consciousness regards as abstract positive Moments of the respective intuitions themselves. Quite as beauty and ugliness, or goodness and wickedness, are judged to be inherent characteristics of external *things*, so also twoness, threeness, etc., are judged to be inherent characteristics of external *groups*.

classified in terms of more and less. Without fear of paradox we can say: the concepts 1, 2, 3, . . . as the species of the general concept of multiplicity, as specifications of the "how many," first came to a more determinate consciousness in the conceptual signification of number signs on the fingers.

In our conceptual domain word language came after sign language, as one recognizes from many examples of number words whose original signification is to be characterized as a mere translation of finger numbers into word language. In any case the finger signs also continued to be employed, for obvious reasons, <247> as means of enumeration alongside the word signs, even at a much higher level of development.

It corresponds to the lower mental level, with which we are here concerned, that even the enumeration of the smallest of groups occasioned no little effort (reports about counting among savage peoples confirms this). And therefore, to safeguard the process, it was required that the enumeration ensue in a step by step manner, in the course of which to each member of the group, in sequence, a raised finger would be coordinated. There thus successively arose the signs for $1, 1+1 = 2, 2+1 = 1+1+1 = 3$, and so on. In this way the number sequence as such was in the process of development.

It is clear that this mode of enumerating was not confined to the narrow domain of *authentic* representations of groups and numbers, but rather in the same manner (and without there being occasion to attend to the distinction) could be continued out beyond it; not, of course, without certain natural hindrances, and specific devices for overcoming them. A first halting point in enumeration was already presented by five, where one had used up the fingers of one hand. Because of this, in several languages the number word "five" signifies the same as "one hand." With the assistance of the fingers on the second hand one could then (in the form $5 + 1, 5 + 2, . . .$) continue counting up to where ten set a new halting point, no longer to be surpassed in the same manner. (Only peoples who also used their toes in counting went further; and only at twenty did they meet the corresponding limit.) If now, nonetheless, larger groups were to be enumerated, there obviously remained nothing left but to make note – on the side, by a sensible sign – of the fact that the fingers (and perhaps the toes also) had been numbered through once, and then to count off the objects yet remaining by means of the fingers (and toes) again. After counting through them again, that sign for ten (or twenty) would have to be repeated, etc.

Thus one enumerated larger groups by partitioning them into groups of ten and one group of less than ten. But then still a second enumeration was necessitated: namely, of the groups of signs for ten. If the number of them was greater than ten, then for their enumeration the ten fingers did not suffice. Again there was required, to indicate that all fingers had been once used, the introduction of a sign; and, indeed, a new sign, in order to prevent confusion of the simple ten with a ten-times-ten. <248> If for convenience of exposition we express this sign as the "hundred," then enumeration of hundreds forced, in the same way, the introduction of a new sign, the "thousand," for $10 \cdot 100$, and so on.

In this way one was led to a general procedure for the enumeration of groups, through which each larger number is already constructed in the form of a polynomial function of powers of ten. Still today there are peoples who follow such a mode of enumeration, which makes the origin [*Ursprung*] of the decimal number system wholly intelligible. The natives on south-sea islands calculate, as *Tylor* tells us, by using in counting a small stone for the ones. When ten of these are together, they are replaced by a small piece of coconut shell. When ten of these latter are together, then a larger piece of coconut shell is used, etc.[3]

Word language followed, in its denominations, the conceptual formations of the number systematic. Of course, besides names for numbers up through ten, only names for the power terms were needed. All the rest could be formed through the mere combination of these. A general employment of the number words in counting would have to facilitate and simplify counting itself, in that certain obvious and practical modifications of the mode of enumeration presented themselves. Through these modifications enumeration would become more cohesive and systematic, and

---

[3] E. B. Tylor, *Einleitung in das Studium der Anthropologie und Civilisation* (translated by G. Siebert), Braunschweig 1883, p. 376.

simultaneously independent of sense perceptible instruments other than words. Instead, namely, of symbolizing each enumeration carried through to ten by new signs to one side, (and then always beginning with one again, only at last to count up the number of tens, in which a similar procedure again would be adopted) one certainly could – by means of the number words, and without other sensible signs – count in the following manner and order:

$1, 2, 3, \ldots, 10; \; 10+1, 10+2, 10+3, \ldots, 10+10 = 2 \cdot 10;$
$2 \cdot 10 + 1, 2 \cdot 10 + 2, \ldots, 3 \cdot 10; \ldots, 10 \cdot 10 = 100;$
$100 + 1, 100 + 2, \ldots, 100 + 10;$
$100 + 10 + 1, 100 + 10 + 2, \ldots, 100 + 2 \cdot 10; \ldots .$

One thus remained within a uniform pattern of enumeration suited to continue the number sequence in conformity with the same principle. <249> Each new step, consisting of the addition of one unit to the number just previously formed, would yield a new number. But it is not this simple sequential principle alone that determines the new formations. The employment of the auxiliary operations of multiplication and exponentiation permits the repeated use of the old concept and name formations, so that each number and each number name appears as a systematic construction from a very few elemental numbers and elemental names (the two running rigorously parallel).

According to our presentation, the natural evolution from the primitive method of enumeration by means of "finger numbers" perfectly explains the origination of the systematic method of enumeration. It is only in the selection of the base number that different paths remained open. For the sake of convenience we developed our line of thought by taking ten as the base number. There were several peoples, however, who had already held to five, which stands out as the number of fingers on one hand, as the constantly fixed base number. Others again, utilizing their toes as well in counting, could develop a coherent vigesimal number system, like the ancient Mexicans. In most cases, however, it was ten that served as base number, whereas the quinary formations under ten $(5+1, 5+2, \ldots, 5+5)$ – after the actual enumeration on fingers had given way to enumerating with words – disappeared, because systematic significance could not longer be attributed to them, so that enumerating up to ten came to be understood as simple sequential enumeration. (The more primitive point of view is still often enough marked in word and inscription, e.g., in the Roman numerals V, VI, VII, VIII, and the like.) In any case the *principle* of the formation of the system is explained in a wholly unequivocal manner by that path of natural evolution.

Our view could still give rise to a difficulty. In his profound fragments, *Zur Geschichte der Mathematik*, H. Hankel supposes, namely, that enumeration beyond 10 · 10 would have been able to progress in a twofold manner, and nevertheless consistently. Upon coming to 10 · 10 + 10, one could take this either as one hundred ten (in a consistently decimal manner of speaking: as 'ten-ty' ten), or else as eleven times ten ('eleven-ty'), and count onward in accordance with that. "In a fortunate move, the first schema was adopted, wherewith the base number X was regarded, similarly to 1, as a unit, but of another . . . order, <250> which must not be repeated beyond the Xth time."[4]

To the contrary, it can easily be shown that the path actually adopted was not merely a "fortunate move," but rather was a necessary consequence of the further development of counting with fingers. The external occasion for that singling out of 10 · 10, through which it from then on had to become a unit of higher order, was simply the necessity of introducing a new sign for ten tens (and, accordingly, a new and distinctive word denomination). But with that the view of (10 · 10) + 10 as 11 · 10 was ruled out, as one sees from the linguistic formation "hundred ten." It no longer lay along the path of natural enumeration. And it obviously is similar with respect to the decision which direction to take at the further turning points of enumeration in the subsequent power terms.

The evolution of the number sequence made available a system of symbolic concepts, which can be continued without limitation, and by means of which any task of enumeration has become feasible. The enumeration of an arbitrarily given group resulted

---

[4] *Op. cit.*, p. 11.

from progressing along this periodically articulated sequence of concepts; or, more precisely, along the chain of number definitions corresponding to it. That sequence provided the univocally determined number concept appertaining to the group, which (disregarding the smallest groups) could be established only in a symbolic form and only with the aid of such an articulated serial procedure. But if any process, this one had the aptitude and tendency to become totally mechanical and external. And this is so precisely in virtue of that two-sidedness which we have stressed. To the progression along the sequence of concepts there corresponds, in rigorous parallelism, a progression along the sequence of names. And the system of names taken by itself is every bit as coherent as that of the concepts. It is clear that as soon as the systematic was mastered through practice, the mental process of concept formation automatically had to vacate the field to the external reproduction mechanism of name formation. Originally one counted by a mental action, picking out of the group one member after another: one, one and one is two, two and one is three, and so on. In this the names were reproduced step by step, <251> structured systematically beyond ten, and at all events supportive of the continuity of the conceptual process through their rigorous succession. But after long practice it automatically came to pass that one counted mindlessly, so to speak, or mechanically, by following out the sequence of names and sentences – which in part were firmly imprinted upon the memory and in part produced mechanically following the principle of the systematic – without any reflexion upon their conceptual signification. From a conceptual procedure for the generation of the number concept there arose in a natural manner, by the falling away of the psychical correlates of the individual steps, a symbolic and exterior procedure for the systematic derivation of the number name corresponding to the concept, and only thereby of the concept itself. For all of this one merely required mnemonic mastery of the sequence of numbers and number definitions up to ten, and familiarity with the decimal system of number word formation. But full mastery of the extended number sequence through memory soon reaches in the trained calculator much farther than ten. Within the limits for which this holds true there is no need to apply the systematic of number words – which always requires thought, even without the accompanying conceptual processes. And thus the *praxis*, intent on simplifications, soon found a yet further, and considerably more abbreviatory device. Instead of the sequence of number definitions, one surely could follow the sequence of number words alone. The sequence of sentences $1+1=2, 2+1=3, 3+1=4, \ldots$ was abbreviated into the sequence of terms $1, 2, 3, 4, \ldots$. By coordinating step by step the names of this sequence (and, of course, each only once) with the members of the group, until all members were exhausted, one attained with the final name produced to the name of the number sought, and therewith to that number itself. This mechanical procedure had to lead to the correct result because the number of the names of the sequence of numbers up to and including $\underline{n}$ is expressed by "n" itself. Within the framework of ordinary practice this is the usual procedure. But for larger numbers, which go beyond that framework, the procedure previously dealt with comes into play. In enumerating beyond the common limits one enumerates by continuing the sequence of names, following the principle of decimal number word formation.

In our explanation of the natural evolution of the number <252> systematic we took only the *word signs* into consideration. We had to do this, for only the number *word* systems accompany the natural course of concept formation, whereas the *numeral* systems – insofar as they can lay claim to the title of *systems* of number designations at all – are later, sometimes crudely superficial, and sometimes exact, copies of the word system. It is in the ingenious Indic system that they first assume the character of a logically perfect instrument of arithmetic, but also of an instrument which originated through scientific reflection. The continuation of logical developments in the next chapter will show that the choice of means of designation is in no way an unimportant matter. The logical differences between these means depend upon the varying degrees of their aptitude for a technique of calculating. This is the reason why we have not yet here drawn them into the field of consideration.⁵

Appraisal of Number through Figural Moments

In Chapter XI we treated in detail the problem of how immediate appraisal of groups comes about without the actual carrying out of the relevant psychical activities – those of individual apprehension and collection. A unitary intuition is given to us, and in *one* glance we judge: a group of balls, coins, and so on. To explain this peculiar fact we referred to the figural Moments of the unitary group intuitions which enter into an association with the name and the symbolic concept of the multiplicity – mediating the reproduction of the latter, and thereby making possible the immediate appraisal of the phenomenon [*Erscheinung*] as a group. Immediate *number* estimation presents a quite similar problem and the means referred to completely suffice for its solution.

The matter stands forth most clearly in examples, as is abundantly illustrated in play at dice, dominos and cards. Each surface <253> of a die possesses a characteristic fixed configuration of dots which enters into an association with the number name (or with the symbolic concept of a certain number named by it). If several dice are thrown simultaneously, then *either* there occurs a rapid quasi-summation utilizing the tables of addition – in which, of course, the mere number words intervene – *or else*, given long practice, the number word corresponding to the sum of dots is reproduced immediately by means of the figural character of the total complex phenomenon. The number of configurations to be impressed upon us for this is in fact only a limited one. The same holds true for play with dominos, and it is well known what a knack experienced players have for instantaneous estimation of numbers. They often can count up to forty dots in one glance.

In the examples considered up to now the configurations were of a fixed type, or even more so, were closely related in type. In order to explain the latter case it should be pointed out that a die

---

⁵ Profuse illustrations supporting the presentations of this section are found in the celebrated anthropological and linguistic works of *Tylor*, *Lubbock*, and *Pott* among others.

surface, for example, in each change of position through rotation, receives another figural character, and that it therefore must basically be the corresponding *generic* character that establishes the association [with the number]. This observation makes it clear that the difference between the cases considered and others where *wholly arbitrary* distributions of objects are estimated as to number is not so great as it might at first appear.

However three cleanly separated objects may be distributed in the field of vision, they together form a characteristic configuration – presupposing that they can in general fuse into an intuitively unitary appearance of a group. The various three-point configurations which arise, depending upon the varying relative positions of the objects, are indeed well-distinguished in intuition. But they possess so much striking analogy that the character common to them all can mediate with certainty the reproduction of the number three (or, more precisely, of the name "three," along with the symbolic concept of a specific number named by it). A somewhat more essential difference is exhibited by the figural character only in cases where the three objects come to lie in a straight or approximately straight line: a boundary case whose quite noticeable special character makes possible the association of the number. It is similar with groups of four objects. Here the configuration exhibits either the familiar <254> quadralateral type, or else other characteristic types show up – as when all four or any three of the objects lie in a row, or when one object falls within a triangular figure formed by the three remaining ones. And so on.

The more objects the group includes, the greater is the respective number of intuitively distinct figural types, and thus it becomes understandable why in reliable number estimation we usually do not get past groups of five members – unless by means of constant, methodical practice. *Preyer*, who did experiments on this, is of the opinion that in the latter case the attainable limits may lie, on the average, at twenty. Nevertheless, the famous calculator *Dahse* could instantaneously estimate some thirty arbitrarily distributed objects.

We see from the foregoing considerations that one has no occasion to borrow from the ever obliging "unconscious" in order to explain instantaneous estimates of number. Accordingly, I also cannot go along with *Preyer*'s theory of "unconscious enumeration,"[6] and I can only wonder that this outstanding physiologist could regard his arguments as compelling. From the mere fact that we instantaneously estimate small groups of objects without there being time for deliberate enumerating, be it ever so rapid, he concludes: "Thus, not only is unconscious enumerating nothing inherently self-contradictory, it is, rather, an everyday occurrence." For more support he further adds: "One must not object that *that* is no longer enumeration. For if someone can definitely state whether three, four or five objects are before him, then he must be able to distinguish the numbers; and it is certain that one who cannot enumerate is also not capable of answering that question." Even if the circumstance upon which the last argument is based were incontestable, it would prove nothing, since the way precisely still stands open to say that, after repeated enumeration of many types of object distributions, the number names enter into fixed associations with their typical figural characters. <255> Moreover, I would hold it to be quite well possible that even someone completely ignorant of enumerating could bring the number names into association with those figural characters, and develop into a skillful domino player, for example.

---

[6] *W. Th. Preyer*'s treatise, "Über unbewusstes Zählen," first appeared in the *Gartenlaube*, 1886 (See pp.15 & 36), and is reprinted again in the collection of his scientific essays that has recently appeared. How seriously *Preyer* takes the hypothesis of the unconscious is shown by the accompanying physiological explanations. According to these, the movements corresponding to the intellectual activity of enumerating ". . . ultimately run unconsciously along the nerve fibers and cells in the brain that are used very often" and "approximate to reflex movements."

Chapter XIII

THE LOGICAL SOURCES OF ARITHMETIC

Calculation, Calculational Technique and Arithmetic

The concept of calculational technique is customarily taken to be closely associated with that of arithmetic. Indeed, the two are often identified. *Arithmetic* is usually defined as the science of numbers. This definition is not sufficiently clear. The individual numbers, considered by themselves, give no occasion for treatment with a view to knowledge of them; and where we are concerned with a scientific grounding of particular characteristics of individual numbers, it is always a matter of properties that accrue to them in virtue of certain relations that bind them to other individual numbers (or whole classes thereof). Only out of the relationships of numbers to one another do there arise problems that require a logical treatment. Accordingly, it would be better to define arithmetic as the science of the relations among numbers. In any case, its essential task consists in finding other numbers from given numbers by means of certain known relationships that obtain between them.

Let us now consider the concept of *calculational technique* [*Rechenkunst*]. It is given to us when we possess that of *calculation*. The concept of calculation, however, admits of various wider and narrower significations. By calculation in the *broadest sense* one can understand *any mode of derivation of numbers sought starting from numbers given*. Accordingly, we certainly would already have to call the unification of the numbers 2 and 3 to form 5, on the basis of the authentic representation of the concepts themselves, a calculation. Likewise for the construction of the concepts of systematic numbers, whether we <257> adopt the path of concept formation or that of mechanical-exterior sign formation. Any arithmetical method would *eo ipso* be a calculational method. The calculational technique would be the technique of arithmetical cognitions, and arithmetic only the systematically arranged entirety of them.

Now, as to the *method of derivation* of sought numbers from given numbers, there are two conceivable cases: *either* this derivation is an *essentially conceptual operation*, in which case the designations play only a subordinate role, *or* it is an *essentially sense perceptible* operation which, utilizing the system of number signs, derives sign from sign according to fixed rules, only claiming the final result as the designation of a certain concept, the one sought.

Which of these methods is the logically more perfect can only be a question of what they are capable of accomplishing. By anticipation we can already say here that the latter, under all circumstances, has the preference in our domain, and also abundantly deserves it. The method of concepts is highly abstract, limited, and, even with the most extensive practice, laborious. That of signs is concrete, sense perceptible, all-inclusive, and it is, already with a modest degree of practice, convenient to work with. I said all-inclusive. In fact there is no conceivable problem which it would not be capable of solving. Thus, it makes the conceptual method entirely superfluous, its use being no longer suited to the scientific state of mind, but only to a childishly backward one instead. There certainly is much yet lacking for a general acknowledgment of this view; and it is surely bound up with the lack of a logic of symbolic methods of knowledge (and, above all, of those of arithmetic) that most researchers – guided by the general prejudice that every scientific methodology operates with the respective intended concepts – have also held the arithmetical operations to be abstract-conceptual, in spite of all clear indications. We will later pay special attention to illuminating the true state of affairs.

The method of sensible signs is, therefore, *the* logical method of arithmetic. Here, then, is presented that concept of calculating which (with regard to the extent of its application) we can designate as the one most commonly in use. It encompasses <258> *any symbolic derivation of numbers from numbers which is substantially based on rule-governed operations with sense perceptible signs*. It follows from the preceding remarks that, in spite of this restriction on the concept of calculating, the relationship between the concepts *arithmetic* and *calculational technique* has undergone no essential modification, since modes of procedure other than calculational (in the present sense) simply do not come into consideration in the former.

But, when we take into account that the mechanism of the symbolic methodology can break completely free of the conceptual substrata of its employment, yet another concept of calculation compels our attention: one which in comparison with the foregoing is more restricted in one sense, but in another, by contrast, is broader.

One can, namely, conceive of calculation as *any rule-governed mode of derivation of signs from signs within any algorithmic sign-system according to the "laws" – or better: the conventions – for combination, separation, and transformation peculiar to that system*.

There are higher logical interests than those of *arithmetica numerosa* (with which we currently have to do) which require this delimitation of the concept. It is a fact highly significant for the deeper understanding of mathematics that one and the same system of symbols can serve in *several* conceptual systems which, different as to their content, exhibit analogies solely in their structural form. They are, then, as we say, governed by the same calculational system.

But this new formulation of the concept of calculating also recommends itself still further in that it places in our hands a logically clear separation of the different stages required by problem-solving in those domains that are susceptible to treatment by technique. Each solution obviously decomposes into one calculational part and two conceptual parts: *Conversion of the initial thoughts into signs – calculation – and conversion of the resulting signs back into thoughts*. In the domain of numbers, where the conceptualization and separation of the concepts (disregarding, of course, the few "authentic" ones) rests upon the symbolization running parallel to it as its indispensable <259> support, that first step consists merely in this: that in the complexes of concepts and names that are given in each case, one abstracts from the former and only holds to the latter.

The relationship between arithmetic and calculational technique, with this new concept of calculating (which from now on is the only one we wish to use), has certainly now been changed. If we loose the number signs from their conceptual correlates, and work out, totally unconcerned with conceptual application, the technical methods which the sign system permits, then we have extracted the pure calculational mechanism that underlies arithmetic and constitutes the technical aspect of its methodology. Obviously, then, the calculational technique is no longer identical with the technique of *arithmetical* cognition.

The Calculational Methods of Arithmetic and the Number Concepts

If a discipline at the height of its development needs no other than calculational means for the solution of its problems, it still certainly cannot dispense with that which is conceptual in its beginnings, where it is a matter of providing a logical foundation for calculation. Only the systematic combination of the concepts and their interrelationships, which underlie the calculation, can account for the fact that the corresponding *designations* interlock to form a coherently developed system, and that thereby we have certainty that to any derivation of signs and sign-relations from given ones, which is valid in the sense prescribed by the rules for the *symbolism*, there must correspond a derivation of concepts and conceptual relations from *concepts* given, valid in the sense that *thoughts* are. Accordingly, for the grounding of the *calculational methods in arithmetic* we will also have to go back to the *number concepts* and to their *forms of combination*. The former are already given to us in the systematic forms of the sequence of natural numbers, or in the higher forms of the number "system" (in the specific sense of the term). But the combinations through which, from given numbers, new numbers are obtained are the *number operations*. <260>

The Systematic Numbers as Surrogates for the Numbers in Themselves

The number systematic offers, as we saw, a uniform and (given Idealizing abstraction from certain limitations on our capabilities) inexhaustible method for the continuation of the number domain beyond any point. Since by means of its rigorously distinct conceptual formations any conceivable multiplicity is denumerable, there also can be no actual [*wirkliche*] number which would not find its symbolic correlate in the system; and, indeed, in each case only one, since different systematic number symbols necessarily refer to different actual numbers. A number system (as, for example, our decimal system) can accordingly be regarded as the most perfect mirror reflection of the domain of the numbers in themselves, i.e., of the actual numbers, which are in general inaccessible to us. And this is also true with respect to their ordering, which – with the symbolic as with the authentic concepts – is that of a simple sequence. Thus we may justifiably regard the indirect formations of the system as the symbolic surrogates of the numbers in themselves.

The Symbolic Number Formations that Fall Outside the System, Viewed as Arithmetical Problems

If, now, the system also includes symbolic correlates of all conceivable numbers, it nevertheless does not contain all conceivable number symbolizations whatever. Any number can be univocally characterized by manifold relations to other numbers, whether these are actually [*wirklich*] represented or are already symbolically represented. And each such characterization provides a new symbolic representation of precisely that number.

Thus, outside the "system" there are still infinitely many symbolic number forms. May we then regard these as equivalent to the rule-governed formations of the system, as legitimate Representatives of the actual [*wirklichen*] number concepts? May we acknowledge them as given in the same sense, and count the reduction of a problem to a numerical result characterized by means of them as that problem's definitive solution?

Simple reflections teach us that we must reply to these questions in the negative. By virtue of the order in the sequence we can immediately decide, for two <261> systematic numbers, whether they Represent the same or different numbers; and in the latter case we also can immediately state which is greater and which smaller. A mere glance at their relative positions suffices. It is quite otherwise with the nonsystematic symbolizations of numbers. Formations of a type such as 18 + 49, 7 × 36, and the like, provide us with symbolic number forms that are no less determinate than the decimal forms corresponding to them. But how does one decide whether any two such numbers are equal or not, and in the latter case which is the greater and which smaller? If we already possess a number system, then the answer is clear: simply through reduction to the corresponding systematic numbers. For this much is certain, that to each nonsystematic number there corresponds a univocally determinate systematic number that is equal to it, i.e., one which symbolizes the same authentic number concept.

Thus it happens that we regard each nonsystematic formation not as something completed and given, but rather as something *problematic*. It poses a problem that demands a solution: namely, the problem of finding the systematic number which corresponds to it and determines its classification. How much is 18 + 48? We answer: 66. And with this we have located this sum-number within the number sequence. For these reasons, then, the systematic numbers – although themselves only symbolic surrogates for other concepts that are inaccessible to us – are regarded in arithmetic as the ultimate number concepts, which all other number forms only lead back to and therefore can, in addition, be reconstructed starting from. But in truth they only function as *normative numbers* [*Normalzahlen*] – fixed standards, as it were – which all other number forms are referred back to for the purposes of an exact comparison among themselves with respect to more or less. And thus we see that *by means of the number sequence the Ideal of a general and exact classification of numbers* is realized in the most perfect way. But that it accomplishes this entirely corresponds to our original intentions. The need for order and classificatory distinction in the jumble of symbolic number forms was in fact the original impulse that forced our logical development to expand the domain of authentic numbers given at the outset precisely into its systematic forms.[1] <262>

The First Basic Task of Arithmetic

Our last investigation led to a *general postulate of arithmetic*: the symbolic formations that are different from the systematic numbers must, wherever they turn up, be reduced to the systematic numbers equivalent to them, as their normative forms. Accordingly there arises, as the *first basic task of Arithmetic, to separate all conceivable symbolic modes of formation of numbers into their distinct types, and to discover for each type the methods that are reliable and as simple as possible for carrying out that reduction.*

The Elemental Arithmetical Operations

That task leads us to the so-called "four species," the most elemental of arithmetical operations. Already in Chapter X we labored at length over their sense and signification – without exactly succeeding. We strayed into the byway of a deleterious skepticism, which negates instead of illuminates, and thus came into conflict with the *fact* of a science so highly successful and so evolved as arithmetic. This was but a rigorous consequence of the standpoint which we adopted there: namely, that of the widespread prejudice to the effect that arithmetic has to do with true and authentic number concepts and the laws of their combinations or "operations." We have only since then first achieved the insight that its true underlying substance consists of symbolic number formations, and ultimately recognized that its first and essential task lies in the discovery of general rules for the reduction of the diverse forms of number to certain normative forms.

Already at this point it is to be foreseen that by the so-called

---

[1] Cf. ch. XII, pp. 238ff.

"*arithmetical operations*" nothing other is to be understood than *methods for carrying out this reduction*. Of course with this the concept of numerical operation, as bound up with the authentic concept of number, is modified in a quite essential way from the
5 original signification. We obtained our first concept of addition, for example, by reflexion upon the way in which <263> several totalities – and accordingly several numbers – are led over into a single one that encompasses all of their units. But now when we speak of an addition, e.g., of 7 + 5, we have in mind a certain
10 systematic (natural or decimal) number that is equivalent to this sum.

Obviously the two concepts of "operation" still remain in an easily recognizable affiliation. Since the authentic number concepts are not accessible to us – not to speak of our classifying,
15 adding and partitioning them – we therefore, in their place, operate with rigorously defined, symbolic surrogate-concepts, which we classify by taking as our basis a sequence of normative concepts, and which we add or partition by locating in that sequence the concepts which exactly correspond to the concepts of combi-
20 nation and partition. And just as the individual symbolic number stands in for a definite authentic one, so also each symbolic operation of combination stands in for a definite (although not actually executable) authentic one.

Then the question will arise: can one really – or rather, *how* can
25 one – carry out those tasks of reduction (the "operations")? The soundness of our new concept of operation will depend upon the answer. If, further, we are to attain the desired and necessary agreement with the arithmetical way of speaking, then we must also show how addition and partition, as the conceptual sources
30 of all operating, lead in the domain of the symbolic concepts to a plurality of autonomous calculational operations: but, first of all, to the *four species*.

We need to regard the *total* number system as given. Or, more precisely, we conceive of the primitive domain of numbers
35 as so far developed, in one of the systematic forms, that now all enumerations in any way required can be treated as actually executable, and accordingly as something already given. All other forms of number formation are, as we straightforwardly say,

combinations of systematic numbers. Only small numbers can be given to us as simple and nonsystematic. But positioning them within the system is so easy that we can, in each case, regard it as already accomplished, and therefore can also take combinations of such small <264> numbers already to be combinations of systematic ones. The forms of operations which the number concept permits are addition and partition. Let us now consider the special turns which they undergo in arithmetic.

Addition

We begin with addition. To add several numbers $\underline{a}$, $\underline{b}$, $\underline{c}$, ... meant, in the original sense, to combine the units in those numbers to form a new number $\underline{s}$. There could be no talk of a special rule for carrying this out in the domain of authentic number concepts. Things stand quite otherwise, now, in our domain of symbolic number forms. Here, to add several numbers $\underline{a}$, $\underline{b}$, $\underline{c}$, ... signifies the finding of the *systematic* number corresponding to their sum. Since the sum is, in concept, independent of a particular order of the summands, we can undertake the sequential execution of it – and indeed, in an arbitrary succession – through, step by step, first picking out a number, perhaps $\underline{a}$, and adding $\underline{b}$ to it, and then to the result of this adding a new number, perhaps $\underline{c}$, and so on. (In all of this it is customary and appropriate to indicate the succession of the operations by means of the sequence in which the summands are spoken or written.) The problem of the addition of arbitrarily many numbers is consequently reduced to that of the addition of any two. The solution itself is then easy. In order to evaluate a sum a + b we need merely to start out from $\underline{a}$ in our number system, and, following the fixed order of that system, count further by $\underline{b}$ members. In fact, each new step provides a number greater than $\underline{a}$ by one, and thus $\underline{b}$ steps a number greater by $\underline{b}$; i.e., a number which contains, besides the units of $\underline{a}$, $\underline{b}$ more units (and no others), thus a + b. But we also obtain the same number by means of enumeration of $\underline{a}$ members beyond $\underline{b}$. There results, namely, the number which contains, besides the units of $\underline{b}$, $\underline{a}$ more units (and no others); and thus, with regard to the sum concept, identically the same number as before.

With this method it is, moreover, a matter of indifference whether we take as the basis of arithmetic merely the sequence of natural numbers or the system numbers. And it will obviously in either case <265> have no effect at all on the trustworthiness of the result whether in carrying out the operation we think of the *concepts* themselves or adhere to the *mere signs*. The sign $\underline{a}$ is an absolutely reliable surrogate for its concept. Each new step supplies the sign of a number which is greater by one, and therefore $\underline{b}$ steps the sign of a number greater by $\underline{b}$. Consequently, from the sign $\underline{a}$ we have only to enumerate by $\underline{b}$ signs further – which enumeration, again, like any enumeration at all, can be carried out in a purely mechanical fashion.

Those untrained at enumerating often prefer another and still more crude procedure, in that they base their enumerations upon sense perceptible groups – e.g., of little stones, tally pieces, dots on a sheet of paper – as supports. They accompany the enumeration from 1 to $\underline{a}$ by the successive singling out of a sensible group of $\underline{a}$ members. In similar manner they count out a second group of $\underline{b}$ members, in order then, finally, either to enumerate onwards from the former in the form $a + 1, a + 2, \ldots, a + b$, or from the latter in the form $b + 1, b + 2, \ldots, b + a$. — The logical justification of this mechanical procedure poses no difficulty whatever. Since the result of the enumeration is independent of the concrete nature of the objects counted, each result attained in the concrete example can immediately be claimed for any conceivable type of units counted, whether, moreover, the counting was authentic or symbolic, provided only it was correct. If, thus, the untrained person finds in the enumerative generation of a sum – whether of numbers in general, or of numbers of units quite inaccessible or inconvenient to him – a difficulty that does not arise for him in the corresponding enumeration of available sense perceptible objects, then he does quite well to solve the corresponding problem for groups of precisely such objects and to appropriately generalize or transfer the result.

In the case of larger numbers these methods of addition, although they can be carried out purely mechanically, are extremely cumbersome and time consuming, and therefore are very limited in their applicability. In the case of the sequence of natural numbers there is certainly no better method, but there surely is a better one in that of the "system numbers" in the narrower sense of the term.

Each number belonging to it is an aggregate of numbers of different types of units: X^0, X^1, X^2. If, then, the sum of two <266> such numbers is to be evaluated, we can in the first place add up the corresponding partial numbers of like units, and then form the sum of these sums. In symbols:

$$(a_0 + a_1X + a_2X^2 + \ldots) + (b_0 + b_1X + b_2X^2 + \ldots)$$
$$= (a_0 + b_0) + (a_1 + b_1)X + (a_2 + b_2)X^2 + \ldots$$

Now since the a_n as well as the b_n represent numbers between 0 and X, the coefficients of the sums obtained will then in general be greater than X, and the sum itself will therefore no longer be a symbolic number. However, the evaluation of the coefficients yields numbers which are either smaller than X – we wish once and for all to designate such numbers by an α with or without an index – or have the form $\alpha + X$. Accordingly, the sum is very easily reduced. One forms $a_0 + b_0$ and obtains α_0 or $\alpha_0 + X$. In the first case one already has the lowest member of the number sought. In the latter case, one retains the mere α_0 and adds the X as one unit of the next higher degree in the sum $a_1 + b_1$ now to be formed. For $a_1 + b_1$, or else $a_1 + b_1 + 1$, one again obtains the number form α_1 or $\alpha_1 + X$. Since X units of X_1 yield one unit X^2 of the next higher degree, one again in the latter case retains the mere α_1 and counts the X^2 as one unit in the degree immediately following. For this one there is $a_2 + b_2$, or $a_2 + b_2 + 1$, to be rendered as α_2 or as $\alpha_2 + X$. In the latter case again one has to add one unit to the following degree, and so on. Thus there results a law governed procedure that successively yields the members $\alpha_0, \alpha_1X, \alpha_2X^2$, the sum of which is the systematic number sought.

The immense advantages of this method are obvious. The addition of two numbers, however large, which according to the previous methods required at least as many individual steps as units in the smaller number, is now reduced in essentials to a small number of pairwise additions of numbers lower than X – a number which is at most as large as the highest degree of the terms involved. But we can economize even on these additions. If, namely, we think of all conceivable sums of any two numbers under X as calculated once and for all (utilizing a more primitive method, of course), and as indicated in tables or imprinted on the memory, then we are spared all their calculations, which are tedious and, even with small numbers, complicated. By means of the familiar and always available propositions of the addition tables up to $(X - 1)(X - 1)$ – in our decimal system it is $9 \cdot 9 = 81$ – <267> each addition problem can be solved in this way, directly or indirectly.

One immediately perceives that in the practical sphere this process of addition must spontaneously become a purely external form of calculation. But whether this is legitimate, whether the truth value of the result actually is necessarily independent of whether we operate with the concepts or simply with their signs: that still requires a logical investigation.

The rule derived above shows how an arbitrary addition of two systematic numbers is to be reduced to a sequence of elemental additions, totally determinate in each case. Now to each systematic number there corresponds a determinate systematic sign in such a way that, not merely the number itself as a whole, but also its mode of formation out of partial numbers, finds its symbolic expression in the sign. But through these same partial numbers (namely, the coefficients of the power units: a_0, b_0; a_1, b_1; ... as designated above), those elemental additions (e.g., $a_0 + b_0$) are univocally determined. Consequently, the linguistic expressions for them are also univocally determined through those of the same partial numbers. Further, the elemental additions are univocal as to their result. They are carried out, perhaps by a determinate process of computation, or perhaps by referring to the table of truths for addition. But the quasi-additions of the signs corresponding to them are also univocal as to their result, whether carried out through the parallel external process of enumeration, or through referring to the table of *sign equivalences* for addition. Thereby is proven, all steps taken one by one, the rigorously univocal correspondence between the method of addition by thinking in concepts and the method of addition by calculating in signs;

and we can place our complete trust in the latter. The huge savings in psychical labor which blindly mechanical calculating makes possible lies here so clearly in view that any more detailed discussion becomes unnecessary.

One easily sees how the method of addition can be expanded to cover the simultaneous addition of arbitrarily many systematic numbers, without changing anything essential or requiring further resources beyond the tables of addition. <268>

Multiplication

The method undergoes a considerable modification in the special case where a number of equal summands are to be added. Therefore one here justifiably speaks of a new calculational operation – multiplication. A major abbreviation is already brought about through the multiplicative mode of representation and designation, in that the number of the summands is introduced as a means of symbolization; and this mode simplifies to the same extent the discovery of the sum sought, or, as one here appropriately expresses it, of the "product." The problem which multiplication solves consists in this: to calculate the product (that sum value) solely from the multiplicand (the numerical value common to all the summands) and the multiplier (their number), without having to actually carry out the addition, or even to begin it. Considerations of a kind similar to the previous ones with addition show that the multiplication of two numbers can be reduced to mere additions and multiplications of numbers between 1 and X. The conceptualizations which essentially come into play here can be symbolically expressed by means of the following chain of equations:

$$(a_0 + a_1X + a_2X^2 + \ldots)(b_0 + b_1X + b_2X^2 + \ldots)$$
$$= b_0(a_0 + a_1X + a_2X^2 + \ldots) + b_1X(a_0 + a_1X + a_2X^2 + \ldots)$$
$$+ b_2X^2(a_0 + a_1X + a_2X^2 + \ldots) + \ldots$$
$$= b_0a_0 + b_0a_1X + b_0a_2X^2 + \ldots$$
$$+ b_1a_0X + b_1a_1X^2 + \ldots$$
$$+ b_2a_0X^2 + \ldots$$
$$+ \ldots$$
$$= b_0a_0 + (b_0a_1 + b_1a_0)X + (b_0a_2 + b_1a_1 + b_2a_0)X^2 + \ldots.$$

If one again thinks of all possible elemental multiplications up to $(X-1) \cdot (X-1)$ as calculated out once and for all (through actual addition), and indicated in tables or imprinted on memory, then the execution of each multiplication reduces to a rule governed process that, in its individual steps, requires only the reproduction of propositions from the addition tables as well as from the multiplication tables. And again, by inferences analogous to those above, one is convinced that the trustworthiness of the result can suffer no harm when the multiplication operating conceptually gives way to a calculating that is purely mechanical. <269>

From the concept of multiplicative combination it follows (presupposing the meaning of the arithmetical notation to be familiar) that the products $a \cdot b$ and $b \cdot a$ correspond to identically the same numerical value. So if the value of the one product is to be determined, then, in case it is more convenient, one will be permitted, in carrying out the multiplication, also to use the other product as a basis.

Subtraction and Division

We now pass on to the operations of partition. From the standpoint of the authentic number concept, partition was the only basic operation apart from addition. General rules for its execution could not be given. More specialized tasks based on the concept of partition were subtraction and division. For those in possession of the authentic number concepts, here too hardly any technical rules were required. For us, who are essentially limited to symbolic number formations alone, these problems modify their sense, and the solutions suited to them do indeed permit and require a rule governed method. However, the general problem of partition remains excluded here too, whereas its two specializations actually lead to calculational operations. We commence with *subtraction*.

To subtract b from a meant, in the original sense (thus, for the authentic [*eigentlichen*] number concepts), to separate a number b of units from the units in a, and unite the remaining ones to form a new number, the "difference" number c sought. In other words, a is to be partitioned into a sum b + x, of which one member b is known and of which a second member x is sought. Obviously the problem does not always have a sense and solution. It must, precisely, be possible to take a as an additive whole of which b is a part. But it can be, conversely, that the number b is a whole of which one part is a – a case that is incompatible with the previous one. It is only the same condition clothed in other words if one says that b must be smaller than a, but not greater than a. If we are dealing with symbolic, and specifically with the systematic number concepts, then the problem to be solved by the subtraction operation is to find from the systematic numbers a and b the systematic number corresponding to their difference. The <270> condition stated, upon which depends whether the operation can be performed or not, can, with the symbolic formations – which are ordered in the number sequence – be decided merely by their relative positions, without relapsing into the respective authentic concepts and the attempt at actual partition. And this stated condition can, consequently, be replaced by one equivalent to it: that the number b must have a lower position than the number a.

As to the solution of the problem, then, it can result from a mere operation on the completed number sequence (whether it be that of the natural numbers, or that of the systematic numbers in the narrower sense of the word), by either starting from b and proceeding one by one, counting through the number of steps up to a, or by starting from a and, progressing backwards (i.e., in the direction of the beginning of the sequence), retaining the number settled on after b steps. The latter way of doing it supplies the number sought by means of a process which corresponds to the successive taking away of b units from the whole a, the former through a process which corresponds to its direct construction as the number which added to b yields a.

In the special case where a and b are members of a "number system," a different and far superior solution presents itself, which in its manner and orientation follows precisely the corresponding method of addition. One first attempts (similarly as on p. 281) to replace a - b by

$$(a_0 - b_0) + (a_1 - b_1)X + (a_2 - b_2)X^2 + \ldots$$

But in this it can happen, as one easily sees, that, even if not all, yet some of the subtractions indicated have no sense. If, now, this is so at the position $(a_k - b_k)X^k$, for example, then one "borrows" one power unit from the next higher position, and thus transforms the member into $((a_k + X) - b_k)X^k$, while the next position becomes $((a_{k+1} - 1) - b_{k+1})X^{k+1}$. The execution of the subtraction problem posed is reduced to that of straightforward elemental subtractions of the form $\alpha - \beta$ or $(\alpha + X) - \beta$, in which α, and β represent numbers between 1 and X (in the latter case excluding the values $\beta = 1$, $\alpha = X - 1$).

It is clear that the addition table of the "one plus ones" is also labor saving in these easy enumerations. Each of its propositions has the form $\alpha + \beta = \gamma$, where α and β can again represent one of the numbers <271> 1, 2, . . . X, and γ one of the numbers 2, 3, . . ., 2X - 2. In addition problems α and β are given to us, and their sum γ is to be selected from the table. In our present case, however, γ and α are given, and their difference β is to be taken from the table.

Subtraction is called the "inverse" operation to addition because, according to its concept, $(a + b) - b = a$. Thus, what the attachment of b accomplished is cancelled out again by means of the subtraction of b.

Similar observations would now also have to be made about *division*. It too is limited as to conditions where it can be carried out inasmuch as it is possible to partition into a equal parts only such numbers b as are multiples of a. The procedure for finding the systematic number corresponding to the quotient a : b can once more – in the case where it is possible to carry it out – be reduced to a sequence of elemental divisions, in the course of which, similarly as with the "one plus one" tables in subtraction, now those of the "one times one" come into play. Certainly that same table does not suffice here. Basically, a table is needed of all the elemental divisions of the numbers of the second degree by those of the first, and, if the divisions are not "even," with an exact indication of the remainder. For our purposes it is not necessary to go into the details, already quite complicated, which bear upon the justification of the calculational procedure in the case of division.

What was established for addition and multiplication holds true, once again, for subtraction and division: The operating with concepts that was necessary to lay a logical foundation for the procedure becomes superfluous in its application. Its place is taken by *mechanical calculation*, the logical soundness of which is guaranteed by means of the rigorous parallelism between the systematic of the numbers and number relations, on the one hand, and that of the number signs and relations of number signs (equivalences of symbols), on the other.

After these discussions we can, therefore, designate the four species as operations of *arithmetical calculation*. They are operations of *calculation*, since they work with mere signs; and are *arithmetical* operations, since they serve in the derivation of numbers. They Represent logical methods for the evaluating of symbolic number compositions (sums, products, differences, and quotients); <272> i.e., for determining the symbolic normative forms corresponding to those compositions as the logically qualified surrogates for the actual [*wirklichen*] number concepts. They are indirect methods for the classificatory subsumption of those number-compositions under the correlated, surrogate number concept.

All of the difficulties and doubts encountered in Chapter X with regard to the understanding of the calculational operations and the arithmetic which treats of them, we may already at this point regard as resolved. With the modified sense which the operations acquire in the domain of symbolic number formations, it has become fully intelligible why scientifically elaborated methods for carrying out the operations are here required, which seemed pointless there. And the broader investigations which we add on below will significantly expand and deepen this knowledge. They will bring to our understanding the true sense of arithmetic through the development of the logical needs from which it arises and which it satisfies.

Before we follow up further on these systematic guidelines, we first wish to supplement the presentations to this point with some additional remarks concerning other types of calculational methods, as well as concerning the significant influence of external circumstances upon the individualized character of such methods.

Methods of Calculation with the Abacus and in Columns. The Natural Origination of the Indic Numeral Calculation.

The logical discussions through which we have arrived at the methods last developed presuppose, on the one hand, the completely and coherently elaborated number system, and, on the other hand, a system of number designation completely and coherently mirroring it. We therefore have set out from a logically Ideal case, which could only find its realization in a highly developed arithmetical understanding. Through the ages of history this requirement went long unsatisfied. Although the principle of systematic number formation essentially achieved a breakthrough [*Durchbruch*] with most peoples who rose above the level of savagery, numerous residues hung on as vestiges of the older, non-systematic modes of number formation, <273> and so complete systematic consistency was lacking in the number systems – and therewith also in the parallel systems of number terminology. To a far higher degree this still holds true of the *numeral* designations. They come down from more ancient times and, stubbornly clung to, lagged far behind the systematic of the word signs; so that, even with a people so highly civilized as the Romans, they hardly deserve the name of "systematic" designations. Only the Chinese and the Hindus had already in ancient times produced a rigorous consistency in the respects mentioned.

Accordingly, peoples other than those last mentioned also could not, for the purposes of practical reckoning, make use of any type of mechanical method that presupposes coherent word and numeral designations for the systematic numbers. Nevertheless, there was no lack of mechanical methods before the invention of such designations, indeed, not even of those with a principle essentially identical to that of the method logically justified above. I refer to the *methods of calculation with abacus* or in *columns*. These Represent transformations of that natural mode of enumeration which we have from time to time invoked in the explanation of the number systems.² Within spatial areas delimited in a fixed manner (in columns or on rods) moveable tokens, perhaps small stones, spheres, etc., can be placed in sufficient amounts. These areas have a stable order. The tokens in the first row are signs for ones, those in the second row are signs for tens, and so forth. Each distribution of tokens in the columns thus presents us with a decimal number in a rigorously articulated form.

Thereby the lack of word and numeral designations is already essentially eliminated. Through these simple arrangements there is procured an artificial and rigorously systematic decimal terminology, the invention of which of course presupposes a clear insight into the principle of the number system. A crucial step forward consists in introducing tokens, not merely for the power units, but also for the power terms. For a written calculation within columns formed of parallel lines, one then writes the signs required in each case <274> from the domain 1, 2 . . ., 9 in the appropriate columns, e.g.:

| 1 | 8 | 9 | 1 | for 1891,
| 3 | – | 7 | – | for 3070,

instead of, as with the cruder method, modelling each power term by corresponding groups of sense perceptible objects which were to be counted only later. After the discovery of these arrangements for rigorously systematic decimal number formation and designation, the rules of calculation – at least for addition, multiplication and subtraction – could correspond exactly to those which we developed above in a general form. This surely requires

---

² Cp. ch. XII, p. 263.

no further exposition. Only in the procedure of the first mentioned and more primitive formation, that of the abacus, are there minor deviations, due to the fact that constant enumerations of the unit tokens in the individual columns are required.

It is highly probable that the Indic numeral system arose from this manner of writing the numbers – articulated in the rigorously decimal manner – in columns, since, with the horizontal entry in the columns, a meaningless sign 0 was entered to mark the absence of a power unit, thus clearly maintaining the linear continuity; just as today the shopkeeper usually puts a dash in the columns of his account book with a similar aim. From there it was only one step to the realization that the column lines are superfluous, given a consistent use of that sign, and by introducing the null sign the position system was completed.

Influence of the Means of Designation upon the Formation of the Methods of Calculation

But the methods of carrying out the four species, logically developed above and currently used in the practice of numeral calculation, are not the only conceivable ones. It does not correspond to my intentions to describe and logically justify other methods invented for the same purpose, which are likewise based on the double systematic of the decimal number concepts and number designations. <275> My primary concern was to clearly show, by using a typical example, the essential character of such purely mechanical systems of rules, their logical-arithmetical signification, and their logical origin. In all mechanisms of this type the fundamental principle is the same.

More important for us is the observation that the diversity among symbolic methods of calculation is by no means conditioned, solely or mainly, upon an accidental tendency in their inventors' interests. It is a fact worthy of attention, from the logical point of view, that even the external means of designation which each culture imposes upon the arithmetician can be of the most essential influence upon the elaboration of algorithms. Thus the specialized methodology, and thereby the entire mode of operation [*Habitus*] of the science, seems conditioned upon Moments the importance of which logicians, who are mostly attuned to the highest of abstractions, might be only too much inclined to underestimate. In illustration of these observations I want to append here some sentences from the repeatedly cited fragments, *Zur Geschichte der Mathematik* . . . by *H. Hankel*, whose insightfulness first brought this fact to light.

Discussing the Indic methods of calculation with numerals he comments as follows:

"If these methods often very significantly differ from the ones familiar to us, we must keep in mind the fact that the Hindus did not calculate, as we do, on paper with pen and ink. Rather, they calculated with a reed pen on a black wooden tablet, using a thin white fluid that produces easily erasable signs; or on a white board less than a foot square, coated with a red powder, upon which they inscribed the numerals with a stylus, so that they appear white on a red ground. Since the numerals, in order to be clearly readable, must be written rather large, the space on the board is therefore very limited. So the Hindus must consider how to save the most space possible in their operations. They achieve this by immediately erasing all numerals in a calculation as soon as they have served their purpose, and putting others in their place."

"This requirement, occasioned by the external circumstances that the calculation is to take up a smallest possible space, <276> together with the possibility of successively putting different numerals in the same place, must lead, as is clear, to algorithms essentially different from those we execute on paper, where, in place of that requirement, we have the one of the greatest possible surveyability of the calculation, but where alteration of a numeral is not permitted."

Hankel provides a fuller justification for these propositions, based on the methods for addition and multiplication among the Hindus. (*Op. cit.*, pp. 187ff.)

It is in general quite interesting to observe how methods of calculation can also in other respects come to have the most far-reaching differences by virtue of the different cultural conditions out of which they arise and the specific requirements which they must satisfy in those conditions. Thus *Hankel* observes that the strange-seeming methods of division by *Gerbert* (stemming from the 10th Century) are fully explained when one posits the following requirements: "1) The use of the multiplication tables is to be restricted as far as possible – especially, division of a two-place number by a one-place number shall at no time be required to be done 'in the head'; 2) Subtraction is to be avoided as far as possible and replaced by addition; 3) The operation is to proceed according to a fully mechanical method, without any trial and error. — Our current methods of division satisfy none of these requirements; they are, on the other hand, all fulfilled, in the best way possible, by *Gerbert*'s antique method of division, and can . . . even explain his selection of methods for each special case, which sometimes appears to be capricious." (*Op. cit.*, p. 322)

The Higher Operations

The four species, which we have dwelt on up to now, by no means exhaust the range of conceivable operations in calculation. Still many other forms of symbolic number compositions are conceivable, and for each of them a problem must be posed that is correlative to those which the four species solve for their forms of number compositions. E.g., a sum of equal addends (thus the cumulative iteration of one and the same number) has yielded, through the counting up of the occurrences of its repeated term, a new means of symbolic number formation: $\underline{b}$ times $\underline{a}$. We have obtained the <277> *product* representation. But a product of equal factors (thus the multiplicative iteration of one and the same number) then provides, once again, through the counting up of the occurrences of the repeated factor, a means of abbreviated and indirect number characterization. We obtain the *power* concept, a^b. And one easily sees that this new type of symbolic number formation has a sense for any pair of numbers $\underline{a}$ and $\underline{b}$, i.e., it characterizes a wholly determinate number. In the same way we can continue on: through counting how often a number has been iteratively raised to a power there arises a new type of symbolic number characterization, that of *elevation*; through the counting up of iterated elevations, again a new one; and so on *in infinitum*.

THE LOGICAL SOURCES OF ARITHMETIC 293

But with this we still are not finished. As the concept of product led to the inverse concept of the quotient, so also each of these new forms leads to corresponding inverses. If, for example, the power concept is established, then the symbolic formation a^b points to a certain number c, where $a^b = c$. But now, in virtue of precisely the same relation, b also in a certain way is characterized by a and c, and likewise a by b and c. b is characterized as the number of multiplicative iterations of a which is equivalent to the number c; and a as the number which multiplicatively iterated b times yields the number c. We therefore have here acquired two new ways of indirectly symbolizing number formations (in symbols, $log_b a$ and $^b\sqrt{c}$), through the inversion of the relationship defining the concept of power. And in the same manner each further member in the above sequence of number characterizations obviously supplies, through inversion of its definition, a new pair of characterizations.

Again the problem poses itself of evaluating number compositions of all these types, i.e., of determining the systematic number corresponding to them; and once again one will want to be able to achieve this without doing the actual calculating, which with larger numbers certainly surpasses our time and strength. We will therefore do our utmost, e.g., with exponentiation, to reduce the operation to a relatively small number of steps that are easy to master, whether of elemental exponentiations (which are to be carried out as simple multiplications) or of other calculational operations of a lower level, in which convenient tables can intervene to support us and to spare us labor in calculating.

Whether, now, the problems here characterized have a signification under all circumstances, i.e., for any arbitrary pairs of numbers a, b, <278> and further, whether or not they are affected by interchanging the numerical values without changing the form of the combination – these are questions which, depending on the degree of complication of the concepts, could no longer be so directly decided without deeper analysis. But in any case, they would require a preliminary response: the first mentioned because already, prior to beginning the calculation, one must know whether it could yield any result at all, whether the number problem posed (and consequently also the problem pertaining to magnitude requirements which imposes it) does not involve an impossibility *apriori*; and, on the other hand, also because the progression of the partial operations which one carries out (similarly as with subtraction and division) would undergo modifications depending on the possibility or impossibility of those initially attempted. But as to the questions mentioned second – concerning the interchangeability of the numerical values while retaining the form of combination – the knowledge of propositions such as, for example, $a \cdot b = b \cdot a$, will under certain circumstances spare us double labor in calculation, and can also lead in other cases, by selection of the more convenient from among equivalent forms, to welcome abbreviations.

Mixing of Operations

But with the number compositions and the corresponding operations taken into consideration up to now, the totality of those that are in general conceivable is still not exhausted. There is added the entire manifold of new forms that arise from combination of the ones already formed by using them as their basic elements. There arise problems such as, for example, multiplying a sum by a number, dividing a product by a sum, raising a quotient to a power, etc. Every such demand provides *eo ipso* a type of symbolic number representation which is to be evaluated in the sense indicated above, i.e., reduced to a determinate systematic number. Now the evaluation can certainly result from simply following, step by step, the progression of the elemental operations indicated. But logical interests once again require that, on the one hand, one must take thought to guarantee beforehand the possibility of execution (thus, that the problem posed is free of contradiction); and <279> on the other hand – with regard to the limitations of our time and strength – one must take thought for the greatest possible simplification. As to the first part, the rules for whether the individual fundamental operations can be executed are quite sufficient, and one will decide according to them from case to case. But in the latter respect, a precise *study of the reciprocal relations in which the various elemental operations*

THE LOGICAL SOURCES OF ARITHMETIC   295

stand to one another is indispensable. Thereby it becomes possible to replace one sequence of operations with an equivalent sequence of other operations that is either simpler or at least is easier to carry out; and perhaps to replace this latter again with another, until one finally arrives at a sequence that exhibits an irreducible minimum of complications or difficulties, and then is evaluated through actual execution of the basic operations indicated. In fact, it often enough happens that entire sequences of interconnected operations cancel each other entirely in their formations and inverse formations.

Examples easily present themselves. If a number is defined as the result of the multiplication of b with a and the subsequent division by b, then the determination of that number by executing as indicated would be silly. One immediately sees that a would have to result. It is no longer so directly that one sees the same result with the problem of multiplying a by b + c, dividing by c, then again dividing by b + c, and multiplying by c, although the corresponding line of thought is still easy to carry out. In the complicated cases, however, the thinking can be highly subtle and laborious, and in general can be impossible to carry out for those who do not have the most thorough familiarity with the laws of relations between operations; anyone experienced in arithmetic certainly knows this. Considerations given above – those which have led to practical methods for carrying out calculations of the four species using systematic numbers in the narrower sense of the word – can also serve us as more simple examples of the beneficial reduction of more difficult to easier operations. Thus the product of two such numbers is *eo ipso* the product of two sum formations. It was, then, knowledge of the law governed relations between addition and multiplication that made possible the undertaking of that decomposition through <280> which the operation, originally so complicated, could be reduced to a small number of very simple ones.

It is obvious that the construction of logically complete methods for the execution of the higher operations (the concepts of which we have indicated above) would also have to adopt similar paths. It would require deliberations that strive to profit from the knowledge of relationships between the higher and lower operations for a most advantageous reduction possible of the former to the latter. With this we see at the same time that it is not just the need for the most perfect execution possible for more complicated compound operations that confers importance upon those reduction problems, but also the need just validated for the most perfect methods possible for the evaluation of the basic operations themselves. And thus this point of view also commends a scientific study of the reciprocal relationships between the basic operations.

Certainly nothing would prevent us from undertaking the relevant reflections on a case by case basis. But would this likewise be adequate for the goals of arithmetical knowledge? Certainly not! Rules formulated once and for all spare us the ever-renewed labor of difficult deliberations, and also permit here the substitution of a purely mechanical operation in the place of actual thinking. *Also, the rules for the combination, arrangement and transformation of operations unite to form a seamless calculational mechanism*, as do the more specialized rules for the evaluation of the individual operations. Now if these rules are established and inculcated in advance, then for the execution of any type of operational complexes the shortest and most advantageous path of calculation can be chosen in every circumstance – without at any time having to recur to the signification of the signs.

The scientific deliberations by means of which the concepts of the various symbolic forms of number construction are defined, along with the knowledge of their reciprocal relations used in the setting up of a mechanism of calculation-rules which govern the equivalent combination, arrangement and transformation of those forms, thus form the *domain of general arithmetic*. <281>

The Indirect Characterization of Numbers by Means of Equations

To this point we have only come to know of a series of problems that have given rise to a logical need for the scientific study of the various forms of symbolic number formation and of the laws of their reciprocal connection. We have all along had in mind only the cases where a number is defined symbolically by means of a structure of numbers that are, without exception, known, and are connected and formed by means of those basic forms of symbolic combinations which we have above come to know as sums, differences, products, quotients, powers, etc. But yet other cases are conceivable. A number can also be symbolically defined as an unknown constituent of such a precisely characterized structure of numbers whose value is already known or is to be calculated by means of an operation structure built up completely out of known numbers. In one word, numbers can also be defined by *equations*. What we know of such a number is the fact that its value, *in case* it were known, would (being bound to certain given numbers by these and those operations) yield that directly or indirectly known result. Thus, whereas in the first case it is always a question of *the mere* (and the simplest possible) *execution* of a sequence of operations with known numbers, here we now have before us a far more difficult problem: namely, that of *unravelling complicated number interrelationships into which the unknown number itself is interwoven*.

Finally, there is yet to be mentioned the possibility that a number is defined by a *system of equations*, rather than by a single equation. This is a case which, in spite of the greater degree of complication, offers nothing essentially new from the logical perspective.

One immediately sees that in all of these cases we have a generalization of that type of indirect characterization of number which we have observed in every "inverse" number formation. In the first sequence of number formations a number $\underline{x}$ was defined by means of the combinations:

$a + b, a \cdot b, a^b$, etc.; <282>

and in the second sequence by means of the conditions:

$a + x = b; a \cdot x = b; a^x = b; x^a = b$; etc.

The numbers which satisfy these conditions are thought of as ultimate and elemental formations, since they are reducible neither to one another nor to those of the first sequence.[3]   But now through combinatorial linkage of these there are also to be constructed other complicated conditions of the same character: e.g., $ax \pm b = c$, $ax^2 + b^x = c$, and the like. Whether, now, such problems as these also lead to essentially new number forms – i.e., to ones that are ultimate and not further reducible – and whether what they yield is free of contradiction, under all circumstances or only under certain ones: these matters require separate investigations. In any case it immediately occurs to us to ask whether all or just determinately characterized classes of these number forms can be reduced to those elemental ones. It may be manageable to unravel the complex of operations in which the unknown number is entangled, through a series of equivalence transformations, in such a way that it finally stands forth as a number of the character first considered, one which has its equivalent in an operational structure founded upon numbers all of which are clearly known. It is obvious that these "unravelling" transformations could be based only upon an exact scientific knowledge of the relations between the various elemental types of number formations and the forms of their complication. And thus this second great class of problems – which is treated by a special branch of number theory of the highest importance, that of algebra – also evokes *the need for a general arithmetic in the sense defined above of a general theory of operations.*

Result: The Logical Sources of General Arithmetic

With this we have characterized the two vast groups of problems which require for their solution a general arithmetic – and <283> logically demand it. The *first* has to do with an indirect determination of number by means of an equivalent complex of given conjunctions of known numbers, and the task here consists in reducing to a minimum the difficulties and complica-

---

[3] We are here at a standpoint where the "negative," "imaginary," "fractional," and "irrational numbers" have not yet been introduced. Through them there occurs in our number domain a calculational/formal – although by no means a conceptual – reduction of the inverse number forms to the direct ones.

tions involved in the actual execution. The *second* has to do with a number determination which is indirect to a yet much higher degree, by means of a complex of operations that are only incompletely given, inasmuch as the unknown number itself functions as one term in the conjunctions. And the task here consists in determining the unknown – whether completely, or at least by means of an equivalent complex of the first type, which then (presupposing that the relevant methods of evaluation are already sufficiently elaborated) can be calculated out at any time, and consequently can be regarded as the Representative of a known number.

With the last problems considered, all conceivable types of symbolic number determinations are exhausted. Thus, we can also express our *result* in abbreviated form as follows:

The fact that in the overwhelming majority of cases we are restricted to *symbolic number formations* forces us to a rule governed elaboration of the number domain in the form of a *number system* (whether that of the natural number sequence, or that of the "system" in the narrower sense of the word). According to a fixed principle a number system always selects one from among the totality of the symbolic formations corresponding to each actual number concept and equivalent to it, and simultaneously assigns that one symbolic formation a systematic position. For every other conceivable number form there then arises the problem of evaluation: i.e., of classificatory reduction to the system number equivalent to it. But a survey of the conceivable forms of number formation taught us that the invention of appropriate methods of evaluation is dependent upon the elaboration of a *general arithmetic*, in the sense of a general theory of operations.

[End of *Philosophy of Arithmetic*]

SELBSTANZEIGE – PHILOSOPHIE DER ARITHMETIK[1]

For a deeper philosophical understanding of arithmetic two things currently are necessary: on the one hand, an analysis of its basic concepts; on the other hand, a logical illumination of its symbolic methods. In this twofold respect the author seeks to lay the most secure foundations possible, but not to construct a completed system of a philosophy of arithmetic. According to his view, all preconditions for that are still lacking.

The first volume, which has just appeared, is divided into two Parts. The First Part includes, in the main, psychological investigations concerning the concepts *multiplicity, unity*, and ⟨cardinal⟩ *number*, to the extent that these are not given to us in symbolic (indirect) forms. The Second Part considers the symbolic representations of multiplicity and number, and seeks to detect, in the fact that we are almost always confined to symbolic representations of number, the logical origin of a general arithmetic. In the course of this investigation there are already clarified the most elemental symbolic methods of number arithmetic, which are based upon the rigorous parallelism between the concepts and the signs, and between the rules for compounding concepts into judgements and the rules for compounding symbols into formulae. The higher level symbolic methods, quite different in nature, which constitute the essence of the general arithmetic <288> of cardinal numbers, are reserved to the second volume, where that arithmetic will appear as one member of a whole class of arithmetics, unified in virtue of the homogeneous character of identically the same algorithm.

---

[1] From pages 287-8 of "Husserliana" Vol. XII; originally appearing in *Vierteljahrsschrift für wissenschaftliche Philosophie*, Vol. 15, 1891, pp. 360-361. A "Selbstanzeige" was, at the time, an author's invited descriptive/promotional statement for the advertisement of his own book. {DW}

SUPPLEMENTARY TEXTS (1887–1901)

A. ORIGINAL VERSION OF THE TEXT THROUGH CHAPTER IV

ON THE CONCEPT OF NUMBER:
PSYCHOLOGICAL ANALYSES[1]

Introduction

From Antiquity – in fact, for millennia – there have been repeated attempts at the analysis of the concepts upon which mathematics is based, of the elementary truths on which it is built up, and of the methods owing to which it has always stood as the model of rigorously scientific deduction. And this endeavor has not been exclusively one of mathematicians. Rather, it has mainly been metaphysicians and logicians who, out of the plentitude of problems present here, have taken up now this matter, now that, depending upon the particular interest moving them, and have made it the object of special investigation. In fact, these are not problems which are either solely or mainly the concern of mathematicians. A fleeting glance at the history of philosophy teaches one how views with reference to the theoretical character of mathematics have influenced in an essential and often decisive manner the formation of important philosophical *Weltanschauungen*. In mutual opposition, the most diverse of philosophical schools have each thought that they could invoke the testimony of mathematics: the Rationalists as well as the Empiricists, the

---

[1] Husserl's "Habilitationsschrift," presented to the Philosophical Faculty of the United Friedrichs-Universität Halle-Wittenberg when he came there to commence his teaching career. It was first published in Halle a. S.: Heynemannsche Buchdruckerei (F. Beyer), 1887; second edition in *Husserliana* XII, (The Hague: Nijhoff, 1970), pp. 289-338. Much of it was incorporated in the "First Part" of the *Philosophy of Arithmetic* of 1891, translated above. Numbers enclosed in "< >"s throughout this translation are the page numbers of the 1970 edition. On the process of Husserl's "Habilitation" at Halle, see p. 357 below. {DW}

Phenomenalists as well as the Realists. Even Skeptics did not shun this battlefield. Especially <290> since *Kant*, the issues of the philosophy of mathematics have moved ever more forcibly into the foreground. As for *Kant* himself, investigations into the nature of mathematical knowledge form the foundations of his theory of knowledge.

During the most recent times in Germany it has been, in the main, the widespread Neo-*Kant*ianism which, in an effort to secure anew the basic principles of the *Kant*ian critique of reason and to support them against the Empiricism imported from England, has focused its primary attention upon these questions. Not without influence in this connection were the discussions carried on in England – for many years and with great brilliance – between *Whewell* and *Hamilton* (and his students), on the one hand, as representatives of *Kant*ian views, and the thinkers of the Empiricist school, on the other hand, led by *J. Stuart Mill*.

But beyond the narrowly confined circle of questions to which these epistemological controversies originally had reference, there yet lay a number of considerably more difficult questions, which at first were dealt with only by professional mathematicians, but later drew more general attention and presented new matter for philosophical reflection.

The interests which brought mathematicians into such frequent interaction with philosophy had their origins in the state of their own science.

It is well known what great progress mathematics has made over the course of recent centuries. A series of new and very far reaching instruments of research was found, and an almost boundless profusion of important pieces of knowledge was won. It was an exhilaratingly creative period – where the great ideas of a *Newton* and a *Leibniz* were yet to be worked out, and ever new domains of knowledge were still to be impregnated by those ideas. It is easy to understand, in such a case, how reflections concerning the logical nature of all of the puzzling, auxiliary concepts, to the introduction and subsequent application of which mathematicians saw themselves forced, had to be postponed in favor of the quest for results, for discoveries, and for the utilization of all those admirable tools. Only later – when the main or most proximate consequences of the new principles were drawn, and when <291> errors which arose in consequence of the unclarity about the nature of the auxiliary means used, and about the limits of reliability of the operations involved, became more and more numerous – only then there arose the need, more and more vivid and finally inescapable, to logically clarify, survey and secure what had been attained, to analyze the primitive and mediating concepts closely, to gain logical insight into the interdependency of the various mathematical disciplines, which at some points are only loosely connected, but at other points again are inextricably intertwined, and, finally, to develop the whole of mathematics out of the smallest possible number of self-evident principles in a rigorously deductive manner.

Since the beginning of this century the number of works giving such logical analyses of mathematics has increased immeasurably. One promises us a complete and consistent system of mathematics. Others promise an elucidation of the relationship of general arithmetic to geometry. Others again attempt to clarify those obscure auxiliary concepts (seemingly laden with contradiction, but still indispensable for analysis) such as the imaginary and the irrational, the differential and integral, the continuous, and so on. Others again – and their number is legion – deal with the axioms of geometry: especially, *Euclid*'s eleventh axiom. They attempt to prove it, to refute alleged proofs of it, or finally, by means of fictive constructions of geometries *without* that axiom, to show its dispensability and merely inductive certainty, in opposition to assertions of its apriori necessity.

Of course the philosophy of our times has had to take a lively interest in this literature which arose within mathematics; and not merely with regard to the needs of *metaphysics*, but also with regard to the needs of *logic*.

In fact, ever since modern logic came, in contrast to the older logic, to understand its true task to be that of a practical discipline (that of a technique [*Kunstlehre*] of judging correctly), and came to seek, as one of its principal goals, a general theory of the methods of the sciences, it has found many urgent occasions for giving special attention to questions about the character of the mathematical methods, and about the logical nature of their basic concepts and <292> principles. In the context of metaphysical and logical works, therefore, such discussions go on to considerable lengths, while, in addition, a large number of philosophical treatises on special topics deal now with this, now with that question from the fringe area between philosophy and mathematics.

Modern psychology also is not wholly a stranger to this domain; even if it does come in only to subject to separate investigation a few questions which either have been treated in confusion with metaphysical and logical questions, or have as of yet not been clearly raised at all – viz., the questions about the phenomenal character and the psychological origin of the representations of space, time, number, continuity, etc. But that the results of such investigations must also be of significance for metaphysics and logic is perfectly clear to everyone.

After so many attempts, undertaken from different approaches and in different epochs, one should expect that, at least with regard to the main ones among the problems in question, resolution and general agreement would have been attained. Nonetheless, the centuries have passed away and the questions remain. In fact, to the old ones some new ones have even been added. Will our time be more fortunate in this respect? Undoubtedly! There are many indications that in this, as in other respects, it may be permitted to our age to resolve old puzzles. Certainly the conviction that it will be so is justified to us by the great steps of progress made by scientific psychology and logic in recent times. The tools lie ready there for the framing of final judgments. But certainly one must also there seek out those tools. One will never succeed in charming away material difficulties by means of verbal or formalistic tricks.

In the light of the way things have gone in the past, many have come to the opinion that the issues debated in the philosophy of mathematics are nothing more than a hopeless knot of superfluous subtleties, which it is not worthwhile to unravel. Sidestepping these issues, science calmly pushes onward, all unconcerned.

However, this view is in fact false. Even if we were to disregard the fact that the unravelling of those subtleties constitutes an essential interest of philosophy, yet the merest allusion to <293>

the many portentous errors which have been committed within mathematics itself – because of false views of the concept of the differential and other concepts – alone teaches how far such a view goes wrong.

Now as to the reasons for this lack of completely satisfactory solutions, excluding all doubt, to such important problems, they lie, as a more exact critique would prove, partly in crippling metaphysical prejudices, and partly in deficiencies of method.

In the latter respect, failure to interrelate the studies which have been made has, in particular, been a hindrance to progress. The intimate, systematic interconnectedness within the range of problems under discussion would have necessitated a natural sequence of treatment. In actual fact, however, what was followed were special interests dominant from time to time. People thus sought to understand by itself that which could be understood only in its dependency upon other things. An excellent example of this is offered by the well-known *Riemann-Helmholtz* theory of space. The method which that theory holds to be superbly suited to the solution of the questions of principle connected with the axioms of geometry, and which it also uses to that end, is the method of analysis through algebraic calculation. *Helmholtz* repeatedly extolled, as the special advantage of analytic geometry, its characteristic of calculating with pure concepts of magnitude and of needing no intuition in its proofs.[2] In this way it obviates – in contrast with the purely intuitive procedure of *Euclid*ean geometry – "the danger of taking customary facts of intuition for conceptual necessities."[3]

However, serious questions immediately arise on this point. Does not the analytic method in geometry also presuppose certain facts of intuition? Obviously it does. How, otherwise, could one arrive at those general rules, according to which every geometrical form can be algebraically defined by means of an equation, and according to which, then, from every algebraic relation a geometrical relation can be derived? For is not the well-known

---

[2] Cf. "Über die thatsächlichen Grundlagen der Geometrie," *Wissenschaftliche Abhandlungen*, Vol. II, p. 611.
[3] Cf. "Über den Ursprung der geometrischen Axiome," *Vorträge und Reden*, Vol. II, p. 16.

fundamental expedient of analytic geometry, which first makes possible the transposition just mentioned – namely, the univocally <294> characterizing presentation of any spatial point by means of the vectorial numbers of its distances from three fixed "coordinate axes" – based upon the peculiar properties of our representation of space? And could those properties be abstracted from anything other than intuition? What, therefore, are the facts of intuition upon which, in the last analysis, the possibility of applying general arithmetic to geometry is based?

But these and ever so many other questions were not even raised up to now, not to speak of being answered. It is obvious that, so long as the relation of arithmetic to geometry is not completely cleared up, no attempt to answer questions of principle in geometry by reference to numerical analysis offers us a sure guarantee that we are not being led in a circle – as, in my opinion, is actually the case with the *Riemann-Helmholtz* theory.

A definitive removal of the real and imaginary difficulties in the problems which constitute the fringe area between mathematics and philosophy is to be expected only when, in natural sequence, first the concepts and relations which are in themselves simpler and logically earlier, and then, subsequently, the more complicated and more derivative ones – and these taken, indeed, in the order of their degree of derivativeness – are subjected to analysis. And the very first term of this natural sequence is *the concept of number*.

In a sense this point also appears to be generally acknowledged. Today there is a general belief that a rigorous and thoroughgoing development of higher analysis (the total *arithmetica universalis* in *Newton*'s sense), excluding all auxiliary concepts borrowed from geometry, would have to emanate from elementary arithmetic alone, in which higher analysis is grounded. But this elementary arithmetic has, as a matter of fact, its sole foundation in the concept of number; or, more precisely put, in that neverending series of concepts which mathematicians call "positive whole numbers." All of the more complicated and more artificial forms which likewise are called numbers – the fractional and irrational, the negative and complex numbers – have their origin and basis in the elementary number concepts and in the relations uniting them. <295> With these latter elementary concepts also the former (and, in fact, the whole of mathematics) would fall away. Therefore, it is with the analysis of the concept of number that any philosophy of mathematics must begin.

This analysis is the goal which the present treatise sets for itself. The means which it employs to this end belong to psychology, and they must do so if such an investigation is to attain solid results.

Certainly one might ask right off: What, after all, does number have to do with psychology? We would like to reply to this question by putting another: What have space, time, color, intensity, etc., to do with psychology? Is not space the object of the geometer, color the object of the physicist, and so on? And yet what an extensive psychological literature – still growing day by day – has been occasioned by these concepts!

In regard to the concept of number, to be sure, this is not so. But the lack here is highly improper. In truth, not only is psychology indispensable for the analysis of the concept of number, but rather this analysis even *belongs within* psychology. As to the first part of this claim, it must be proven in this work itself. In reference to the second part, note that analyses of elementary concepts, that is, those which present us with only a few levels of complication – and, indeed, the number concepts are of this sort – may nowadays be counted among the more essential tasks of psychology. For how otherwise could it attain insight into the internal structure of that intricately interwoven tissue of thoughts which constitute the substance of our thought-life? The understanding of the first and most simple modes of composition of representations is the key to the understanding of those higher levels of complication with which our consciousness constantly operates as with seamless and fixed formations.

The preceding remarks may serve to justify exclusive engagement with such a specific question as that about the content and origin of the concept of number. They should summarily characterize the significance of that question for philosophy, on the one hand, and for mathematics, on the other; and should, at the same time, indicate the more profound reasons which have led the author into the investigations which follow. <296>

Chapter One

THE ANALYSIS OF THE CONCEPT OF NUMBER AS TO ITS ORIGIN AND CONTENT

Section 1: *The Formation of the Concept of Multiplicity [Vielheit] out of That of the Collective Combination*

Common consciousness finds two sorts of numbers: *cardinal numbers* and *ordinal numbers*. The former are usually intended when simply "numerals" or "numbers" are spoken of. Already in ordinary discourse a special emphasis appears to be laid upon the close relationship between the two sorts of number concepts by the likeness of their denominations: the spoken and written signs for cardinal numbers go over by slight modifications into the signs for the corresponding ordinal numbers (1, 2, 3, 4, . . .; 1st, 2nd, 3rd, 4th, . . .). As cardinal numbers refer to *groups*, so ordinal numbers refer to *series*. But series are ordered groups; and so one may perhaps apriori suppose that the concepts of ordinal numbers proceed merely by a certain delimitation out of those of the cardinal numbers. However, noted scientists, such as *W. R. Hamilton, H. Grassmann, H. v. Helmholtz,* and *L. Kronecker,* among others, have held that the natural starting point is the series. In this way they vindicate the superiority of the ordinal numbers (or related concepts) with respect to generality. The question as to whether the one or the other of these views – or perhaps a third view which completely rejects all logical subordination of the one class of concepts to the other – is worthy of preference must be dealt with later. In now beginning with the analysis of the concept of the cardinal number, we do not aim to prejudge <297> the issues in favor of any of these views. In the light of the relationship of the two sorts of numbers to the representation of groups, in the one case, and to the representation of series, in the other case – as is already impressed upon us by superficial consideration of the matter – the loose characterization of the one as "group numbers," and of the other as "series numbers," appears to be quite adequate.

Now to mention *all* of the authors who ground the concept of number upon that of the group would hardly be possible. Already in *Euclid* we find the definition (in the preamble to Book VII of his *Elements*): "A *unit* is that by virtue of which each of the things that exists is called 'one'. A number is a multitude composed of units."[4] As on other points, so also here, *Euclid* was *the* authority for a long time. *Hobbes* declares: "Number is 1 and 1, or 1, 1 and 1, and so on, which is the same as saying, number is units."[5]

In his chief work, on the human understanding, *Locke* gives extensive descriptions of the psychological process involved in enumerating, but without ever putting his view on the content of the concept of number into the form of a definition. In the context of his discussions, however, numbers are characterized as representations which (as "complex ideas," or "collective ideas") are composed out of units; more precisely, as "ideas for several collections of units, distinguished one from another."[6]

In a letter to *Thomasius*, *Leibniz* gives a definition which reads almost the same as Hobbes: "I define number as one and one and one, etc., or as unities."[7] In the *New Essays*, Book II, Chapter 16, the (whole) number is defined as a multiplicity (multitude) of units.[8] When compared to the earlier statement, there is obviously no difference here worth noting; only there the word "multiplicity" is avoided – but the use of the plural means the same thing.

These outstanding examples should do for now.

The most common definition reads: The number is a multiplicity of units. Instead of "multiplicity," the terms "plurality,"

---

[4] Translation from T. L. Heath, *The Thirteen Books of Euclid's Elements*, (Oxford: 1926), Vol. I, p. 277. {DW}

[5] *De corpore*, ch. VII, sect. 7; cf. Baumann, *Die Lehren von Raum, Zeit und Mathematik*, Vol. I, p. 274.

[6] *Essay on the Human Understanding*, Book II, ch. 16, sect. 5. {Husserl erroneously cited Book II, ch. 4, sect. 5; retained in the 1970 edition. DW}

[7] *Opera philosophica*, ed. J. E. Erdmann (Berlin, 1840), p. 53. {English translation from L. E. Loemker, tr./ed., *G. E. Leibniz: Philosophical Papers and Letters* (Chicago, 1956), Vol I, p. 157. DW}

[8] *Opera philosophica*, p. 243.

"totality," "aggregate," "collection," "group," etc., are also used: all of them expressions which are equivalent or very closely interrelated, although they are not without distinguishable nuances.[9] <298>

But there certainly is very little accomplished with this definition. What is "multiplicity"? And what is "unity"? Most controversies revolve around precisely these questions. Also, "multiplicity" appears to signify almost the same thing as "number." In fact, the name "number" is used in a more extended sense – namely, where it is not supposed to designate a *determinate* number – in which it becomes completely equivalent with "multiplicity." Because of all this, many authors have supposed that they had to abandon this definition (if one wants to call it that). However, it is precisely "number" which here is used in an *extended* sense. And, in any case, this much is clear: The concrete phenomena [*Phänomene*] to which determinate, numerical assertions are applied are concrete multiplicities, i.e., groups of determinately given things; and, hence, they are precisely *the same* phenomena which also fall under the general concept of multiplicity. Precisely herein resides the necessity of setting out from these phenomena, and of observing how it is from them that the more indeterminate and universal concept which underlies that sequence of names, "multiplicity," "plurality," "group," etc., as well as the determinate number concepts, is abstracted.

The first question which we have to answer is the question about the *origin* of the concepts in question.

The concrete phenomena which form the basis for the abstraction of these concepts are, as just noted, totalities [*Inbegriffe*] of determinate objects. But we also add that these totalities are completely arbitrary and optional. In the formation of concrete totalities there is in fact no limitation whatever upon what particular contents are to be included. Any object of representation, whether physical or psychical, abstract or concrete, whether given in sensation or in imagination, and so on, can be united into a

---

[9] For the present we refrain from excluding these nuances by using one of these names alone. However, for reasons which will appear later, the words "totality" [*"Inbegriff"*] and "multiplicity" [*"Vielheit"*] are preferred.

totality with any, and with arbitrarily many, other objects.[10] E.g., a few particular trees; the sun, moon, earth, and Mars; a feeling, an angel, the moon and Italy; and so on. In these examples we can, in each case, speak of a totality, of a plurality, or of <299> a determinate number. The nature of the particular contents makes no difference at all.

But, if this be true, exactly *how* does one succeed in getting from concrete totalities to the general concept of plurality, of totality, or of number? What abstraction process is supposed to yield the concept? What is it that one retains, in abstraction, as the content of the concept? And what is that *from which* abstraction is made?

We assume that concepts originate through a comparison of the specific representations which fall under them. Disregarding the attributes which differ, one retains the ones they have in common; and these latter are the ones which then constitute the general concept. Let us now attempt to follow this guideline in the case at hand.

It is obvious that a comparison of the particular contents which we find before us in given totalities would not straightaway yield to us the concept of multiplicity, of totality, of number. And it would be absurd to expect it, even if it did happen. It is not those particular contents that are, in fact, the basis of the abstraction. Rather, the basis is the concrete totalities *as wholes* in which the particular contents are comprised. But even comparison of the totalities appears not to offer the desired result. The totalities, one might say, consist *merely* of the particular contents. How, then, are the *wholes* to exhibit some common attribute, when the *parts* constituting them may be utterly heterogeneous?

However, this specious difficulty is easily resolved. It is misleading to say that the totalities consist merely of the particular contents. However easy it is to overlook it, there still is present in them something more than the particular contents: a "something

---

[10] Probably there is little need to recall that, where we are dealing with objectively real things, these still must be represented in our consciousness by means of representations. The represented totality is then related to the intended totality of real things as the representation of a single real thing also is related to that thing itself.

more" which can be noticed, and which is necessarily present in all cases where we speak of "totality." This is the *combination* [*Verbindung*] of the particular elements into the whole. And it is here as it is in the case of many other classes of relations: There can be the greatest of differences between the related contents, and yet there be identity of kind with respect to the combining relations. Hence, similarities, gradations [*Steigerungen*], and mediations involving continua are found in wholly heterogeneous domains; and they can occur between sensible as well as between psychical phenomena. It is, therefore, quite possible for two wholes, *as wholes*, to be similar, although <300> the parts constituting the one are completely heterogeneous to those constituting the other.

Those combinations which, always the same in kind, are present in all cases where we speak of multiplicities are, then, the bases for the formation of the general concept of multiplicity.

As to the sort of abstraction process which yields our concept, we can best characterize it by referring to the way in which concepts of other composites (wholes) originate. If we consider, for example, the cohesion of the points on a line, of the moments of a span of time, of the color nuances of a continuous series of colors, of the tonal qualities in a "tone progression," and so on, then we acquire the concept of combination-by-continuity and, from this concept, the concept of the continuum. This latter concept is not contained as a particular, distinguishable, partial content in the representation of any concretely given continuum. What we note in the concrete case is, on the one hand, the points or extended parts, and, on the other hand, their peculiar combinations. These latter, then, are what is always identically present where we speak of continua, however different the absolute contents which they connect (places, times, colors, tones, etc.) may be. Then in reflection upon this characteristic sort of combination of contents there arises the concept of the continuum, as a *whole* the parts of which are united precisely in the manner of continuous combination.

Or, to take another example, consider the quite peculiar way in which, in the case of any arbitrary visual object, spatial extension and color (and color, in turn, and intensity) reciprocally penetrate and connect with each other. With reference to this manner of combination – which, following *F. Brentano*, we shall call the "metaphysical" – we then again can form the concept of a whole, the parts of which are united in just such a manner.

We can say quite generally: Wherever we are presented with a particular class of wholes, the concept of that class has only been able to originate through reflexion upon a well-distinguished manner of combining parts, one which is identical in each whole of the class in question. <301>

How, then, does the matter stand in the case with which we are concerned? We can likewise say that a totality forms a whole. The representation of a totality of given objects is a *unity* in which representations of single objects are contained as partial representations. Certainly the combination of parts as found in any arbitrarily selected totality must be called loose and external, when compared to other cases of combination. So much so, in fact, that one would almost hesitate to speak here of any combination at all. But, however that may be, there *is* a peculiar unification there; and it would, as such, also have to have been noticed, since otherwise the concept of totality, and that of multiplicity, never could have originated. So, if our view is correct, the conception of the multiplicity originates by means of reflexion upon the peculiar and, in its peculiarity, quite noticeable manner of unification of contents, as it shows up in every concrete totality (concrete multiplicity). And it arises in a way analogous to that of the concepts of other sorts of wholes, all of which are come by through reflection upon the modes of combination peculiar to those wholes.

From here on, we shall use the name "collective combination" [*kollektive Verbindung*] to designate that sort of combination which is characteristic of the totality.

Now before we proceed with the development of our subject, it will be good to deflect an apparent objection. We could be charged as follows: If the multiplicity is defined as a whole the parts of which are united by collective combinations, then this definition is circular. For in speaking of "parts" we certainly represent a multiplicity; and, since the parts are not individually determinate, we have a general representation of this multiplicity. That means, we are explaining multiplicity by means of itself.

Nonetheless – and however much plausibility this objection may have – we cannot concede its cogency. First, note that our task is not a *definition* of the concept of multiplicity, but rather a *psychological characterization* of the phenomena upon which the abstraction of that concept is based. All which can serve to this purpose we must therefore regard as welcome. Now the plural term "parts" certainly involves (disregarding its necessary correlation with the concept of the whole) the general representation of a multiplicity; but *that* term does not express what peculiarly characterizes this multiplicity *as* multiplicity. <302> By adding that the parts are collectively combined, we made reference to the point upon which our special interest reposes, and in virtue of which the multiplicity is characterized precisely *as* a multiplicity, in contrast to other sorts of wholes.

Section 2: *Critical Exposition of Certain Theories*

The shortest answer to the question about what kind of unification is present in the totality lies in a direct reference to the phenomena. And here we are genuinely dealing with ultimate facts. By saying that, however, we are not spared the task of considering this kind of combination more carefully, and of bringing out its characteristic differences from other kinds: – especially since false characterizations of it, and confusions of it with other species of relations, have occurred often enough. In order to accomplish this task we shall try out a series of possible theories, some of which have actually been advocated. Each theory characterizes the collective unification in a different way and, in relation thereto, seeks also to explain in a different way the origin of the concepts *multiplicity* and *number*.

I

The combination of representations to make up a totality, someone could say, still hardly deserves the name of a "combination." What is given, then, when we speak of a totality of certain
5 objects? Nothing further than the co-presence of those objects in our consciousness. The unity in representations of the totality consists, thus, only in their belonging to the consciousness which encompasses them. Still, this "belonging" is a fact which can be attended to; and with reflexions upon it there originate, then, those
10 concepts whose analysis is here in issue.

Now this view is obviously in error. Quite a number of phenomena make up, in each moment, the total state of our consciousness. But it is the role of a special interest to lift certain representations out of that plenum and collectively unite them.
15 And this occurs without the disappearance of all of the remaining representations from consciousness. Were this view <303> correct, then in each moment there would be only a single totality, consisting of the whole of the present partial contents of our total consciousness. But at any time, and in any way we choose, we
20 can form various totalities, can expand one already formed by the addition of new contents, and can narrow others down by taking contents away (without necessarily excluding these contents from consciousness). In short, we are conscious of a spontaneity which would be inconceivable on this view.

25 But this view, in its general and indeterminate form, contains in addition an absurdity. In fact, do not continua, with their infinite groups of points, belong to the materials of our consciousness? Who has ever actually represented them in the manner of a totality?

30 It is important to stress that a totality can have as elements only contents of which we are conscious in the manner of things separately and specifically noticed [*als für sich bemerkte*]. All contents which are present only as things incidentally noticed, and which either cannot be separately noticed at all (like the points of
35 continua), or merely cannot for the moment be separately noticed:
– all these cannot yield elements out of which a totality is constituted.

All of this will perhaps be quickly conceded; and the representative of the view just criticized might forthwith restrict his assertion in such a way that by the "encompassing consciousness" which unites representations into a multiplicity a special act of consciousness is to be understood, and not consciousness in the widest sense, as the totality of our psychical phenomena. So it would be, accordingly, a question of unity in an encompassing act of noticing, or of unity of interest, and so on. We intend to come back later to consider more closely the theory as thus corrected.

II

Let us now turn to consideration of a new theory, which argues as follows:

If a totality of contents is present to us, what else are we to notice but that every content is there *simultaneously* with each other one? Temporal co-existence of contents is indispensable <304> for the representation of their multiplicity. Now, indeed, there is required in any composite act of thought the co-existence of its parts. But whereas in other cases there are present, *in addition to* simultaneity, distinctive relations or combinations which unify the parts, it is precisely the distinguishing feature of the representation of the totality that it contains *nothing more* than the simultaneous contents. Hence, multiplicity *in abstracto* signifies nothing other than the simultaneous givenness of certain contents.

This view, as is easily seen, comes under precisely the same objections as does the previous view, and under many others besides. It would be superfluous to repeat the former objections; and, of the latter, it is sufficient to emphasize the fact that to represent contents simultaneously is still not to represent contents *as simultaneous*. For example, in order for the representation of a melody to come about, the single tones which compose it must be brought into relation with one another. But every relation requires the simultaneous presence of the related contents in one act of consciousness. Thus, the tones of the melody must also be simultaneously represented. But they are not at all to be represented *as* simultaneous. Quite to the contrary, they appear to us as situated in a certain temporal succession.

It is no different in the case where we represent a multiplicity of objects. That we must simultaneously represent the objects is certain. But that we do not represent them as simultaneous, and that, rather, special acts of reflexion are required in order to notice the simultaneity in the representing of the objects: this is directly proven by a reference to inner experience.

III

A third view is likewise based upon time as an insuppressible psychological factor. In direct contrast with the foregoing, it argues as follows:

In virtue of the discursive character of our thinking, it is true in general that several contents which are different from each other cannot be thought at the same time. Our consciousness can be employed about only *one* object in each moment. All mental activity of a relational or higher sort becomes possible only in <305> that the objects with which it has to do are given *in temporal succession*. So, then, each complex thought-structure, each whole composed of certain parts, is something which has arisen out of simple factors in succession. In such cases we always have to do with step by step processes and operations which, proceeding through time, intertwine and extend themselves more and more. In particular, therefore, each collection [*Kollektion*] presupposes a collecting [*Kolligieren*]; and each number presupposes an enumeration. And herewith there is necessarily given a temporal arrangement of the collected objects or of the enumerated units. But yet more. The totality is the loosest of ways of combining parts into a whole. Indeed, we speak of a totality or multiplicity there where contents are united by *no further connexions* than by the insuppressible form of intuition, time – where contents, therefore, are present in consciousness *merely* as ordered in the temporal sequence of their entrance into it. Accordingly, it also follows that *multiplicity in abstracto* is nothing more than *succession*: succession of *any sort of* contents separately and specifically noticed. But the number concepts Represent determinate forms of multiplicity or succession *in abstracto*.

Now in order not to dissipate attention through fruitless individual critiques, I have here preferred – instead of criticizing in sequence the authors who have represented such theories or similar ones – rather to state the view itself, which more or less clearly underlies all of those theories, as plainly and as fully as is possible, and to exercise my critique upon that view. And the view which must be combatted here is in fact based upon crude psychological and logical errors.

First, it invokes the psychological fact of the narrowness of consciousness. However, it exaggerates and falsely interprets this narrowness. It is true that the number of distinct contents to which we can turn our attention in any one moment is very restricted. In fact, at the highest concentration of interest, the number shrinks to one. But it is false that we can *never be conscious* of more than *one* content in one and the same moment. Indeed, just the fact that there is thought which relates and connects – as well as, in general, all of the more complicated mental and <306> emotional activities which this very theory invoked – teaches clearly the utter absurdity of this viewpoint. If in every instant only *one* content is present to our consciousness, how should we be able to notice even the simplest of relations. If we represent the one term of the relation, then the other either is not yet in our consciousness, or it is no longer there. We certainly cannot connect a content of which we are not conscious – and which, therefore, does not exist for us at all – with the single content which, supposedly, is present to us and is really given. Hence, reference to the temporal succession of the representations which are to be related can contribute nothing whatever to an explanation of the possibility of relational thinking.

But, then, does not experience teach (so, perhaps, our adversary replies) that as a matter of fact we can always have only *one* present representation, and that it is very well possible to bring it into relation with past representations? In that a representation is past, it by no means therefore ceases to be.

ON THE CONCEPT OF NUMBER: PSYCHOLOGICAL ANALYSES   323

However, it is easily seen that such an answer would rest upon misinterpretations of experience. One must not confuse ⟨temporally⟩ present representations with representations of what is ⟨temporally⟩ present, and past representations with representations of what is past. Not every present representation, as we must emphasize here once again, is a representation of what is present. Precisely all representations which are directed upon things past constitute an exception; for they all are, in truth, ⟨temporally⟩ present representations. If I recall a song which I heard yesterday, for example, then the memory representation is indeed a ⟨temporally⟩ present representation; only it is referred by us to the past. Now, of course, there is no problem in the fact that we are able to bring representations with present contents into relation with representations with past contents. In doing this these representations are all, in fact, simultaneously present in our consciousness. They are *in toto* representations which are ⟨temporally⟩ present. On the other hand, we can relationally unite past representations neither with each other, nor with present representations; for, as past, they cannot be brought back, and are gone forever.

The alleged fact of experience which our adversary has in view reduces, therefore, to the claim that, whenever we represent a plurality of contents, there is always one alone which is a temporally present content, <307> whereas all of the others exhibit greater or smaller temporal differences. Naturally, then, each total representation composed out of distinct (separately and specifically noticed) parts would have had to be originated through *successive* acts of noticing and relating the individual, partial contents, while the total representation itself, as something finished and developed, would contain all of the parts at the same time – only each furnished with a different temporal determination.

Now it is indeed certain that, already with a very modest number of contents, a comprehensive noticing of them is only possible by apprehending and retaining them successively or in very small groups. But, on the other hand, experience does seem to teach with sufficient clarity that we are able to survey two, three, or four contents of a very simple kind with one glance, as it were, and to unite them collectively in one representation, without being conscious of any sort of serial progress from one content to another. (Consider, for example, a small group of sharp dots which are very close to each other upon a blackboard.)

However that may be, we can acknowledge it as a fact that for the formation of representations of groups and numbers (the first ones at most excepted) temporal succession is an indispensable psychological requirement. One is, therefore, quite justified in designating groups and numbers as results of processes, and, insofar as our will is thereby engaged, as results of activities, of *"operations"* of colligating or of enumerating.

But this is absolutely all to which we can agree. Only this one thing, and no more, is proven: that succession in time forms an insuppressible *psychological precondition* for the formation of by far the main part of number concepts and of concrete multiplicities, that is to say, of all of the more complicated concepts in general. These have a temporalized mode of becoming, and thereby each constituent of the completed whole receives a different temporal determination in our representation. But does that also prove that temporal order enters into the *content* of these concepts, or that it perhaps is the special relationship which characterizes pluralities as such, in contrast to concepts of other composites? In fact, people have often been satisfied with such paltry arguments, without considering that time forms, in precisely the same way, the basis for <308> *all* thinking of higher order [*jedes höhere Denken*], and that, for example, one could with equal right infer that the relation of premises to conclusion is identical with their temporal succession. However, such obvious absurdities have already been avoided by the very formulation which we gave the time-theory for our purposes. That formulation asserts solely that the case of the totality (or of the plurality) is distinguished from cases of wholes composed in other ways by the fact that in it *mere* succession of partial contents is present, while with the other wholes there is yet *beyond that* some *other* sort of combination.

So the argument is not simply that, because enumerating requires a temporal succession of representations, number is the comprehensive form of the successive *in abstracto*. Rather, the theory in question invokes the factual distinction between the totality (or collective whole) and all other sorts of wholes, and thus invokes the testimony of inner experience.

However, it does not rightly do so. Again and again an error has been committed on the one side, and censured on the other, with respect to this point: To perceive temporally successive contents is still not necessarily to perceive those contents as temporally successive.

The clock sounds off with its uniform tick-tock. I hear the particular ticks, but it need not occur to me to attend to their temporal sequence. But even if I do notice how one tick sounds *after* the other, that still does not involve lifting out a number of ticks and uniting them, by a comprehensive noticing, to form a totality. Or take another example: Our eyes roam about in various directions, fixing now upon this, now upon that object, providing in this way manifold representations following upon one another. But a *special* interest is necessary if the temporal sequence involved here is to be separately and specifically noticed. And in order to single out to themselves some or all of the objects noticed, to relate them to each other, and to unite them into a totality, here again are required special interests and special acts of noticing directed upon just those contents picked out and on no others. But even if the temporal sequence in which objects are colligated were always attended to, it still would remain incapable of grounding by itself alone the unity of the collective whole. <309> And since we cannot even concede that temporal succession enters into the representation of each concrete totality merely as an invariable constituent which is always attended to, then it is clear that it can even less enter into the corresponding *general concept* (i.e., that of multiplicity or number) in any way. *Herbart* is completely justified in saying that "number has ... no more in common with time than do a hundred other sorts of representations which also can be produced only gradually."[11]

Were it merely a question of describing the phenomenon [*Phänomen*] that is present when we represent a multiplicity, then certainly we would have to mention the temporal modifications which the separate contents had undergone, although those

---

[11] *Psychologie als Wissenschaft*, 1825, Vol. II, p. 162.

modifications were not by themselves given any special notice. But disregarding the fact that the same holds true of every composite whole, we have, in general, to distinguish between the phenomenon as such, on the one hand, and that for which it serves, or which it signifies for us, on the other hand. Accordingly, we must also distinguish between the psychological description of a phenomenon and the statement of its signification. The phenomenon is the foundation of the signification, but is not identical with it.

If a totality of objects, a, b, c, . . . and f, is in our representation, then, in light of the successional process through which the whole representation arises, perhaps only f ⟨the last⟩ will be given as a sense representation, but the remaining contents merely as phantasy representations which are modified temporally and also as regards their content. If, conversely, we pass from f to a, then the phenomenon is obviously a different one. But the logical signification suppresses all such distinctions. The modified contents serve as signs, as deputies, for the unmodified ones which were. In forming the representation of the totality we do not attend to the fact that the contents are changed as the colligation progresses. We suppose we actually keep hold of them and unite them; and, consequently, the logical content of that representation is not, perhaps, f, just-passed e, earlier-passed d, and so on up to a, which is the most strongly modified. Rather, its logical content is nothing other than (a, b, c, d, e, f). The representation takes in every single one of the contents, irregardless of the temporal differences and of the temporal order based on them.

Thus we see that time only plays the role <310> of a psychological *precondition* of our concepts, and that it does so in a two-fold manner:

1) Most – in fact, almost all – of our representations of multiplicities are results of *processes*, are wholes originated gradually out of elements. Insofar as this is so, each element bears in itself a different temporal determination.

2) It is essential that the partial representations which are united in the representation of the multiplicity are ultimately present in our consciousness *simultaneously*.

But we have found that neither the simultaneity nor the successiveness in time enter in any way into the content of the representation of multiplicity; and so, likewise, with that of the representation of number.

As is well-known, *Aristotle* already seemed to bring time and number into a close relation by his definition: "Time is the number of movement in respect to earlier and later." However, it is only since *Kant* that it has become more generally common to stress the temporal "form of intuition" as the foundation of the number concept. To be sure, this happened much more as a consequence of the authority of his name than as a consequence of the weight of his arguments. We do not find in *Kant* a serious attempt at a logical or psychological analysis of the concept of number. Unity, multiplicity and totality are the categories of quantity in his metaphysics. Number is the transcendental schema of quantity. *Kant* fully states his view as follows, in the *Critique of Pure Reason*: "But the pure schema of magnitude (*quantitatis*), as a concept of the understanding, is *number*, which is a representation that comprehends the successive addition of one thing to another thing (of the same kind). Thus, number is nothing other than the unity of the synthesis of the manifold of a homogeneous intuition in general, a synthesis which comes about through the fact that I engender time itself in the apprehension of the intuition."[12]

This passage is obscure and, also, will not exactly agree with the explanations which *Kant* gives of the function of the schema. These explanations themselves certainly are not exactly uniform. For example he says: "We wish to call . . . the formal and pure condition of sensibility, to which the concept of the understanding is restricted in its use, the *schema* of this <311> concept of the understanding."[13] On the other hand we read, a few lines later: "The representation . . . of a general procedure of the imagination in giving a concept its model [*Bild*] I call the schema of that concept."

---

[12] I. Kant, *Kritik der reinen Vernunft, Sämmtliche Werke*, Hartenstein Edition, Vol. III, p. 144.
[13] *Ibid.*, p. 142.

Were we to transfer this last definition to the schema of quantity, then we would have to say that number is the representation of a general procedure of the imagination in giving to the concept of quantity its model. However, by this "procedure" can only be meant the process of enumerating. But is it not clear that "number" and "representation of enumerating" are not the same? Further, it is not very easy to see how, starting out from the category of quantity, we are apriori to arrive, by means of the representation of time (as the common schema of all the categories), at the particular determinate number concepts. Still less intelligible is the necessity which determines us to ascribe to a concrete multiplicity a certain number which is always the same: precisely that number of which we say that it belongs to the concrete multiplicity. The theory of the schematism of the pure concepts of understanding appears here, as elsewhere, to fail in the realization of the purpose for which it was especially created.

We can omit enumeration of all those investigators who, following *Kant*, based the concept of number upon the representation of time. Let us mention here only two famous names. *Sir William Rowan Hamilton* flatly called algebra "the science of pure time," as well as "the science of order in progression."[14] In Germany it is *H. von Helmholtz* who, in a philosophical treatise which recently appeared,[15] has published a detailed investigation into the foundations of arithmetic and into the justification of the application of arithmetic to physical magnitudes. Herein he represents this same *Kant*ian point of view. When we come later to certain other developments (concerning the analysis of the concept of the ordinal number) we will find occasion to deal with this treatise thoroughly.

Finally, it should be noted that, in general, most of the investigators who take the representation of the *series* as basis for the development of the number concepts and the axioms of arithmetic have been essentially influenced by the time-theory. <312>

---

[14] Cf. H. Hankel, *Vorl. ü. d. complexe Zahlensysteme*, p. 17.

[15] In: *Philosophische Aufsätze zu Zeller's Jubiläum*, Vol. I, "Über Zählen und Messen."

IV

Whereas *Kant* put number into an intimate relation with the representation of time, *F. A. Lange* thought that everything which could be done with the representation of time could be derived with far greater simplicity and certainty from the *representation of space*. In the *Logische Studien*[16] he says: "*Baumann* has already shown that number has far greater unison with the representation of space than with that of time . . . . The oldest phrasings of the words for numbers always designate, so far as we can discern their meaning, spatial objects with determinate properties which correspond to the number in question. Thus, for example, rectangularity [*Viereckiges*] corresponds to the number four [*vier*]. From this we also see that number did not originally arise through systematic addition of one to one, and so on; but rather that each of the smaller numbers, upon which the system arising later is based, is formed through a special act of synthesis of intuitions; so that it is only later, then, that the relations of numbers to one another, the possibility of adding, and so on, are recognized." "The algebraic axioms rest, like the geometric axioms, upon spatial intuitions."[17]

"It is peculiar to the representation of space that within the great all-inclusive synthesis of the manifold there can be singled out, with ease and certainty, smaller units of the most various types. Space is, therefore, the archetype, not only of continuous, but also of discrete magnitudes, to the latter of which number belongs; whereas we scarcely can think of time otherwise than as a continuum. To the properties of space belong, further, not only the relations which occur between the lines and surfaces of geometrical figures, but equally the relations of *order* and *position* of discrete magnitudes. If such discrete magnitudes are considered as homogeneous with each other, and if they are united by a new act of synthesis, *then* number arises as sum."

Consider yet one more passage, from the *Geschichte des Materialismus*: "We originally receive each number concept in

---

[16] Pp. 140-141.
[17] *Ibid.*, p. 141.

the form of a sensuously determinate image of a group of objects, whether they are fingers, or the buttons and spheres of an abacus."[18] <313>

Now our critique will certainly not have to look very far for a handhold. The last quotation is especially offensive; for the well-known *general* concept of number appears as an *individual* phenomenon, as the sensuously determinate image of a group of spatial things. However, we may very well have here only an imprecise mode of expression. The view probably is that number is something noticeable *in* such groups – namely, in the manner of a partial phenomenon – something which must be lifted out of them by abstraction. The influence of *J. St. Mill* stands out clearly here. For *Mill*, number is a "*physical* fact," "a visible and tangible phenomenon." It is for him a sensuous property on the same level with color and weight, etc.[19] But whereas *Mill* explicitly declines to state wherein really consists numerical difference (whether because he held this to be too difficult, or, in the light of the elementary nature of the phenomena, held it to be superfluous), *Lange*, by contrast, believes that he can detect its source in the nature and properties of the *representation of space*. If we look at the passages quoted above we find that, in fact, spatial localization of the things enumerated is always emphasized. The spatial relations of order and position of discrete magnitudes considered as homogeneous with each other and united by an act of synthesis – *this* would be the content of the representation of number.

But one immediately wants to raise the question: *Where* are the four cardinal virtues, the two premises of an inference, and so on, located? What spatial order and position serves as the basis for numerical designation in the case of any arbitrary psychical phenomena? This objection certainly would not alarm *Lange*. He simply reduces all logical thinking to spatial intuition. For him everything that is psychical is located. We do not wish here to involve ourselves in criticism of this intrinsically obscure and utterly untenable view. We stress only a few points which especially concern our problem.

---

[18] Book II, p. 26.
[19] *Logik*, Gomperz translation, Vol. II, p. 237.

It is clear that, even if we were to concede *Lange*'s premise, no more would be proven in reference to the representation of space than was earlier admitted in reference to the representation of time. The representation of space would be an insuppressible psychological precondition <314> of the concept of number – and this to no greater extent, and in no other way, than it is for all other concepts. Even if spatial determination did belong to all contents which we unite in thought, it would always remain two different things (i) to represent spatially dispersed contents and (ii) to represent contents in terms of their spatial relations. Now what *does* actually happen when we collectively unite or enumerate certain spatial things? Do we then attend to the relations of order and position? Does the selective interest within which we form the representation of number turn to those relations? Certainly not. There are a great many positions and orders, but the number remains unchanged. Two apples remain two apples whether we group them together or apart, whether we shift them to the right or to the left, up or down. Number has exactly nothing whatsoever to do with relations of spatial position. It *may* be, nevertheless, that relations of order and position are co-represented in the phenomenon (implicitly) when there is a representation of a multiplicity of spatial objects. It is still certain that they do *not* constitute the objects of selective interest in enumeration. Not as separately and specifically noticed, but only as partial representations which are implicitly co-thought, are they then given in the phenomenon. The fact that the oldest phrasings of the words for numbers refer to objects in space with determinate properties which correspond to the numbers "is still no serious counter-instance" to this claim, and has such obvious explanations that we can dispense with discussion of it.

But *Lange* stresses, not merely the spatiality of the numbered; rather, he also speaks of the *acts* of synthesis through which discrete magnitudes are united to form number. For our present investigation, which mainly aims at a more precise characterization of the collective combination, it would be of interest to learn how *Lange* conceived of this synthesis of singulars into the multiplicity. But if we attentively go through the frequent discussions, in relation to the concept of synthesis, which are in the

Logische Studien, a serious confusion shows up. It will have already struck the reader of the above quotations that *Lange*, while he speaks once of acts of synthesis, yet another time calls the representation of space a synthesis.

Already *Kant* used the word "synthesis" (combination) in a double sense: first, in the sense of the unity of the parts <315> of a whole, whether these parts are properties of a thing, parts of an extension, units in a number, and so on; second, in the sense of the mental activity (*actus*) of combining. Both significations are intimately related in *Kant* because, in his view, every whole, of whatever kind it may be, is developed from its parts by means of the spontaneous activity of the mind.

"Synthesis" therefore signifies simultaneously, for him, combining *and* the result of combination. That we presume to observe combinations in the phenomena themselves, and to extract them therefrom by means of abstraction: that is only an illusion. It is we ourselves who have furnished the combinations, and, of course, by means of the "pure concepts of the understanding," the categories.

Lange mounts a polemic against the *Kant*ian concept of synthesis; but, certainly, not where it deserved censure. Rather, in his polemic we find only progress in obscurity and confusion. In opposition to *Kant*, his view is that synthesis is something noticeable in the content of the representation. Synthesis in this sense would signify representation of a relation, and of course – since, according to *Lange*, space is "the intuitional form of the ego with its variable content" – all synthesis would ultimately turn out to be *spatial* combination and relation. But synthesis also is supposed by him to be a *process*, occurring wholly in the unconscious, through which *we as subjects* first originate. And, finally, *Lange* speaks of special (and apparently conscious) acts of synthesis which, for example, yield numbers.

Now, with this multivocal use of the same name, fundamental obscurities are connected. Space is repeatedly designated as the archetype [*Urbild*] of all synthesis – in fact, as the true, objective countertype [*Gegenbild*] to our transcendental ego. The properties of space are supposed to form the norm of all of the functions of our understanding,[20] and so on. Throughout there is presupposed the erroneous view that a psychic act and its content stand to one another in the relation of pictorial resemblance. Not the least part of the source of this absurdity resides, perhaps, in the equivocation of the word "synthesis," in consequence of which it at one time signifies the relational content, and at another time signifies the act of relating.

But *Lange* certainly was also influenced by <316> *Baumann* on this point, whose work[21] he quotes. On the one hand, *Baumann* calls number the result of an activity, of a mental "sketching." But, on the other hand, we find "number again in the external world." According to him, external experience bears the mathematical in itself, independent of our mind; but, on the other hand, we form in ourselves "purely mental" mathematical representations. In this the applicability of mathematics to the external world is supposed to be grounded. With respect to the relations of space and number *Baumann* observes – and this passage is one quoted by *Lange* – "It [number] is together with space and everywhere present in it. It is also because of this that geometry is brought to arithmetical expression."[22]

It is not our task here to criticize in its full range *Baumann*'s theory, according to which, to a certain extent, the mathematical outside of us is known by the mathematical inside of us. (A suspicious similarity with the ancient Empedoclean theory, that "like is known by like," leaps to the eye.) So far as his theory concerns number – and this alone concerns us, dealing with the influence which it exercised on *Lange*'s theories – it is obviously incorrect. It is based upon an erroneous view of that abstraction process which supplies us with the number concepts. Neither are they "purely mental" creations of an "inner intuition," nor can one speak of a rediscovery of these concepts in the external world, or of their being together with space and in space.

Certainly it is true that the formation of numbers, as also of multiplicities *in concreto*, is not a matter of a passive reception, or

---

[20] Lange, *Logische Studien*, pp. 148-149.
[21] J. J. Baumann, *Die Lehren von Raum* . . ., Vol. II, pp. 668ff., 670, 675.
[22] *Ibid.*, p. 670.

a mere selective noticing of a content. If anywhere at all, here are present spontaneous activities which we attach to the contents.

Depending upon whim and interest, we can unite discrete contents, and again take away from, or newly add to, the contents just united. A unifying interest directed upon all the contents, plus with and in it – in that reciprocal interpenetration which is peculiar to psychic acts – a simultaneous act of noticing: these throw <317> the contents into relief. And the intentional object of this act of noticing is precisely the representation of the multiplicity or the totality of those contents. In this manner the contents are simultaneously and together present. They are *one*; and it is with reflexion upon this unification of separate contents by the psychical acts mentioned that the general concepts of multiplicity and (determinate) number arise.

If, now, this is the truth of the matter, then it is clear that designation of numbers as purely mental creations of an inner intuition involves an exaggeration and a distortion of the true state of affairs. Numbers are mental creations insofar as they are results of activities which we exercise on concrete contents. But what these activities create are not new, absolute contents which could then be found again somewhere in space or in the "external world." Rather, they are peculiar, relational concepts, which can only be produced again and again, but which absolutely cannot be simply *found* somewhere already completed.

Also, how are all of the conceivable numbers which we can enumerate by arbitrarily combining spatial contents to be contained in space? That which is intuitively present, which we can find before us in space and can notice, certainly does not consist of numbers in and for themselves, but consists, rather, only of spatial objects and of their spatial relations. But with that no number is yet given. But if a number is given, it is not, and cannot be, identical with the spatial syntheses which enclose the number of spatial objects (or the concrete totality) as the unifying bond. The adjacency of objects in space is still not that collective unification in our representation which is essential to number. That unification is first brought about *by us* through that unified accentuation which is in the psychical act of interest and of noticing. It was by misunderstanding this that *Lange* managed to explain the intuition of space as the "archetype" of all synthesis, and, hence, as the archetype of the synthesis of discrete magnitudes, of numbers, as well. This error was aided and abetted by *Baumann*'s theory of the "rediscovery" of number in space. However, we now wish to discontinue this critique of the views of *Lange* and *Baumann*, especially since they offer no positive suggestions for our further developments. <318>

V

Much more scientific and plausible than all of the theories of the origination of the concepts of multiplicity and number which have been criticized up to this point is the theory to the development of which we now wish to turn. But in order to make completely clear whether or not it does what it promises to do, I shall endeavor to give it as consistent a development as is at all possible, and I shall decline directly to tie my critique down to any one of the forms in which this theory has actually been represented by this or that outstanding author. The following line of argument may be easily admitted:

A totality can be spoken of only where objects which *differ* from each other are present. Were all of the objects identical, then we would in fact have no totality, no multiplicity of objects, but just *one* object alone. But these differences must also be noticed. Otherwise the different objects would form for our apprehension only one unanalyzed whole, and we would again have no possible way of coming to the representation of a multiplicity. Hence, representations of differences essentially belong within the representation of any totality. In that we, further, distinguish each single object within the totality from the others in it, along with the representation of *difference* there also is necessarily given the representation of the *identity* of each object with itself. In the representation of a concrete multiplicity each single object is, therefore, thought of both as an object which is different from all of the others, and as an object which is identical with itself.

Given this, the origination of the general concept of multiplicity also, it seems, lies in the clear. In fact, what common thing could be present in all cases where we speak of multiplicity other than these representations of *difference* and *identity*. All of which fits the fact that, as is well known, in the abstraction of the *general* concept of multiplicity absolutely nothing depends upon the peculiarities of the *individual* contents. Thus, setting out from any one *concrete* multiplicity, we get the determinate *general concept* of multiplicity under which it falls, i.e., its *number*, by relating each content to each other one as <319> different – but this completely in abstraction from the peculiar character of the concretely given contents – and by considering each content merely as something which is identical with itself. In this way there originates the concept of multiplicity as, to a certain extent, the *empty form of difference*. But now the concept of *unity* is also easy to explain. In numbering, i.e., in carrying out the abstraction of numbers, we bring each thing to be counted under the concept of unity. We consider it merely as *one*. That means just this: We consider each thing merely as something which is identical with itself and different from everything else. As distinguishing and identifying are reciprocally conditioning functions which are inseparable from each other, so the general concepts of multiplicity and unity, which are formed through reflexion upon those functions, are also correlative concepts, mutually interdependent.

We especially find ideas of these and similar kinds in the logical works of *W. Stanley Jevons*[23] and *Christoph Sigwart*.[24] Thus we read in *Jevons*: "Number is but another name for diversity. Exact identity is unity, and with difference arises plurality." And, "Plurality arises when and only when we detect difference."[25] Here, as one sees, "number" is taken in the broader sense noted above, where it is synonymous with "plurality." With respect to the kind of abstraction which is here present, this same author remarks: "There will now be little difficulty in forming a clear notion of the nature of numerical abstraction. It consists in abstracting the character of the difference from which plurality arises, retaining merely the fact . . . . Abstract number, then, is the

---

[23] *The Principles of Science*, 2nd Edition, London 1883.
[24] *Logik*, Vol. II, Tübingen 1887.
[25] *The Principles of Science*, p. 156.

empty form of difference; the abstract number *three* asserts the existence of marks without specifying their kind."[26] "Three sounds differ from three colors, or three riders from three horses; but they agree in respect of the variety of marks by which they can be discriminated. The symbols $1 + 1 + 1$ are thus the empty marks asserting the existence of discrimination."[27]

But these statements suffer – presupposing the correctness of <320> their basis – from essential indetermination. Specifically, this is made most apparent when we inquire about the origination and content of the singular, numerical representations, 2, 3, 4, ... They, indeed, are all "empty forms of difference." What differentiates three from two, four from three, and so on? Are we to give the dubious answer: With the number two we notice *one* relation of difference, with three, *two*, and with four, *three* such relations, and so on? The information which the last of the passages quoted gives us is obviously very meager. That phrase, "variety of marks," either signifies the same thing again as "number," or it signifies the same as "form of difference." But what characterizes these "forms" psychologically in contrast to each other, so that they can be grasped through their peculiar determinations, clearly distinguished from each other and, accordingly, also denominated by distinct names?

Let us try to go deeper here. For the sake of simplicity we will consider only a totality of three objects A, B, C. Into the representation of this totality there must enter, according to the view in question, these relations of difference:

$$\widehat{AB}, \widehat{BC}, \widehat{CA}$$

(where the ties indicate the relations). They are given together in our consciousness, and they effect the unification of the objects into the collective whole. Now one may replace A, B, and C with contents of any kind whatsoever, but these differences always remain present as determined in some way. They thus constitute

---

[26] *Ibid.*, p. 158.
[27] *Ibid.*, p. 159.

the "form" of difference which is characteristic of the number three.

However, certain objections to this present themselves: If these relations of difference are together in our representation, then, in case the basic viewpoint of the theory is correct, each of the differences represented must also be in turn perceived as self-identical and as different from any other. For were $\widehat{AB}$ and $\widehat{BC}$, for example, not recognized as different, then they would just blend together as undifferentiated; and then, as one immediately sees, their terms also could not show up in the representation of the totality as distinct from each other. So the sum total of the differences of <321> differences must be also present in our representation, that is:

$$\widehat{\widehat{AB}\ \widehat{BC}}\quad \widehat{\widehat{BC}\ \widehat{CA}}\quad \widehat{\widehat{CA}\ \widehat{AB}}$$

But the same would also be true in respect to them, and so on indefinitely. Hence, in order to get hold of the "form of difference" we would fall into a lovely *regressus in infinitum*.

But there still might be a way of avoiding this consequence. One could say: If we proceed, in our distinguishing, from A to B, and from there to C, then a new distinction of C from A is no longer required. That is to say, in relating the two differences $\widehat{AB}$ and $\widehat{BC}$ (which are connected by the term B) to one another in a higher act of differentiation, the possibility of C and A blending into one is *eo ipso* excluded. So the true schematization would be:

$$\widehat{A\ B}\ C$$

Then, whatever A, B and C may signify, this schematic figure refers us to a process which is everywhere the same. If we therefore abstract from the peculiarities of the particular contents, retaining each only as *somehow* determinate, then we have here the desired form which is common to all multiplicities with three contents, and in virtue of which we also ascribe the number three to such multiplicities.

In such a way one could establish all of the forms of difference which are to form the basis of the numerical denominations. Thus, for example, the schema of the simplest number, two, would be: $\widehat{AB}$.[28] Indeed, as one could say, what is represented in all cases where a duo lies before us but this? — One object is there, and in addition an object *different* from it is there; and the general idea of this fact forms the content of the number concept *two*. If a concrete totality of two contents is given to us, and if we assign to it the number two, then this means that we direct our attention merely upon the fact that one content, and still one other content, is present. Our attention does not come to rest upon the peculiarity of the difference, but rather upon the mere fact of it. <322>

The schematic form for the number four would be:

A B C D

And one now easily grasps the rule for how the forms are further complicated. In all cases, the distinctions are ones which bound one another (i.e., have a term in common), making it possible for all of them ultimately to be grasped together in a single act, by means of higher-order acts of distinguishing.

These schemata would perhaps best be regarded as models of those mental processes which occur in the representation of any totality of two, three, four or more contents. And in reflexion upon those mental processes, whose well-characterized distinction would have to be given to inner observation, the number concepts would arise.

The extremely rapid increase of complication in these forms would also explain why we attain authentic [*eigentliche*] representations only with the very first numbers, whereas we can conceive of larger numbers only symbolically or, so to speak, only indirectly.

Further, one easily comprehends that the independence of number from the *order* of the enumerated objects follows, ac-

---

[28] One must not be confused by the fact that there is only *one* relation of difference involved in the schematic form for the number *two*, or that there are six such relations in the schema for the number four. The theory here holds only that for every number there is, founded in the concrete totalities to which it applies, a distinct and determinate group of relations, relations of relations, etc., which is *its* peculiar "form of difference." The group for the number five would contain ten such relations, for six fifteen, etc. {DW}

cording to this theory, directly from the nature of the concept of number.

Finally, one could also invoke linguistic usage to support this theory. For example, the same thing is usually meant when it is said that A and B are different as is meant in saying that A and B are two things. And so on. So it appears that we have here a well-grounded theory with a claim upon our assent.

However, even if we consider all of the essential supplements which alone work *Jevons'* assertions into a theory (these assertions being of little use in their indeterminate state), the psychological foundation of the theory yet remains, it seems to me, untenable. But before going deeper into this matter, I must reject as misleading the invocation of linguistic usage. More closely considered, usage says much more against this view than it says for it. Only when given a certain emphasis does the statement, "These are *two* things," have the same signification as the statement, "This thing is different from that thing." It is that <323> emphasis, namely, which is given when one wishes to ward off a threatened confusion of things with each other.

Now let us turn to a critique of the psychological foundations of this theory.

It is true that we can speak of a totality only where there are contents present which are different from each other. But the assertion here derived from this truth is false; *viz.* that these differences must be represented *as such*, because otherwise there would be in our representation only an undifferentiated unity, and no multiplicity. It is important to keep distinct: to notice two different contents, and to notice two contents *as different from one another*. In the former case there is, presupposing the simultaneous, unified grasping of the contents, a representation of a totality; in the second case there is a representation of a difference. There where a totality is given, our apprehension primarily goes merely upon absolute contents (namely, those which compose the totality). By contrast, where a representation of a difference is given (or a complex of such representations), our apprehension goes upon *relations* between contents. This much alone is correct: Where a plurality of objects is perceived, we are *always justified*, on the basis of the particular contents, in making Evident judgments to the effect that every one of the contents is different from each other one. But it is not true that we *must* make these judgments.

With regard to the concepts of *distinguishing* and *distinction*, certain obscurities generally prevail which have their origin in equivocations, and which certainly may have contributed not a little to the errors which I have touched upon here:

(1) "Distinction" or "difference" signifies the result of a comparison. A comparison can yield either of two results: that the contents considered are the same, or that they are different, i.e., *not* the same. Thus, difference here signifies something negative, the mere absence of an identity. In this sense one speaks of comparing and distinguishing as correlative, intimately connected *activities*. Indeed, in any case where we have an arbitrary act of comparison, two sorts of results may occur: affirmative judgments which acknowledge sameness are made, or, on the other hand, negative judgments which reject sameness are made. To this affirming of sameness there refers, <324> then, the term "comparing," while the term "distinguishing" refers to the denial of identity, when we combine them to speak of "comparing and distinguishing."

In the case where comparison of contents in a certain respect leads to the result, *non-identity*, it can, nonetheless, happen that at least a similarity, or "gradation," or such like is noticed. These are well-characterized classes of relations, in the case of which, quite as in the case of identity, representation of the relation Represents a real [*reellen*], positive content of the representation in question. Now these relations, too, are called relations of difference; and, in particular, the names "distinction" and "difference" are customary for *intervals* in continua (distinction of place, distinction of time, distinction of pitch in tones, etc.). But then this narrower signification of those terms led again, on the other hand, to cases of *mere* non-identity (since such cases, too, were called distinctions) being thought of as if they were content-relations; i.e., as if in their case too the relation lay in the positive content of the representation; whereas, in fact, nothing further is given than an Evident negative judgment which denies the presence of one such content-relation (*viz.*, the specific relation of identity).

From the practical viewpoint, it may still be useful to classify all of the results to which comparison can lead under the two headings: "Identity" and "Difference." It must not, however, be overlooked that under the latter heading there stand together classes of relations which, as to their phenomenal character, are foreign to each other, while, moreover, a part of them are closely related to the identity relations which have been brought under the other main heading. But from the psychologically scientific point of view, the relations of similarity, identity, metaphysical combination, etc. – in short, all relations which have the character of representation phenomena in the narrower sense (hence, not merely represented psychical phenomena) – belong in one class, that of content-relations. But difference, in the broadest sense, does not belong in that class; for it is not a positive content of representation which is directly noticeable at the same time as the terms are. Rather, it is a negative judgment made, or represented as made, upon the basis of those terms.

(2) But the term "distinguishing" is used in yet another <325> signification, which is connected with analysis. According to this signification, the "distinguished" is that which has been thrown into relief and especially noticed through analysis; and "to distinguish" means the same as "to segregate" or "to analyze."

By investigating the conditions which favor analysis, it is found that a plurality of partial contents are the more easily and certainly segregated the greater, in number and degree (or disparity), are their distinctions amongst themselves and against the environs. Now these reflections about analysis consisted of comparisons and distinctions of contents that were *already analyzed*; and thus they commonly misled people into believing that the activity of distinguishing (in the sense of analyzing) is also such a *judgmental* activity of distinguishing (in the sense of distinguishing compared contents). Then one reasoned: To be able to hold several contents in consciousness *as segregated* – i.e., as analyzed and separately and specifically noticed – they must be thought of as *distinguished* from each other – i.e., as compared and specifically characterized in terms of their distinctions. But this is false. In fact, it is obviously absurd. The *judgmental* activity of distinguishing evidently presupposes contents which are already segregated and separately and specifically noticed. Hence, these contents cannot have *first* become noticeable through their being distinguished from one another ⟨in judgment⟩.

Now it is this error which the theory we are contesting commits by arguing: "The differences between objects of a multiplicity must have been noticed *as such*. Otherwise, in our representation we would never get beyond an unanalyzed unity, and there would be no talk of any multiplicity. Hence, representations of differences must be explicitly contained in the representation of the multiplicity."

It is true that, if the contents were not different from each other, then there would be no multiplicity. Further, it is true that the distinctions must not have been too small. Otherwise no analysis at all would have occurred. But it is not true that every content first becomes a distinct content, i.e., one which is separately and specifically noticed, by means of apprehension of its distinctions from other contents; whereas it is surely Evident that every representation of such a distinction presupposes contents which are already separately and specifically noticed and which, in that sense, are distinguished. <326>

In order for a concrete representation of a totality to originate, all that is necessary is that each of the contents comprised therein should be a content which is noticed separately and specifically, a segregated one. However, there is no absolute necessity that the distinctions of the contents be attended to, even though this frequently will occur – and does so as a rule, where the distinctions are intervals.

Precisely the same thing which has been stated above of the representation of mere distinction holds true also of the representation of *identity*. Here also we have to do with results of reflection upon the content which are later slipped into it as something supposed to have been originally given with and in it. According to *Sigwart*, identifying and distinguishing must be the functions which supply the concept of unity. "For what is posited as identical and is distinguished from another thing, is, *eo ipso*, like this other thing, posited as *one*."[29] However, distinction and

---

[29] Sigwart, *Logik*, Vol. II, p. 37.

identification are judgment activities which pursue a wholly different end than the one here ascribed to them. "A is identical with itself, i.e., A is not non-A, is not B, C . . ., but rather is just A." Such a line of reflection has the aim of staving off confusions of the content A with other contents. This intent is realized by seeking out and throwing into relief the points of distinction of A from B, C, and so on. But while this process develops, A, B, C, etc., are already present to consciousness as contents which are distinct from one another. The task of this process absolutely is not to divide up for the first time what originally is an identical unity, but rather is only this: for the further purposes of thought, to segregate similar things from each other by the use of characterizing marks supplied by distinctions, and to obviate, thus, all future confusion. In this process it is in nowise a question of "constant activities which are repeated in every act of thought," in which "self-consciousness, identical and the same in all acts, is realized." Nor is it a question of "factors which constitute the unity of our self-consciousness." (*loc. cit.*)

So I believe I have shown that representations of identity and of distinction do not explicitly belong among the contents of the representation of multiplicity. Thus they also could not have constituted the basis <327> for the abstraction of this concept. Likewise for the number concepts.

Section 3: *Establishment of the "Psychological"[30] Nature of the Collective Combination*

Now let us review our reflections to this point and their results. We undertook to indicate the origin of the concepts *multiplicity* and *number*. For this purpose it was requisite to get a precise view of the concrete phenomena from which they are abstracted. These

---

[30] Because of the obvious suggestion of this terminology to the contrary, it is necessary to point out that *Husserl*'s conclusion is not that the collective combination is psychological or "mental" in any usual sense of the word. Rather, it is 'psychological' only in the sense that it is a member of a unique class of relations the defining features of which are paradigmatically exemplified by *intentionality*, which *Brentano* used to characterize the psychological or psychical in contrast to the physical. See below. {DW}

phenomena were evident as concrete totalities or groups. However, special difficulties appeared to obstruct the transition from these to the general concepts. We distinguished and discussed a series of views – rejecting them all, however. Our attention especially rested upon the sort of synthesis which unites the objects of a multiplicity into a whole; for in the false characterization of that synthesis lies the source of the main errors. Our results were, briefly, the following: Whenever we represent a totality we are conscious of the contents as separately and specifically noticed. But, in order to characterize the unification of the contents, we may have recourse neither to appertainence to *one* consciousness, nor to the relations of simultaneity, temporal succession, spatial combination, or, finally, difference. Now what possibilities remain?

We have not yet investigated all classes of relations. Is collective combination to find its place among those which yet remain? For obvious reasons, however, we are here exempted from a detailed examination of the various particular species of relations. Since we know that the most heterogeneous of contents, whether physical or psychical, can be united in the collective manner, all relations with a range of applicability restricted by the nature of peculiar contents fall away from the start. Thus it is with similarity, gradation, continuous combination, etc. In fact, it appears that none whatever of the familiar sorts of relations can satisfy the set requirements, after temporal relations and relations of mere distinction are excluded. Possibly relations of resemblance <328> still could be brought in here; for, however much two contents may deviate from each other, it will always be possible to state a respect in which they are similar to one another. In fact it is often thought (indeed, it is the rule) that with regard to the origination of the number concepts recourse to similarity relations must be had. We must take this up later on. Here it is sufficient to point out that, as to concrete totalities, the similarities which it is possible to discover cannot constitute the relations which unite the elements of a totality. The clock and the pen – this is a totality. But in thinking of it I do not need antecedently to bring the two contents under the concepts *colored, extended*, etc.

So there is nothing left to do but to claim for the collective combination a new relation type, quite different from all others. Accordingly, we must say: The totality representation Represents a whole of a special kind, the parts of which are united by certain relations exclusively characteristic of it – precisely those called by us "collective combinations."

Inasmuch as it is now established that we have here to do with a new and original class of relations, we wish to turn to a closer characterization of them, in contrast to other relations. They have in fact peculiar characteristics which very essentially distinguish them in their phenomenal existence from all of the remaining kinds of relations.

Since I am not in a position to base my remarks upon a generally acknowledged theory of relations, I think I must fit in here some general observations concerning this very dark chapter of descriptive psychology.

First, it will be useful to come to agreement on the term "relation." What is the element common to all cases where we speak of a "relation," in virtue of which precisely this name is used? To this question *J. St. Mill* gives us – in a note to his father's book on psychology – an intelligible answer, which, in my opinion, is sufficient: "Objects, whether <329> physical or psychical, are in relation to one another in virtue of one complex state of consciousness into which they both enter: even for that case where the complex state consists of nothing more than thinking of the two together. And they are related to one another in as many different ways – or, in other words, they stand in as many distinct relations to one another – as there are specifically different states of consciousness of which *both* are parts."[31]

For purposes of a classification of relations, one might at first use as a guide-line the character of the phenomena which they interrelate (i.e., of the "terms"). However, such a classification would be superficial. In the most diverse of domains we find relations which have one and the same character. Thus, identity,

---

[31] James Mill, *Analysis of the Phenomena of the Human Mind*, ed. J. St. Mill, London 1897, Vol. II, pp. 7ff. Cf. Meinong, *Hume-Studien*, Vol. II, "Zur Relationstheorie," Vienna, 1882, p. 40.

similarity, etc., occur both in the domain of "physical phenomena" and in that of "psychical phenomena."[32]

But one can also (and here is the more penetrating principle of division) classify relations in terms of their particular phenomenal character. From this vantage point, relations fall into two main classes:

1. Relations which possess the character of "physical phenomena," in the sense defined by Brentano.

Every relation rests upon "terms." It is a complex phenomenon which comprises in a certain way – which cannot be more closely described – partial phenomena. But in no wise does every relation comprise these its terms by intentionality,[33] i.e., in that specifically determinate manner in which a "psychical phenomenon" (an act of noticing, of willing, etc.) comprises its content (what is noticed, willed, etc.). Compare, for example, the way in which the representation called the "similarity" of two contents includes these contents themselves, with any case of "intentional inexistence," and it will have to be acknowledged that we have here two wholly different kinds of inclusion. <330> Precisely because of this, similarity must not be designated as a "psychical," but rather as a "physical" phenomenon. The same is true of other important relations as well, e.g., identity, gradation, continuous combination (i.e., the combination of the parts of a continuum), "metaphysical combination" (i.e., the combination of properties ⟨in individual objects⟩, as with color and spatial extension), logical inclusion (as color is included in red), and so on. Each of these relations Represents a peculiar "physical" phenomenon (in the signification assumed here for this term), and in that regard belongs in the same main class.

I would, in addition, expressly point out that it makes no difference here whether the terms, i.e., the contents, which are

---

[32] In regard to the significations of the terms "physical" and "psychical" phenomenon, and the fundamental distinction underlying them, which also is indispensable for the following reflections, cf. F. Brentano, *Psychologie vom empirischen Standpunkte*, Vol. I, Book 2, ch. 1. [English translation, *Psychology from an Empirical Standpoint*, by Antos C. Rancurello, D. B. Terrell, and Linda L. McAlister, New York, Humanities Press, 1973, pp. 77-100. {DW}]

[33] *Ibid.*, p. 115.

interrelated are themselves physical phenomena, or are some sort of psychical phenomena (represented psychical states). Even such identities, similarities, etc., as we perceive between psychical acts or states (judgments, acts of will, and so on) are physical phenomena. They only show up here on the occasion of psychical phenomena, and are grounded in them.

Relations of this class could most briefly be designated by the name, "*physical relations*." But one would have to guard against the misunderstanding that we here have to do with relations of (or "between") physical contents, whereas, as was just emphasized, it is not a matter of that at all.

2. On the other hand there stands a second main class of relations, which is characterized by the fact that here the relational phenomenon is "*psychical*." If a unified psychical act is directed upon several contents, then, with regard to it, the contents are combined or are related to each other.

Were we to realize such an act, then, of course, we would seek in vain, among the contents of the representation which it includes, for a relation or combination (unless *in addition* a physical relation were there). The contents are, in this case, unified precisely by the act alone; and the unification, therefore, can only be noticed by means of a special reflexion upon the act. Any arbitrary act of representation, judgment, or emotion and will, which is directed upon a plurality of contents, can do as an example. Of any of these psychical acts <331> we can say, in agreement with *Mill*'s definition: It sets the contents into relation with each other. There especially belongs here the relation of "distinctness" in the widest of senses, which has already been discussed, and in the case of which two contents are brought into relation merely by means of an Evident, negative judgment.

The characteristic distinction between the two classes of relations can also be marked by saying that the physical relations belong to the respective contents of representation in the same sense as do their terms, but that this is not so with psychical relations. In reference to this, one could also quite appropriately call physical relations "*content relations*."

After this digression into the theory of relations we return now, once again, to those particular relations upon the characterization

of which we have set our aim; and we put the question: Are the
relations which unify the objects of the totality, and which we
called "collective combinations," content relations, in the sense
just now made precise – as, for example, metaphysical and
continuous combinations are? Or must we perhaps assign them to
the class of psychical relations? More exactly expressed: Are
collective combinations intuitively contained in, and separately
noticeable among, the contents of the representation of the totality
as partial contents – as are, say, metaphysical combinations in the
metaphysical whole? Or is no trace of a combination to be noticed
in the representation contents themselves, but rather only in the
psychical act which unifies the parts in its embrace?

In order to decide this question let us, to begin with, compare
the totality with any represented whole.

In order to note the unifying relations in such a whole, analysis
is necessary. If, for example, we are dealing with the represent-
ated whole which we call "a rose," we get at its various parts
successively, by means of analysis: the leaves, the stem, etc. (the
physical parts); then the colors, their intensities, the odor, etc. (the
properties). Each part is thrown into relief by a distinct act of
noticing, and is retained *together with* those parts already segregated. As the next issue of the analysis there results, as we see, a
totality: namely, the totality of the separately and specifically
noticed parts of the whole. But then by means of a simultaneous
reflexion upon this whole <332> which unifies the parts there also
stand forth the combining relations, as separate, specifically determinate phenomena of representation. In our example we have the
continuous combinations within the leaves; or the combinations of
the properties, like redness and spatial extension, etc., which
combinations are characterized quite differently again from the
continuous. In such a way, therefore, *these* combining relations
present themselves as, so to speak, a certain "more," in contrast to
the mere totality, which appears merely to gather up its parts, but
not really to unify them. What, then, distinguishes a case of
physical combination from a case of collective combination?
Apparently it is this: that in the first case a unification is
intuitively noticeable within the contents of representation, while
this is not so in the latter case. In the totality there is a lack of any

intuitive unity in the way such unity so clearly manifests itself in the metaphysical or continuous whole. And this is so even though a certain unity is present in the totality, and is perceivable with Evidence.

The same thing is also shown by a comparison of the collective combination with the relations of identity, similarity, gradation, etc., (which, within the class of content relations, constitute, like the combining relations, a group of relations that are psychologically well-characterized). Although they do not "combine" the contents upon which, as terms, they are based, yet they constitute perceptible representation phenomena; and in contrast with them, again, the collective combination appears almost as a case of lack of relatedness. And so one also speaks of "disjoined" or "unrelated" contents when it is a matter of emphasizing the absence of any *content* relation whatever, or of content relations upon which the current governing interest is directed. In such cases the contents are just simply thought "together," i.e., as a totality. But in no wise are they actually disjoined or unrelated. To the contrary, they are joined by means of the psychical act grasping them together. It is only within the content of that act that all perceptible unification is lacking.[34] <333>

The following circumstance also shows that between the collective combination and all of the elementary content relations which are known to us there is an essential distinction which can make sense only upon the assumption that the former really is not to be counted among the content relations. Every relation rests upon terms and, in a certain sense, is dependent upon them. But whereas, with all content relations, the variability of terms which is admissible without a change in the species of relation is limited,

---

[34] Therefore *Mill* is right in expressly stressing that objects already stand in relation to each other even if we only think of them together. Precisely with respect to the psychical act which thinks of them together, they constitute parts of a psychical whole; and they also can, at any time, be recognized as joined by means of reflexion upon that whole. This whole constitutes their "relation." And only if one were to restrict this term to what we have called "content relations" could there, of course, be no more talk of relation in the case of a psychical combination. On the one hand, this certainly is a terminological matter. But, on the other hand, there is *de facto* so much in common between the content relation and the psychical relation, as to their main moment, that I fail to see why a common term would not be justified here.

with the collective combination, any term can be varied completely without restriction and arbitrarily, while the relation yet remains. The same also holds true of the relation of distinctness in that widest sense discussed above. Not every content can be conceived of as similar to, continuously joined to, etc., every other content. But each *can* be conceived of as different from, and also as collectively united with, every other. These two latter cases are, precisely, cases where the relation does not immediately reside in the phenomena themselves, but, so to speak, is external to them.

So testimony from various sources – and, above all, inner experience itself – tells us that we must decide in favor of the second view mentioned above, according to which collective unification is not intuitively given *within* the representation content, but has its existence only in certain psychical acts which unifyingly embrace the contents. And obviously these acts can only be those elemental acts which are capable of enclosing any and all contents, be they ever so disparate. So, then, an attentive inspection of the phenomena teaches the following:

A totality originates in that a unified interest – and, simultaneously in and with it, a unified noticing – throws into relief and encompasses various contents by themselves. Hence, the collective combination also can only be observed by means of reflexion upon that psychical act through which the totality comes about.
And this also is positively confirmed by inner experience. <334> Wherein, for example, consists the combination when I think of a number of such disparate things as redness and the moon? Obviously only in this: that I think of them "together," think of them in *one* act.

Collective combination plays a highly important role in our mental life as a whole. Every complex phenomenon which presupposes parts that are separately and specifically noticed, every higher mental and emotional activity, requires, in order to be able to occur at all, collective combinations of partial phenomena. There could never even be a representation of one of the more simple relations (e.g., identity, similarity, etc.) if a unified interest and, simultaneously, an act of noticing did not accentuate together and unifiedly seize upon the terms of the relation. This psychical relation called "collective combination" is, thus, an indispensable psychological precondition of every other relation and combination whatsoever.

The abstraction which provides the general concept of the collective combination requires, then, no further special discussion. In any case, in virtue of its elementary nature this concept found its expression in language very early. A mere collective combination is expressed in language by the occurrence of the conjunction "and" between the names of particular things mentioned.

Section 4: *The Analysis of the Concept of Number as to its Origin and Content*

Since we have established the 'psychological' nature of the collective combination, we can bring to completion the solution of our problem, which was the exhibition of the origin and content of the concepts *multiplicity* and *number*.

We stated that the abstraction which yields the concept of multiplicity or totality requires reflexion upon the collective mode of combination, similarly as, for example, the abstraction of the concept *metaphysical whole* requires reflection upon the metaphysical mode of combination. In order to render such abstraction possible, all that is necessary is that the combining relations between the elements of the totality always be perceptible as what they are in essence, well-distinguished from all other relations; and it is in this respect unimportant whether these combining relations are given among the contents of the representation itself, or merely in the psychical act which represents the totality. <335> Now we have decided in favor of the latter. In reflexion upon that elementary act of emphatic interest and noticing which has for its content the totality representation, we attain to the abstract representation of the collective combination; and, by means of this abstract representation, we form the general concept of the multiplicity as a whole the parts of which are united merely in the collective mode. However it is better to avoid the terms "whole" and "part." They involuntarily evoke the thought of a more intimate unification of contents, such as is not present here at all. Hence we prefer to say that a representation which is occupied with contents merely as "collectively" united, this all thought *in abstracto*: – such is the concept of multiplicity.

But with this we still have only a paraphrase. What is the actual conceptual content when we think the concept *multiplicity*? The contents which can be colligated to form totalities are, as we know, utterly without restriction. There can also, therefore, enter into the general concept of the multiplicity no peculiarities of content. However, since this concept is a relational concept, parts must somehow be thought of in it. And, without any difficulty, this also is what takes place in a suitable manner. The particular contents are thought of, not as determinate, but rather as totally indeterminate, as *any sort of* content: each one as *anything*, as *any one* thing. If, now, we dispense with the scientific term, "collective combination," and hold ourselves merely to the little word "and," which designates or indicates the same thing in a completely clear and intelligible manner, then we can quite simply say without any circumlocution: totality or multiplicity *in abstracto* is nothing other than "anything" and "anything" and "anything," etc.; or, any one thing, and any one thing, and any one thing, etc.; or more briefly, one, and one, and one, etc.

Thus we see that the concept of the multiplicity contains, besides the concept of collective combination, only the concept of *something*.

Now this most general of all concepts is, as to its origin and content, easily analyzed.

"Something" is a name which is proper for any conceivable content. Any real or conceptual being is a "something." But we also <336> can give this name to a judgment, an act of will, a concept, an impossibility, a contradiction, and so on. Of course the concept *something* is not to be obtained by any conceivable comparison of contents which takes in all objects, both physical and psychical. Such a comparison would simply remain without a result. In fact, "something" is no partial content. That wherein all objects – actual and possible, real and unreal, physical and psychical, etc. – agree, is this alone: They either are contents of representations, or are represented in our consciousness by means of contents of representations. Obviously the concept *something* owes its origination to reflexion upon the psychical act of representing, as the content of which just any determinate object may be given. Hence, the "something" belongs to the content of any concrete object only in that external and non-literal fashion common to any sort of relative or negative attribute (such as, for example, with similar-to-B, non-C, etc.). In fact, it itself must be designated as a relative determination. Of course the concept *something* never can be thought unless some sort of content is present, on the basis of which that reflexion mentioned above is carried out. Yet for this purpose any content is as well suited as another: even the mere name "something."

Let us turn back, now, to the concept of the multiplicity. We explained it as: something and something and something, etc.; or one thing and one thing and one thing, etc. This "etc." indicates an indetermination which is essential to the concept. It does not, of course, mean that we should continue on *in infinitum*. Rather it means only that no limit to our continuation has been set. *De facto*, to be sure, in thinking out "multiplicity" a boundary is speedily found. But it is always with the consciousness that it is an arbitrary one, which is of no significance at all. This gives us the concept of *multiplicity in the widest of senses*.

Through elimination of that indetermination just noted there originate from the general concept the determinate multiplicity concepts or *numbers*. The more general concept of multiplicity encompasses all concepts of the same sort as *one and one, one and one and one, one and one and one and one*, etc., as its special cases. These special cases are, in their determinate delimitation from each other, well distinguished; and accordingly they would receive separate names: "two," "three," "four," etc.

Each concrete totality falls under one – and, of course, a <337> determinate one – of these concepts. To each such totality there "belongs a certain number." It is easy to characterize the abstraction which must be exercised upon a concretely given multiplicity in order to attain the number concept under which it falls. One considers each of the particular objects merely insofar as it is a "something" or "one," simultaneously retaining the collective combination; and, in this manner, there is obtained the corresponding general form of multiplicity, one and one and . . . and one, with which a determinate number name is associated. In this process there is total abstraction *from* the specific characters of the particular objects. But this neither means nor implies that the concrete objects have to disappear from our consciousness. To "abstract" from something merely means to pay no special attention to it. Thus, also in our case at hand, no special interest is directed upon the peculiarities of content in the separate individuals, while those peculiarities, nonetheless, do constitute the precondition of the acts of reflexion which yield the "units" of the respective number, and are the ground of the distinctness of those units.

Let us look once more, then, at the psychological foundation of the number concepts.

According to our view two things make up the concept of number: 1) the concept of "collective unification" and 2) the concept of "something."

The abstraction of the former concept becomes possible in virtue of the fact that, in all cases where discrete contents are thought together, i.e., in a totality, there is present one and the same, constantly uniform act of combining interest and noticing, which unifyingly gathers each of the particular contents, separated off to itself (i.e., as separately and specifically noticed) and at the same time in a union with the others. It is with reference to this unifying act that we win the abstract representation of collective combination.

As to the subsumption of any content under the concept of "something," that requires reflexion upon the act in which that content is represented. <338>

The two psychological constituents of the concept of number obviously are not independent of each other. We cannot conceive of a collective unification without united contents; and, if we wish to represent them *in abstracto*, then they must be thought of as "any something." But if this is so, what, then, constitutes the distinction between the concepts *collective unification* and *multiplicity*?

The answer is obvious. In the first case, interest rests exclusively upon the combination of the arbitrarily conceived contents; but in the latter, it rests upon the totality of those contents as a whole, i.e., the elements considered in this their unification. So both concepts are equally essential to the concept of the multiplicity – the concept of "something," and that of "collective unification."

It is clear that the concept *something* is related to a concrete content in exactly the same way as the concept *number* is related to a totality of concretely given contents. However, the concept *something* is the more primitive one. Without it there would be no number. The elementary fact which originally manifests itself in it, and essentially conditions it, is that which makes possible the concept of the collective unification.[35]

---

[35] These pages contain the first chapter of a book which will shortly be published by C. E. M. Pfeffer (R. Stricker) of Halle. [The book in question developed into Husserl's *Philosophy of Arithmetic*. {DW}]

APPENDIX TO "ON THE CONCEPT OF NUMBER: PSYCHOLOGICAL ANALYSES"

THESES[36]

I. Every law of nature is an hypothesis.
II. In maintaining the intuitability of the non-*Euclid*ean forms of space, *Helmholtz* essentially modifies the concept of intuition.
III. With regard to the demonstration of the law of causality, *Helmholtz* contradicts himself at numerous points.
IV. The concept of time is not included in the concept of number.
V. One can hardly count beyond three in the authentic sense.
VI. *Hankel*'s "Principle of the permanence of formal laws" in arithmetic is neither a "metaphysical" nor an "pedagogical" ["*hodegetisches*"] principle.
VII. The logical justification of the use of irrational and imaginary numbers in all mathematical domains has not been demonstrated up to now.
VIII. Color phenomena constitute a bounded, space-like continuum of *four* dimensions.

---

[36] Theses which the author defended in a Disputation at the University of Halle in the year 1887, first published in "Husserliana XII," p. 339. They were printed for the first time on the back side of an invitation: "Dr. phil. E. G. Husserl takes the liberty herewith, upon the consent of the honorable philosophical faculty of the United Friedrich University of Halle-Wittenberg, to invite you most respectfully to the Disputation occurring in The Great Hall at 12 noon on Friday the first of July, for the purpose of the habilitation of his doctorate. The adversaries: Dr. Wiener, Privat-docent, and Mr. H. Schwarz, scholar in mathematics." {LE} <340>

B. ESSAYS

ESSAY I [1]

⟨ ON THE THEORY OF THE TOTALITY ⟩

⟨ I. The Definition of the Totality ⟩

Let A, B, C, . . . designate any objects whatever, whether intuited or thought, existing or imaginary, so long as they are compatible with each other – i.e., are such that the being of the one does not exclude the being of the other. Then the expression "A and B and C and . . . ," taken in its general sense, yields a definition of the term "totality of the objects A, B, C, . . . ". The objects totalized are also called the members of the totality. If they are individually given, then the collective act which combines them to form a conceptual unit can occasionally be carried out in the authentic [*eigentlichem*] sense. From such cases the concept of the totality took its psychological origin. Reflection upon the collocating act supplies the intuition which the concept of the totality presupposes. In the majority of cases, the collective unification cannot be carried out in actuality [*wirklich*], and it then is merely intended. For logical purposes this is no hinderance. It does not matter whether the objects which are to be grasped together are intuited separately, along with their individual peculiarities, or are only Represented [*repräsentiert*]. It does not matter whether the objects separately given come to synthesis in one act of actual unification. Important epistemic interests are already satisfied if we have the inauthentic representation of a totality that refers exclusively to the objects intended. But this

---

[1] This corresponds to "IV. Abhandlungen" in *"Husserliana XII"* (pp. 385ff), dated by the editor {LE} as from 1891. These are not essays in the polished literary sense, but in the sense of an "assay," often tentative exploratory notes. Pp. 343-384 of *"Husserliana XII"* were translated in an earlier volume of the series, "Edmund Husserl Collected Works," titled *Early Writings in the Philosophy of Logic and Mathematics*, Dordrecht, Kluwer, 1994. {DW}

exclusive reference does not presuppose separate representations of the respective objects. Indeed, conceptual determinations can be given that decide in a general manner which objects are and which are not to belong to the intended unity. The conceptual representation of a certain collective unity <386> that links all the objects together is then a completely determinate one. This is the case, for example, when we speak of a totality of objects that fall under a concept C. We then have the conceptual representation of a certain collective combination, with the appended determination that whatever possesses the distinguishing marks C is to be united in it, while all else is to remain excluded.

We here take note of an *Axiom* of which we must make use later:

Axiom I. *For every totality, with the single exception of that one which includes everything representable in the widest sense of the word, there is a possible further object which is not contained in it.*

There are operations through which, from given totalities, new ones can be obtained. These are: augmentation and diminishment, combination and partition.

A totality is *augmented* when the objects collocated in it are united with any new objects to form a totality.

Axiom II. *It is Evident that every totality can be augmented by an arbitrarily selected object not contained within it.*

With reference to Axiom I we deduce the *Theorem:*

There is for every totality – excepting only the totality of all possible objects – a possible object by which it can be augmented.

A totality is *diminished* if one takes away from it any of the objects collocated in it and unites the remaining objects into a totality.

Axiom III. *To augment (or diminish) a totality by certain objects, and to diminish (or augment) the resulting totality by identically the same objects restores the original totality. In other words, augmentation and diminishment are inverse operations.*

A totality T can be augmented by a single object T', or by the objects of a totality T' disjoined from T. In each case one says that T' is *added* to T, and indicates the result of this operation by the symbol T + T'. The resulting totality T_s is called the *sum* of T and T'. <387>

Axiom IV. *If the symbol "≡" is read as "is identical with," it is Evident that* $T + T' \equiv T' + T$; and, further,

Axiom V. $(T + T') + T'' \equiv T + (T' + T'')$.

If, when all the letters designate individual objects, one understands the sign "+" as the "and" of collective combination, then the Axioms remain meaningful. In the light of that, we dispense with treating as exceptions the cases where the signs do not designate totalities, and treat individual objects entirely as totalities.

Instead of adding T' to T, or T to T', one can unite the members of the two totalities to form a new totality without in any way favoring the one totality over the other. If we designate the operation thus defined by Σ(T, T'), then it is Evident that: Σ(T, T') ≡ T + T' ≡ T' + T. Because of this identity, one need not distinguish the operations designated by "Σ" and "+" from each other. They differ only inessentially. Whereas there correspond to the symbols "T + T'" and "T' + T" well segregated concepts, "Σ(T, T')" and "Σ(T', T)" are only different inscriptions for identically the same thought. Thereby Axiom IV loses its significance. At the most it may express the fact that no importance whatever attaches to the unavoidable succession of the signs T and T' in the written expression of the Σ-operation, and thus that any reordering of the signs is permissible.

Similarly for Axiom V. The operation Σ can be carried out at one stroke with arbitrarily many totalities, T, T', T" . . . . That is, one can simply think the units of arbitrarily many totalities as united into a new totality instead of composing successively, as in the addition { . . . (((T + T') + T") + T''') + . . . }

That any operation Σ(T, T', T", T''', . . .) is equivalent to such a successive composition, requires no proof. Likewise that the operation Σ no longer has need of an associative Axiom, since the different totalities $T^{(k)}$ have no conceptually distinct status in Σ, and the operation is a simple one in spite of the multiplicity of the members.

Because of the greater simplicity attained through the operation Σ, it merits in and for itself a decisive preference over the operation of addition equivalent to it. We will discuss later the reasons <388> why it is nonetheless preferred to define addition, and to regard (thus, to symbolize) any composition of several totalities as a successive unification of two-member additions.

If T' is a partial totality of T, then by T - T' we understand the totality which results from T through diminishment by the objects of T'. One then speaks of a *subtraction* of T' from T and ⟨calls⟩ T - T' the *difference* between T and T'.

We can then write *Axiom III* as follows:

$$(T + T') - T' \equiv T$$
$$(T - T') + T' \equiv T,$$

whereby the first identity has a sense only if T' is disjoined from T, and the second only if T' is contained in T as a part.

Axiom VI. *Any totality admits of being diminished by one unit.*

The diminishing rests upon a partitioning. One part is distinguished and suppressed in thought, and the remaining part is constituted as a totality. One can in this way also segregate and suppress in thought several parts. The resulting totality can then always be regarded as the result of a successive subtraction, ((T - T') - T'') - T''' . . . . The partitioning could also directly serve to obtain new totalities out of a totality T, and would then be considered the inverse operation of Σ, above. As through unification of T, T', T'' I obtain the new totality Σ(T, T', T''), so I can, by working backward from this, obtain once again T, T', T'' through partitioning.

Within the unity of the totality the members are taken up, or to be thought of as taken up, as that which they are. That is the intention. Each member has its determinations, partly those which it has in common with others, and partly those which distinguish it from all others and thus individuate it. It of course depends on the degree of authenticity with which we represent the totality how many of the determinations of the members actually fall within our representation. We will for now content ourselves with a minimum.² We want to stipulate that, with whatever <389> fullness of determinacy we have represented the members of a given totality T(A, B, C, . . .), now abstraction is to be made from all determinacy of the content of those members and each member is to be considered only as a "something," more precisely as a "something" *in some way* distinct from the remaining members.³ It is Evident that this abstraction can be carried out on any totality, however given. To any arbitrary T_A(A, B, C, . . .) there thus belongs a T_1(1, 1, 1, . . .), which proceeds from the former by each member A, B, C, . . . being replaced by "something" or "one" (1). Obviously nothing prevents our conceptualizing the representation of a T(1, 1, 1, . . .) without going through the abstraction described, on the basis of a T(A, B, C, . . .); for we have the concept of the totality and that of the "something."

In the formation of T(1, 1, 1, . . .), i.e., "something and something and something, etc . . .," each "something" means one that is different from every other "something." That this can be what is meant, in spite of the fact that "something" and "something" allows no distinction in terms of content, is clear. Conceptual objects which are given only as such can, *eo ipso*, not be distinguished in content. But nothing prevents our *thinking* of them as in some way distinct, and representing concrete objects through them. We thus understand "one and one" as "one lion and one lion."

We call a totality of ones (units) a *(pure) number*. To every totality of however determinate objects "belongs a certain number," which is to be found in the manner just described.

We obtain concepts analogous to pure numbers, only less abstract, through formation of totalities whose members are taken

---

² Marginal note: Yet this is of no importance, but rather how much of each content we intend as belonging to the object.

³ Marginal note: It is no doubt more useful, with Bolzano, to put in place of "something" "something of kind K," but to leave K undetermined – and invariant nevertheless, so that it would not have to be brought into consideration.

exclusively from the extension of a certain concept C, e.g., T(C, C, C, ...). We then call each C as such *a unit of kind C*. That unit results from the pure 1 through determination by C. Each pure number passes over into a "concrete" ["*benannte*"] number by means of such determination.

If we represent to ourselves the unit C as undetermined, but as arbitrarily determinable, if we therefore form in complete generality the concept of a collective combination whose members all fall under one and <390> the same concept, then it is clear that the concept of the pure number also can be regarded as a special case of that concept, since the concept of *one* presents itself, after all, as a special case of the C, of the concept in general. In considering the matter from this point of view, the pure number is regarded as that special concrete number whose unit is the abstract *one*. This of course in no way conflicts with the fact that given another – namely, our earlier – point of view, any concrete number falls under the concept of the pure number, so long as we precisely bring each unit C under the concept *one*. Where totalities are of interest to us only with reference to the fact that they are *one* and *one* and ... etc., they fall as objects under the concept of the pure number.

⟨ II. Comparison of Numbers ⟩

Let us limit ourselves for the moment to totalities of objects of a determinate kind K. In two such totalities, now, we always find a certain sameness: Both are, precisely, totalities of nothing but A's. But it is easily seen that there is something peculiar about the case when the two totalities are equal with respect to their concrete number. In a totality of A's we grasp the concrete number when we indeed do take an interest in *each* member in it, but in each one *only* as an A.[4] Given this orientation of interest, the other distinctive marks of any one A – in particular also those that individuate it – are inconsequential to us, except for the one quite general circumstance that there are some individuating determinations that

---

[4] Marginal note: Take an interest? Subsumption under A is the essential point.

distinguish each A from all the others in the totality.⁵ If we therefore consider some member or other of the totality to be arbitrarily modified, but in such a way that it remains an A and becomes identical with none of the remaining A's of the totality, then for our interest *nothing* is changed. The infinite manifold of modifications which are conceivable in this manner Represent [*repräsentieren*] no modifications for our interest – or, if you wish, only non-essential modifications. Any two totalities derivable from each other through such variations <391> are, in essentials, "the same." Given the orientation of interest, they are *equal* to each other,⁶ i.e., equal with respect to the concrete number.

Conversely, it is Evident that any modification that does not remain within the limitations stated is an essential one, and thus leads to a totality different as to number. In other words, if two totalities are equal in number, then it must also be possible to transform the one into the other through modifications of the type characterized. From this the following criterion results: The totality T_A is *equal* to the totality T'_A with respect to the concrete number of the units A if and only if T'_A can be derived from T_A through considering the A's of the latter as transformed into the A's of the former; however, in such a way that **1)** each A remains one A and **2)** different A's are each time transformed into different A's. If, thus, one changes A_κ into A'_κ, A_λ into A'_λ, then A'_κ and A'_λ must not be identical. Likewise, it lies in the sense of the first condition, that A_κ, for example, is not transformed into A_μ *and* A_ν; for A_μ *and* A_ν does not fall under the concept A (one cannot say: A_μ *and* A_ν is an A).

That in this way an *equality* is in fact "defined" or – more accurately stated – a criterion of equality is obtained, follows from the validity of the theorems: (i) If $T_A = T'_A$, then also $T'_A = T_A$, and (ii) if $T_A = T'_A$ and $T'_A = T''_A$, then also $T_A = T''_A$ – theorems which are manifestly Evident on the basis of the above criterion: the latter indirectly, the first directly.

---

⁵ Marginal note: It is not a matter of the interest, but of the concept!

⁶ Marginal note: Conceptually!

Exactly what we have established carries over without further ado to the comparison of *arbitrary* totalities with respect to their pure numbers or, what is in essentials the same, to the comparison of pure numbers themselves. We grasp the (pure) number in an arbitrary totality when our interest bears indeed upon each member in it, but on each only as a *one*. We have, therefore, in the foregoing line of thought merely to replace the concept A by that of the pure unit, that of the "something." The criterion of equality will then read: A totality T is equal to the totality T' with respect to number if and only if T' can be derived from T through considering the members of the second as transformed into the members of the first; however, in such a way that <392> **1)** each member remains a *one*, and **2)** different members are each time transformed into different members. In other words we can also say: A number N is equal to a number N' if and only if N' can be derived from N through considering each unit of the second as replaced by one unit of the first – and, indeed, different units of the second each time by different units of the first.

These criteria also admit of yet another equivalent reformulation. Instead of *replacing*, according to the latter criterion, the units of N (with the precaution stated) by corresponding units of N', it suffices to consider them as *coordinated with* the corresponding units. The limiting condition then reads: At no time are two different units of N to be coordinated with identically the same unit of N'. In fact, if there is the possibility of that replacement, then *eo ipso* there is also the possibility of such a coordination: Through the replacement, the corresponding units are automatically coordinated in a determinate manner, and in a permissible manner if the replacement was a permissible one. And conversely: If the number N can be put into a "univocally one-to-one" correspondence[7] – i.e., in such a way that to each unit of N there corresponds one from N', and never two different units of N to one and the same from N', then we need only to consider the corresponding units as replaced by one another and the

---

[7] Note: I understand that currently "gegenseitig eindeutige korrespondenz" may also be read in English as "one-to-one onto correspondence." I have stayed with "univocally one-to-one" because that is still, it seems to me, the common usage for philosophers.

ESSAY I 367

equality of the two numbers comes to Evidence in accordance with our original criterion. We can therefore say:

The number N is equal to the number N' if and only if the units of the latter can be univocally coordinated one-to-one with those of the former. One can verify the soundness of this criterion once again by proving from it the theorems: (i) If N = N' then N' = N, and (ii) if N = N' and N' = N" then N = N". From the relationship between equality and one-to-one univocal correlation there also immediately follow, if we indicate the latter by the sign $\cong$, the theorems:

If $N \cong N'$, then $N' \cong N$.
If $N \cong N'$ and $N' \cong N''$, then $N \cong N''$.

No special discussion is required to show that a correspondingly modified criterion results for concrete numbers, which we will disregard in this chapter.

Many will prefer to call the criteria brought to light <393> "definitions of equality." However, equality in terms of number or concrete number is just as little a matter of arbitrary definition as is the equality in terms of redness or blueness. The significance of these criteria lies in the fact that they offer the irreplaceable means of making classifications in the domain of numbers, as we will later demonstrate.

In the case of inequality one speaks with justification of numbers being more or less, or greater or smaller, but only under certain presuppositions. At first one is tempted to say quite in general: If two totalities T and T' are unequal, then obviously either T is equal to a part of T' or T' is equal to a part of T. In the second case of inequality one says that T is greater than T' (namely, by the complementary part) or that it is more than T' (namely, by the members of that part); or again, that T' is smaller or less than T. The reverse holds in the first case. However the following must be considered: **1)** The obviousness of the presupposed situation is not at all beyond doubt; **2)** it lies in the concept of the greater and the smaller that one excludes the other. It would thus also be presupposed – what is altogether *not* obvious – that if T is equal to a part of T', T' is not simultaneously equal to a part of T, unless also T = T'. This would, therefore, have to be demonstrated beforehand, which is not a very simple matter. The definition: T is greater than T' if the latter is equal to a part of the former, would be radically mistaken. For then being equal and being greater would no longer exclude each other. The correct place for the discussion of the new distinction is in the domain of finite numbers.

⟨ III. Addenda ⟩

⟨ 1. Addendum to p. 367: *Identity and Equality* ⟩

Finally, the relationships between identity and equality shall briefly be touched upon. For identity there of course hold true the same formal relations as for equality.

1) If $A \equiv B$, then $B \equiv A$.
2) If $A \equiv B$ and $B \equiv C$, then $A \equiv C$.

The different letters on the two sides of each relation signify different conceptual determinations (among which we also <394> must count different designations) of one and the same object.

If in 2) we replace the "$\equiv$" sign with the "$=$" sign in the first or second premise, it then is necessary to write "$=$" also in the conclusion.

We therefore make note of the theorem: If ⟨in 2)⟩ an equality is interlinked with an identity or conversely, the first and the third terms ⟨obviously *not* premises!⟩ stand in a relationship of equality.

For reasons that are discussed later, it is not necessary in arithmetic to continually maintain the distinction between identity and equality. If the "$=$" sign is read as "identical with or equal to," then, in cases where the distinction between them is superfluous, one can simply invoke the laws of equality, 1) and 2), which, thus understood, are universally valid.

⟨ 2. *On the Definition of Number* ⟩

The concept of cardinal number, above all as here defined, is an invariant feature of the defined group in question, only if the group is equivalent to itself in every arbitrary mode of correspondence (= if every permutation of all elements of the group is equivalent to every other permutation) = if the cardinal number remains the same for all permutations. Bernstein has demonstrated that a sufficient condition for this is (without it necessarily occurring in actuality) that the group of the permutations of the given group "exists." There are, we should mention, groups where the group of the permutations contains a contradiction. On this point we still must have an exchange with Bernstein.

⟨ IV. The Classification of the Cardinal Numbers ⟩

The concept of number equality in no way excludes the possibility that a number might be equal to one of its partial numbers, i.e., to a number which we obtain by leaving aside certain units in the number before us and colligating the remaining ones. An example will suffice to confirm this. The number of the familiar endless series of signs <395>:

$$1, 2, 3, 4, 5, \ldots$$

is, as one may convince oneself through application of our criteria, equal to the number of its partial series:

$$1, 3, 5, 7, 9, \ldots$$

On the other hand there are also cases, as arbitrarily many examples show, where a number can be equal to none of its partial numbers. To mark out this fundamental distinction we state the definition:

A number is said to be infinite if among its partial numbers there is one that is equal to it. A number for which this is not true is finite. From its units, therefore, no partial number can be formed that is equal to it.

One can now demonstrate numerous interesting theorems. For example, that the part of an infinite number that is equal to it would also have to be infinite. That *no infinite number whatsoever could ever be equal to a finite one.*[8] That, therefore, all of the numbers equal to a finite number form a class of finite numbers. That to this class no number can belong which proceeds from any among them through diminishment or augmentation. Etc. But here the following special theorem suffices:

If a is a finite number, then a + 1 is also a finite number.

Proof: If a + 1 were infinite, then it would have to be true that a + 1 = θ(a + 1), where θ(a + 1) designates a certain partial group from the units of a + 1. There is then at least one determinate, univocally one-to-one correlation of the two numbers. We indicate this by writing:

(1) $a + 1 \cong \theta(a + 1)$.

Let us assume that the totality of units of θ, which hereby stand in correspondence with a, is Θ, whereas 1 is coordinated with a certain 1_0. We thus have:

$a \cong \Theta, 1 \cong 1_0$
$\Theta + 1_0 \equiv \theta(a + 1)$ <396>

in which, again, "≅" indicates the univocal one-to-one correlation and "≡" indicates identity.

It is easy to convince oneself that Θ cannot exclusively contain units of a. For otherwise it would have to contain *all* units of a: since a as a finite number can, in fact, not be equal to one of its partial numbers. If Θ contained all units of a, then necessarily Θ = a and, if 1 is to have any correspondent at all, also $1_0 \equiv 1$. Thus:

---

[8] Marginal note: This theorem will certainly be made use of in what follows. <See Addendum 1, p. 379>

$$\Theta + 1_0 \equiv a + 1 \equiv \theta(a + 1).$$

Contrary to the hypothesis θ would not be part of a + 1. We come to the same conclusion by assuming that, in (1),

$$1 \cong 1,$$

i.e., that 1 is coordinated with itself. Consequently in Θ there must be, besides units from a, also the unit 1, and 1_0 can only be one of the a units. This presupposed, we have only to exchange in Θ + 1_0 the 1 in Θ with 1_0, to coordinate the 1 with itself and 1_0 with the earlier correspondent of the 1 in a, and we again have the case acknowledged to be impossible. The modified Θ, call it Θ', would contain only units of a. We thus would have:

$$\Theta' \equiv a, \Theta' + 1 \equiv a + 1;$$

but on the other hand:

$$\Theta' + 1 \equiv \Theta + 1_0 \equiv \theta(a + 1),$$

and consequently under no circumstance is θ(a + 1) a part of (a + 1), which is contrary to the hypothesis.

We can now turn to the formation of the series of natural numbers. If "1" is a general sign for any "something" or "one," then by successive augmentation by one, step by step, the numbers

$$1 + 1, (1 + 1) + 1, ((1 + 1) + 1) + 1, \ldots$$

can be formed, and be designated in succession by

2, 3, 4, . . .

We have therefore the chain of definitions: <397>

$$2 = 1 + 1, \ 3 = 2 + 1, \ 4 = 3 + 1, \ 5 = 4 + 1, \ldots$$

The series of numbers thus defined we denominate "the series of natural numbers."

Theorem: *The natural ⟨number series⟩ has a beginning, but no end.* The first point is clear. The latter point follows from Axiom I, according to which every totality (disregarding that of the conceivable in general) can be augmented by one object, and thus every number by one unit.

Theorem: *1 + 1 is a finite number.*

Theorem: *The natural number series contains finite numbers only.* According to the theorem, a + 1 is finite if a̲ is finite. Now $1 + 1 \equiv 2$ is finite. Thus every natural number is also finite.

Theorem: *The number series contains unequal numbers only.* For the units of each number N_0 are contained in each number N_1 later in the series. If N_1 were equal to N_0, then N_1 would be equal to $\theta(N_1)$ and therefore infinite. — If number formation in the natural series is understood in such a way that m + 1 is formed from m̲ by adding one new unit to m̲ or to m' = m, then the demonstration is only inessentially modified. N_1 contains, then, a part N_0' that is either identical with or equal to N_0. If $N_1 = N_0 = N_0'$, then once again $N_1 = \theta(N_1)$.

Theorem: *The number of the natural numbers is an infinite one.* If from the series we select the first, third, fifth, . . . , by always skipping one step in the sequence, we then obtain a partial series that can be set into univocal one-to-one correlation with the entire number series, as any series that is bounded at one end and is at all points bounded dense. The serial principle of the partial series consists in the fact that from each member *a* the successor arises by the univocal operation (a + 1), which, as is easy to see, always remains repeatable.

Theorem: *To each finite number there corresponds in the series of natural numbers one, but also only one, equal to it.*
<398>

In order to demonstrate this for an arbitrary finite number N, we select some pair or other of its units and we form $1 + 1 \equiv 2$. Then we continue with $2 + 1 \equiv 3, 3 + 1 \equiv 4, 4 + 1 \equiv 5$ etc., so long as we still find new units which we can introduce into the process of formation. If all units are exhausted, then obviously the last natural number attained is equal to N. But it is also certain that this process has an end, and thus leads to a last natural number. Otherwise the entire infinite number series would in fact appear in univocal one-to-one coordination with our *finite* number, which is impossible. Were there, further, several natural numbers equal to the given number, then they would be equal to one another, which contradicts the theorem proven above. It follows that it is irrelevant in what order I run through the units of a finite number when counting: the resulting natural number must be identically the same. In order to grasp this one does not first have to establish the theorem that, if two finite totalities coincide (are equivalent) under one mode of coordination, they do so under every mode – a theorem which, as great as is the importance usually placed upon it, is certainly superfluous for a rigorous system of arithmetic.

We therefore obtain for arithmetic the fundamental result:

The series of natural numbers secures the exhaustive classification of the entire domain of finite numbers. Each natural number Represents a class, namely, the entirety of the numbers equal to it. To each finite number there corresponds one and only one natural number, and thus each belongs in one and only one class.

Through this classification it is simultaneously demonstrated that the concept of the *finite* number divides itself up into an infinite number of species concepts, which come together in their entirety to form an ordered manifold in which each member is followed by one member in univocal determination and only one member has none before it.

This order is not something externally imposed upon the number species, something that can be established only by recourse to consideration of concrete data. It arises, rather, from the pure concepts, through immediate and mediate Evidence, and consequently is apriori. It can be regarded as the fundamental fact of <399> arithmetic, inasmuch as it univocally determines as to their form the totality of arithmetical propositions. In every case where a general concept C can be analyzed into such an ordered manifold of species concepts, there hold true, as we will demonstrate, operations, basic laws, and accordingly also corollaries which are equiform (or, if you prefer: formally identical) with the arithmetical.[9]

⟨ V. Remark ⟩

It is totally inadmissible to begin the systematic treatment of arithmetic with the series of natural numbers, considering it as self-evident or even as having been demonstrated that every number is to be univocally counted off by means of a natural number (or by means of a fragment of the natural series that terminates in the latter), and consequently that for every number a determinate natural number can be substituted as an equivalent. In so doing tacit use is made of a basic presupposition that one limits oneself to *finite* numbers alone, constructs an arithmetic only for them, as then it is also only for them that the basic laws to be established hold true in their entirety. A rigorous system of arithmetic must therefore begin with the precise distinction of the numbers into the finite and the infinite, and then on the basis of that distinction it must provide proof of the complete classification of the domain of the finite numbers by means of the series of natural numbers.[10] But for this purpose there is, it seems to me, no shorter and more appropriate route than the one forged above. Of course I do not suppose that it is the only possible one. The most obvious one is, for example, to give the definition of finitude by means of the temporal series. A number is finite if we, traversing

---

[9] Marginal note: Read to *Cantor* when he spoke to me of a treatise of *Schröder*'s for the *Leopoldina*.

[10] Marginal note: This reproach falls, for example, upon the estimable works by Stolz (*Allgemeine Arithmetik I*, 10ff) and by Schröder (*Lehrbuch der Arithmetik und Algebra I*). The latter indeed does subsequently mention (*op. cit.*, p. 20) the presupposition of finiteness, without, however, establishing the definition of this important concept and without making systematic use of it. The first volume of this present work (Husserl, *Philosophie der Arithmetik*) is also deficient in this respect.

its units successively, encounter a last one. Precisely formulated, the definition would have to be much more complicated. If we supply temporal determinations to all the units in a number – and indeed different units with <400> different temporal determinations – then they form, if the number is to be termed finite, a temporal series closely compacted at all points and bounded in both directions, i.e., a series in which each point has an immediate predecessor and each point has an immediate successor. This in such a way, however, that in the first respect an exception must hold for one point which has no predecessor at all, and in the latter respect for one point which has no successor at all. Instead of the segments of an endless temporal series there can serve, as is easily to be seen, segments from any arbitrary endless series that is closely compacted at all points, whether it be one given in particular or a series in general, given in general representation. From the first perspective one can choose, for example, the series of natural number signs formed by one-strokes. The introduction of a particular series tacks onto the progression of Ideas a contingent character that is decidedly reprehensible in the establishment of a general theory. But the recourse to the general concept of series is also to be rejected because it introduces into the theory an element foreign to it. One must not invoke the fact that with the number series the concept of the series automatically makes its way into arithmetic. It is correct that the natural numbers form a series, but it does not follow that arithmetic would have to reason from the concept of the series. That concept is, to the contrary, entirely avoidable; and if we have recourse to it, that happens only to set the theory of the numbers into relationship with concepts that are formally related to it – a perspective which is important for the logic of arithmetic, but which to the theory of numbers, however, is a matter of indifference.

⟨ VI. Corrections ⟩

If we have authentic representations of numbers, then we can directly distinguish the different number types – that is, the species – e.g., 4 and 5. But with symbolic representations of numbers, how do we arrive at distinctions between species? To begin with, how do we recognize numbers as equal? Number is now equal to: totality of certain units. Various ways offer themselves:

1) We form the "series of natural numbers": symbolically, 1, 1 + 1 = 2, 2 + 1 = 3, etc., possibly within the artificial context of a deductive system.

But how do we know that any arbitrary number occurs in that series? *Which* numbers occur in it is clear: all of those for which <401> we finally come to a determinate formation as we enumerate step by step along this chain of ordered formations. "Finally" – that is, after all the units are exhausted. But what is the source of my knowledge that the units enumerated in every sequence yield *the same* number? Perhaps from the fact that it is absurd that the same multiplicity should have different numbers? No. For I in fact absolutely do not yet know of arbitrary totalities of units whether they admit of specific distinctions as I discover them in the case of authentically represented totalities. Apriori that is absolutely not Evident. We need therefore the **theorem:** ***A multiplicity which under one enumeration yields a number n from the number series yields the same number under every enumeration***. And this theorem is a demonstrable consequence of the other and likewise demonstrable theorem, that if a multiplicity can be coordinated in *one* arrangement to the members of a series segment thought of as closed in both directions, it can be coordinated to that same segment under any arbitrary arrangement. *If, now, one defines finite numbers as those whose units can be univocally* (not necessarily univocally one-to-one) *coordinated to the members of a bounded series* (a segment), then the **theorem:** ***All finite numbers can be enumerated by means of the number series***, holds true. Two numbers are equal if they correspond to the same natural number. All equal numbers constitute a class, and get their class name from the natural number.

2) Not essentially different is the classification of numbers by means of the series of the signs for the natural numbers. That I can follow this method in practice results from 1) by an easy line of thought. But I also can ground this classification independently of 1) – through considerations that are analogous, only more specific than those under 1).

3) Instead of relating all numbers to the formations of a particular series, or to a series of signs defined in some peculiar manner, I can also proceed in a different way:

First, the concept of a finite number can be explained. This is either done as above, or, equivalently to that, by reference to the temporal series. (A number is finite if, traversing its units in a sequence, I come upon a final unit. It follows then that the same holds true for any serial sequence, as shown above.)

Or finally, I do not define finitude first <402> but start with equivalence (or whatever one may call it). Two numbers are equivalent or coincide if they can be univocally coordinated one-to-one. If the numbers are finite in the sense of the definition above, then numbers coincide regardless of the mode of coordination. But instead of giving this definition in advance, one can also give the following definition subsequently: *Numbers are finite if they coincide with themselves under every conceivable mode of coordination (if there is no mode of coordination under which they do not coincide with themselves)*. This implies that two finite numbers which coincide under one mode of coordination do so under every mode. For if NN_1 coincide under mode A and A_1, and if A' is another mode of N, then A' and A coincide, and therefore also A' and A_1. If the modes of A_1 and A_1' are simultaneously modified, then the two still coincide, and then A' and A_1' are also in coincidence.

That there are numbers which are finite is shown by reference to authentic numbers or by reference to numbers that are coordinated to a segment of a series (But then I would need the corresponding theorems for series!). Two numbers are said to be equal, then, if they coincide under every ordering.

Theorem: *Two finite numbers are equal if they coincide under one arrangement.* One can then prove: $a + b \neq a$ and $a \neq b$. That only presupposes the following **Theorem:** *If b is finite, then every part of b is finite*. For otherwise I would be able to coordinate b with itself in such a way that ($b = a + \alpha$ assumed) a remains unchanged, but the units of α are reordered in such a way that now

α' does not coincide with α. Then b also would no longer coincide with itself. — If a and b are finite, so also is a + b. For I can arbitrarily reorder a, and b also; and likewise I can interchange each unit of b, one after the other, with each arbitrary unit of a. If
5 a is finite, so also is a' — i.e., the number which I obtain from a by putting an arbitrary unit in place of one of its units. (It is obviously the same number.) It then also follows that a + b = b + a. For both numbers are finite and coincide under one arrangement, and therefore under all. Or better: In both cases I have the same units
10 in differing order. Since a + b is finite, I then have "the same" number. And (a + b) + c = a + (b + c) (in the same way).

One then proceeds further: Since a + b is finite if a and b <403> are, then *every sum* is also finite. Now I form the number series: 1, 1 + 1 = 2, etc., and thus *the number series contains nothing but*
15 *finite numbers. It itself is infinite.*

For every number there is one in the number series that is equal to it, for otherwise in enumerating (in whatever arrangement) I would advance onward in the number series without end. Thus the number or a part thereof would come into univocal one-to-one
20 coordination with the number series itself, and would therefore be infinite, as the series is. So the number-series classifies the number domain of the finite numbers.

This is sufficient as a foundation. It is now clear that all further operations operate entirely with finite numbers and, when they are
25 possible, yield finite numbers.

In my chapter on number definitions by means of equivalences I have therefore been decisively mistaken.[11] I was there dealing with authentic number representations, to be sure, and for them such definitions are not required. But I overlooked the fact that
30 the inauthentic number representations require a principle of classification that is not given apriori, so that we cannot know apriori whether a classification is in general possible. Now the concept of equivalence serves admirably for that purpose. Certainly an important point is overlooked by arithmeticians most of the time:
35 Namely, that the distinction between finite and infinite numbers must show up at the outset of the investigations. Without it one

---

[11] Footnote: Cp. E. G. Husserl, *Philosophy of Arithmetic*, Chapter VI, pp. 101ff.

arrives at no classification. Schröder ⟨in his *Lehrbuch der Arithmetik und Algebra I*⟩, for example, who states all presuppositions in detail, omits that of finitude, which he designates as "tacit" only very much later (p. 20). Also, he omits to give a
5 definition of finiteness, which is absolutely essential, and to bring it into the correct systematic relationship with all the other definitions.

Nota: In the above definitions of continuity we speak of series bounded in both directions (series segments with two ends). That
10 is not adequate. The series must be closely compact at all points, i.e., for each member there is an immediately adjoining member (one that is no longer mediated through a member falling between them). To compare a definition <404> of the "natural series" that seems to me to move toward something similar, see Veronese
15 ⟨*Grundzüge der Geometrie von mehreren Dimensionen und mehreren Arten gradliniger Einheiten in elementarer Form entwickelt*, Leipzig 1894⟩ p. 15.

That two arbitrary numbers a̲ and b̲ can unite into one number that includes the units of both and them alone is Evident. But that
20 two finite numbers determine again a finite number, and indeed univocally so, is not Evident at all, but rather must be demonstrated.

⟨ VII. Addenda ⟩

⟨ 1. *Addendum to p. 369* ⟩

25 The *Theorem*, **An infinite number can in no case be equal to a finite one**, must be demonstrated.

If $N = N'$, and N is finite while N' is infinite, then $N' = \theta(N')$. Corresponding to the equations there must be determinate, univocal one-to-one correlations. Thus we have:

30 (1). $N \cong N'$ and $N' \cong \theta(N')$.

In the first congruence the units of $\theta(N')$ are one-to-one univocally coordinated to only some of the units of N, say to $\Theta(N)$, in such a way therefore that $\Theta(N) \cong \theta(N')$. But from (1) it follows that $N \cong \theta(N')$, thus also $N \cong \Theta(N)$, contrary to the hypothesis.

The following *theorems* have, moreover, to be demonstrated:

If $a = \theta(b)$ and $b = \theta(a)$, then also $a = b$.

If, thus, $a \neq b$, then, in the case where $a = \theta(b)$, $b \neq \theta(a)$. We therefore can then distinguish the case where $a > b$ from the case where $a < b$.

But is it true that, if $a \neq b$, then certainly either $a = \theta(b)$ or $b = \theta(a)$ must be so? That is highly doubtful. How am I to say whether the totality of possible squares in general is equal to or greater or smaller than the totality of possible lions? Where there is no general principle of univocal one-to-one coordination, there a comparison is also impossible. For comparison of number it does not suffice that I know for each number which units belong to it, and hence what is to be enumerated. I also must have a principle that makes a comparison apriori conceivable. We can therefore attach the distinction between greater and smaller only to the finite domain.

⟨ *2. Addendum to p. 377* ⟩

Theorem: *A finite number can in no case be equal to an infinite one.*

If $\underline{a}$ is finite and $b = a$, then $\underline{b}$ too is finite. If, therefore, $\underline{a}$ is a finite number, then the entirety of the numbers which are equal to $\underline{a}$ is a class of finite numbers. If $\underline{m}$ belongs to this class, then $\theta(m)$ cannot ⟨405⟩ belong to it, and likewise also not $m + n$, where $\underline{n}$ is a unit or a number.

<As *Proof:*> *If $N = N'$ and $N' = \theta(N')$, then $N = \theta(N')$.*

According to the first equation the coincidence is total; according to the second, partial. To every U_N there corresponds on the one hand a determinate $U_{N'}$ (univocally one-to-one), to every $U_{N'}$ a determinate $U_{\theta(N')}$ (ditto), thus to every U_N a determinate $U_{\theta(N')}$; on the other hand, all $U_{\theta(N')}$ together form one part of the $U(N')$. In the first coordination, to this, the $\theta(N')$, only some of the U_Z were coordinated, the $\Theta(N)$. But now all $U_{(N)}$ are coordinated to them. Therefore indirectly $N = \Theta(N)$.

More briefly: *$N = N'$, $N' = \theta(N')$, therefore $N = \theta(N')$.*

Now in the arrangement $N = N'$ in $\theta(N')$ only some of the units of N were coordinated, the $\Theta(N)$. Therefore N can be thought of as partitioned $\equiv \Theta(N) + R$, where $\Theta(N) = \theta(N')$.

Since, now, $N = \theta(N')$, then also $N = \Theta(N)$, which is impossible.

*Corollary: **If a number N is infinite and $\theta(N)$ is a part equal to it, then $\theta(N)$ also is infinite.*** (Every part of an infinite number that is equal to it is infinite.)

*Further Corollary: **If a is finite, then $a + b \neq a$***, for otherwise $a + b$ would be infinite, and since it would be equal to a, an infinite number would be equal to a finite one.

Theorem: *If N is finite, then every part of N is finite.*

According to the foregoing proposition, the converse also holds true. For every infinite number has infinite parts. Therefore, if a number has only finite parts, it also is finite.

⟨As proof of the theorem:⟩ Assume P to be such a part which is infinite. Then $P = \theta(P)$ and simultaneously $P \equiv \theta(P) + R$.

Further: $N \equiv P + U \equiv \theta(P) + R + U$
$= \theta(P) + U$

It would therefore be possible to coordinate N to only a part of its units (the R would be omitted).

Theorem: *If a and b are finite numbers, then $a + b$ also is finite.*

Theorem: *If a and b are finite numbers, then $a + b \neq \theta(a + b)$.*

If this were so, then the units of a (or b) would coincide with certain units Θ (or Λ), so that $\Theta + \Lambda = \theta(a + b)$, thus $a = \Theta$, $b = \Lambda$.

Then it is clear that Θ cannot contain exclusively units of a, or Λ exclusively units of b. For then Θ would have to contain *all* of the units of a (respectively, Λ all of the units of b). Otherwise a and b would not be finite. But if Θ were identical with a, then Λ would either be identical with b, which would contradict our hypothesis, or Λ would be identical with some part of b, which also would be impossible. Similarly one sees that Θ cannot be identical with b and Λ cannot be identical with a. It therefore only remains that either Θ and Λ contain a mixture of the units from both numbers or that Θ contains a part of the units <406> from b, but none of a, and for Λ the corresponding inverse situation. Let us now think of each unit of b in Θ as replaced by a corresponding unit of a. That is always possible. Then the units of a are:

$$U^{(a)}_\lambda \quad U^{(a)}_\mu \quad U^{(a)}_\nu \quad \text{(I)}$$
$$U^{(a)}_{\lambda'} \quad U^{(a)}_{\mu'} \quad U^{(a)}_{\nu'} \quad \text{(II)}$$

And if these latter are in part put into coordination with themselves – i.e., but in part with units from b – we then would need only modify series (II) through a valid internal transformation of a, and utilize the units which are not put into coordination with those of the series (I), in substitution for the units of b. Such units can never be lacking. For if a coincides with Θ, and thus with units of the groups $U^{(a)}$ and $U^{(b)}$ which are present in Θ, and if it also coincides with some internal transformation, then I would need only choose the transformation in such a way that the $U^{(a)}$ remain unchanged and only the units corresponding to the $U^{(b)}$ are imagined, in order to see that I can find the substitute in every case. I replace, then, each $U^{(b)}$ with its correspondent. Through this alteration of Θ, Θ' will emerge, which is equal to Θ. We proceed the same way with Λ. (See the following remark.)

In this way we obtain Θ + Λ = Θ' + Λ' = θ(a + b), where Θ' contains only a and Λ' only b, while simultaneously a = Θ' and b = Λ'. Thereby the impossibility is already demonstrated. What we said just previously finds application. If a = Θ' and b = Λ', then Θ' must contain all of a, and Λ' all of b. Their sum thus cannot be θ(a + b).

The demonstration is perhaps easier if one utilizes the (logical) corollary: If a is finite, it can be that a + b ≠ a.

⟨Remark.⟩ Θ thus consists of units from b, or in part of units from b and in part of units from a.

For example: $\Theta = \zeta(U^{(a)}_\lambda) + \eta(U^{(b)}_\mu) = a$

Since a = Θ, it is clear that the units of a are coordinated with these units. If $\zeta(U^{(a)}_\lambda)$ are coordinated with themselves, then the remaining units of a must correspond univocally one-to-one to $\eta(U^{(b)}_\mu)$. Or if any other internal transformation of a is coordinated with the two groups, then ζ' is to be coordinated with ζ (a part of the a thus internally transformed), and the remaining units of a with the η. In order to carry out the substitution I therefore need each time only to have recourse to the units coordinated.

Theorem: *a + b ≠ a, a + b ≠ b, if a is finite.*

What does this mean: a + b = a? That there is then a univocal coordination of the units of a by themselves with the units of a + b together. The units of a are then coordinated with a part of themselves.

⟨Proof:⟩ a + b ≠ θ(a + b)
          θ(a) + b
          a + θ(b)   θ(a) + θ(b)   <407>

If a + b = θ(a) + b, i.e., the units of a + b (1) can be brought into coincidence with those of (θ(a) + b) (2), then the units of b in (1) coincide with certain units Θ in (2). Thus, b = Θ. The leftover R = a. But I can also bring the same units of b into coincidence with the units of the constituent part of b in θ(a) + b. Thus I can coordinate b in the latter sum with the part Θ.

If a is finite, then a + b ≠ a. For if a + b = a were true, then a + b would be infinite, and thus an infinite number would be equal to a finite one.

ESSAY II [1]

⟨ ON THE CONCEPT OF THE OPERATION ⟩

⟨ I. Arithmetical Determinations of Number ⟩

There are various general forms of number determination that make it possible to determine new numbers by means of arbitrary numbers. In such concepts of determination are grounded all of the lawlike regularities that prevail in the domain of number, to investigate which is the goal of arithmetic.

Accordingly, not every determination of number is to be called an arithmetical one. With such a determination as "the number of flowers in this garden," arithmetic has nothing to do. Only determinations of numbers by means of numbers and numbers alone come into consideration for it. But, where the task is to discover the most general laws of the domain of numbers, it must also disregard every specific determination of numbers. In order to investigate what can be asserted apriori of numbers *as such* we must inquire: Under what forms, starting from arbitrary numbers, however they might be given, as species or concretely, can we determine new numbers? If we have found such forms, we then think of "any given" numbers as united by them, and then consider which general laws result from the concepts of these forms of construction. The determining numbers remain necessarily indeterminate in the course of these investigations. They are only – in totally inauthentic representation – thought of as in some way determinate or determinable. Therefore the number signs which are used here are not numerals, but rather are arbitrarily chosen letters. Each of these – if nothing else is stipulated – has the

---

[1] This corresponds to "V Abhandlung" in "*Husserliana XII*," pp. 408-429, dated by the editor {LE} as from 1891.

function of designating some number, identical within the same assertion or context of inquiry, but no further specified. If we call the numbers by means of which <409>, in an arithmetical determination of number, the conceptual representation of the new number is realized, the "members" of this "combination," then each of the $\underline{a}$, $\underline{b}$, $\underline{c}$, . . . serving as signs for the members of the combination would amount to "a certain number" – an expression which in its indeterminacy is suited for any number, so that in the domain of application each can be regarded as "this certain number."

All arithmetical determinations of number rest in the final analysis upon certain modes of activity that can be brought to bear, as upon totalities in general, so upon totalities of units, upon numbers. We recognized union [*Zusammenfügen*] and partition [*Teilen*] as such activities. The units of two or of arbitrarily many separate numbers can obviously be united into one number which includes within itself all of the units of those numbers and no others. (In a similar manner, numbers and isolated units, or isolated units alone, can be united.) Conversely, one can partition (any) number (at least into units, but often) into numbers (or into numbers and units). This too is recognizable apriori. For to any union of several numbers there corresponds a division of the resulting number into precisely those numbers. Union and division are, obviously, inverse operations.

As a result, then, there are the following possibilities for determining new numbers by means of given numbers – we limit ourselves at the outset to two – or, as it is also said, of "combining" given numbers into new numbers. We begin with union:

1) The two disjunct numbers $\underline{a}$ and $\underline{b}$ are given; they determine the concept of a number $\underline{c}$ in which the units of both, and they alone, are united.

2) But here $\underline{c}$ and $\underline{a}$ (or $\underline{b}$) can also be regarded as the determining numbers, and $\underline{b}$ (or $\underline{a}$) as the one determined. There arises the concept of a number which when unified with the disjunct number $\underline{a}$ (or $\underline{b}$) yields $\underline{c}$.

We now introduce the concept [*Gedanken*] of division:

1') The disjunct numbers a and b are given; they conceptually determine a number which can be (exclusively) partitioned [*zerstückt*] into these two numbers.

2') Similarly as with 2), there results here the concept of a number <410> which united with a (or b) presents us with the two disjunct parts into which c can be partitioned.

But with this not all conceivable problems or forms of determination are exhausted. In the case of union we until now thought of the units of the two composing numbers being unified without any favoring of one of the numbers over the other. Accordingly, to unite a and b means identically the same thing as to unite b and a. But the operation can also be intended in such a way that the two numbers are assigned a conceptually distinct status. Often we understand, namely, the uniting as an adjoining of the units of b to those of a, or conversely. To the form of determination 1) there then is joined a new one, 1"), in which the thought of adjoining (increasing) replaces that of uniting in the first sense. There result, certainly, two different determinations, depending on the status of a and b, but two that are distinct only by means of the particular content of the determining numbers, and thus present the same general type of determination. 1''') would be the problem of finding a number from which I can first sever a part a, and then a part b, with no remainder left over. Similarly with the form 2), alongside of which there now comes a new one 2"). To the adjoining (or increasing) corresponds the inverse thought of detaching (or diminishing). From it there arises a new form of the determinations belonging to the second group. We designate it by 2'''); namely, the concept of a number which proceeds from c in that we set aside the units of its partial number a ("take the units of a away from c") and reassemble the units then remaining into a number.

Further concepts of combination are obtained by means of some easily executed modifications and combinations of the ones previously defined; and in the first place by arbitrarily increasing the number of the members in combination without modifying the concept of combination in other respects, as if we, for example, were to define a number as the result of the union of arbitrarily many given numbers a, b, . . . , p. The combination retains in this case the character of a simple combination. Other modifications result from the fact that we also utilize in the determination the relations of equality and inequality, and, for example, specify that instead of the numbers a, b, c, . . . themselves, the combination is to be undertaken with numbers <411> that are *equal* to them. To this then is added the infinite manifold of composite combinations, the possibility of which is immediately evident: the results of arbitrarily defined combinations can serve as members of certain new combinations, whose results in turn . . . , and so forth.

Now is it actually required to distinguish all of these forms of combination from each other, or even to separately define them, in laying the foundation of arithmetic? As the reader knows, such a thing has never entered the minds of mathematicians. They are satisfied to take into consideration merely two from the manifold of elementary concepts of combination enumerated above, which they define as "addition" and "subtraction." Of the rest, no use is made. They are non-existent for the science. Certainly we will trust the scientific tact of the mathematician and can assume that this restriction can be logically justified. But there is a wide-ranging interest in attaining to clarity concerning this matter. Let us for this purpose return to the group of concepts of combination which first result (assuming combinations of two members) from the activities of uniting and dividing. They split, as is easily seen, into two groups, each of which consists of concepts that are immediately, or as well as immediately, equivalent with one another, and are for this reason designated by the same numbers. The determinations, "result of the union of a and b," "result of the attachment of a and b," "whole which is decomposable into a and b," refer collectively to the same number. Likewise for the determinations of the second group (2), 2'), 2"), 2''')). And to grasp this with Evidence does not first require proof. In view of the immediate equivalence of these concepts, the desired justification seems to be already there. The psychological distinctiveness of the concepts of combination in question is still no reason for their logical distinction. Indeed, for purposes of knowledge the psychologically distinct can, in general, be fully equivalent. That holds true, one may say, in every case of direct logical "equivalence" – and indeed just as well with concepts as with judgments or bits of knowledge. What logical benefits are we ever to reap from the separation of concepts for which it is immediately Evident that that which corresponds to the one <412> concept must also correspond to the other, and conversely (as the definition of the immediate equivalence of concepts states)?[2] For then what holds true in general of the objects of the one concept must reciprocally have validity for those of the other also. If the knowledge of the equivalence of the concepts is an immediate one, so also is the knowledge of the equivalence of the judgments with reference to them. It therefore requires, as we can say, no logical work to derive the one from the other. Precisely the same seems to be clear for immediately equivalent states of affairs (or for those to be so judged); i.e., those which, with respect to validity, mutually condition each other with immediate Evidence. In fact we often enough speak – where not psychological, but logical interests are decisive – of "the same" concepts and "the same" judgments, where merely equivalent ones are present. And it is, by the way, quite intelligible how out of this circumstance – given the habitual dominance of the logical (objective [*sachlich*]) way of deliberating, which outside of psychology we always and everywhere follow – a great danger arises for psychology: one which unfortunately all too often (especially in the psychology of knowledge) succumbs to the temptation to regard what is logically "the same" as also psychologically identical. No one considers it to be even a rational task to work out all of the transformations that convert the concepts and judgments under consideration into their immediately equivalent ones. Through such a project, so it is said, knowledge is not "expanded." It always remains fixed at the same point instead of progressing. Although this interpretation has some justification, it demands essential restrictions. The logical separation of equivalents, as well as the transformation into immediate equivalents, is, although certainly not always useful, still not always useless. And the former is not such because the latter is not either.

Here we do not need to engage in the controversy over whether immediate equivalences – in particular, transformations into immediate equivalents – considered in and for themselves must be

---

[2] Marginal note: The immediacy of the equivalence can certainly be disregarded.

regarded as extensions of knowledge or not. In any case they can serve as extremely important instruments for unquestionably extending our knowledge <413>, namely, for leading it out beyond the domain of immediate equivalents. This will prove true in all cases where a proposition – through immediate equivalent transformation – first receives that form which makes it appear as an appropriate premise for certain inferences. The possibility of an inference can be immediately evident or impose itself if the premises are of a certain form. But it can just as soon be remote if the one or other premise is subjected to an immediately equivalent transformation. And conversely: such a transformation can at once impose and bring to Evidence an inference which otherwise would be remote. The equivalent transformations that show up under the "immediate inferences" of the traditional logic can serve as examples. Their advantage for making syllogisms possible is well known. A similar role is played by the transformations of relationships into their correlatives for the class of non-symmetrical relations that are characterized by "transitivity," i.e., where two interlinked relations of the same sense ground in turn a relation of the same sense between the first and third term.

What holds true for the immediately equivalent transformations of judgments is similarly valid for those of concepts. In fact, the two are intimately enough connected. Such a transformation can supply a concept that vividly suggests and renders immediately obvious the applicability of a law that otherwise would have remained unnoticed or unnoticeable. In the knowledge so won the transformation of particular concepts can become beneficial in a similar manner, or the knowledge can take on a form which makes it combinable with other already available knowledge to yield a valuable inference, etc. It is clear that by such means we can far remove ourselves from the original starting point or domain of equivalence.

One must not here object that, given all of this, the obviousness which resides in the transition to the immediate equivalent makes superfluous a rigorous separation of concepts and judgments that are equivalent, unless the interests of logic (in contrast to knowledge interests of the subject domain to be cultivated – the expression "logical interest" is here annoyingly ambiguous) were guiding us and required a fully developed analysis of the mediating steps. For in the cases discussed <414> the equivalent transformations play a role so essential that they, for example, force the one demonstrating to explicitly mention the helpful equivalences, if he really intends to evoke persuasion based on insight.

Finally, we still have to take the following perspective into consideration. Wherever the sense-perceptible signs do not serve simply for the expression of thought, as in the case of language, wherever they, instead, in the form of algorithmic methods, acquire new functions in knowledge (which we will investigate in detail), there it can be in the interest of those functions that immediate equivalences be expressed as special laws and then be fixed in special formulas. Whoever has, at least through sample cases, gotten clear about the algorithmic methods, which run precisely parallel to the, for short, objective [*sachlichen*] methods (operating on the concepts themselves which are to be developed) will see that all immediate equivalences that intervene in the objective methods must have their counterpart in formulas that intervene in the algorithmic methods. Without an appropriate separation of equivalents it is not possible here to set up the algorithm.

From these general considerations there results for our case the practical application that the mere reference to the immediate equivalence of the operation concepts in question does not suffice to make their segregation appear as arithmetically worthless. That it nevertheless is of no value we recognize through the following line of thought, which we put in a sufficiently general form to have the decision for future analogous cases ready, on generally valid grounds (in the domain of arithmetic or other deductive disciplines).

If $C_1, C_2, \ldots, C_n$ are certain concepts of combination within a domain D, then the units of knowledge to be obtained exclusively on the basis of them can be separated into the following groups:

1) Knowledge units which are founded upon a single concept of combination. If there are (basic) propositions that are valid for one combination C with arbitrary terms – regardless of whether they are founded apriori in the concepts, or are accepted aposteriori on the basis of experiences [*Erfahrungen*], or, finally, are

accepted purely conventionally for objects formally conceived – the application of those propositions then leads to conceptions that are exclusively constructed from combinations of the form C, and to a limitless <415> manifold of units of knowledge. The members of a combination of the form C can, namely, again be combinations of that form, the members of these latter again, and so *in infinitum*. Every such composition offers multiple possibilities for the application of laws, depending on whether it is the main combination or the combinations occurring (individually or together) within the members that are subjected to the laws. We thus obtain infinitely many derived propositions, derived from a certain number of propositions which are not derivable from one another and which we call the axioms of this combination.

It is immediately clear that the entirety of the units of knowledge thus demarcated passes over into an equivalent entirety as soon as we start out, not from the concept C, but from the concept C' (immediately) equivalent to it. The units of knowledge on each side correspond to one another univocally one-to-one, in such a way that the corresponding units are (immediately) equivalent. Accordingly, it is superfluous to separately carry out the respective investigations for each equivalent concept of combination. It suffices to take a single one into consideration, perhaps C. By the mere conversion of C into C' the units of knowledge derivable from the axioms of C pass over into those derivable from the axioms of C'.

2) Units of knowledge which are simultaneously founded in several concepts of combination. Here there are the following possibilities:

a) The respective group of concepts of combination contains only those that are non-equivalent to each other. The combinations interweave in that the members of one combination can be themselves in turn combinations, and so on. However, in this case not all of these combinations are to be of the same form (stand under the same concept C) – otherwise we would have case 1) – but rather are to belong to different and non-equivalent forms. The infinite manifold of generally characterizable combinations founds an infinite manifold of truths by means of the laws, of which certain ones have the character of relationships that place the combinations, partly of the same and partly of different forms, into determinate relation with one another (or to particular terms), while the others have the character of forms of inference that permit us to derive new relationships out of ones given.

b) The group of concepts of combination concerned <416> contains only those that are equivalent to one another. According to the considerations advanced under 1), it is clear without further thought that every unit of knowledge belonging here is identical or equivalent to one that is derivable in the domain of a single combination – and indeed an arbitrary one – of this group (e.g., C). If therefore we limit ourselves to a single combination of this group and investigate the corresponding truths, then the interest of knowledge is in essentials satisfied. All that remains is merely a matter of interchanging the concept C with some or several equivalents C', C", ...

One doubt remains still possible. Could not the transition from C to C' in the middle of a deduction begun in the domain of C lead to results that would remain unattainable without such a transition? If this were so, then there would be truths which, concerning C alone, would nevertheless not be captured in the axioms respecting C alone, but rather would first be captured in the unification of the axioms proper to C and C'. The transition to C' in the middle of a C-deduction could, in fact, only extend our reach if we brought the laws valid for C' into play, in order only subsequently to turn back to the domain of the mere C. But clearly such truths are inconceivable. Any valid deduction that has recourse to the laws of C and C', of however many steps it may consist, must through the mere conversion of C' into C yield a valid deduction that confines itself exclusively within the domain of the C; and the two deductions are composed of nothing but equivalent steps and issue in equivalent results. This is a necessary consequence of the circumstance that to each axiom, and especially to each form of inference, in the domain C' there corresponds an equivalent one in the domain of C, and conversely.[3] There is thus no deduction in the domain C and C' to

---

[3] Marginal note: Moreover, there are general forms of inference that are valid for arbitrary figures. Naturally there then correspond again to each such inference, the figures of which

which there would not correspond an (immediately) equivalent correlate in the domain of C. In particular, we deduce the impossibility of C-knowledge units for the proof of which it would be necessary (or even only beneficial) to pass over into the equivalent domain.

c) The group concerned contains combinations which <417> are partially equivalent to one another and partially not. One becomes convinced that through restriction of the group to combinations that are not equivalent to one another there results no essential loss in units of knowledge. To each unit of knowledge from the broader group corresponds one from the narrower group that is either identical or (immediately) equivalent. Through considerations analogous to those under 1) it further results that no truth with respect to the narrower group can be provable only through means that presuppose a transition to equivalent combinations, and, simultaneously, that the requisitioning of these latter can promote no proof in any way.

These considerations make it Evident that the entirety of truths which are founded in all of the types of elementary combinations $C_1, C_2, \ldots, C_n$ of a domain D undergo no diminution worthy of attention if we restrict ourselves to the combinations not equivalent to one another, and thus retain one from each group of equivalent combinations. Through merely carrying out immediately equivalent transformations the missing truths can at any time be supplied, and so they lose any theoretical interest. But with them also the excluded types of combination, especially since we have proven that they also can accomplish nothing for the purposes of the deduction. But it follows that the application of equivalent types can contribute neither to (essentially) new truths nor to new methods, so it is not merely not beneficial, but rather is positively harmful. Only through concentration of our limited powers of knowledge upon the absolutely necessary can we victoriously penetrate into the kingdom of truth. We can therefore not afford to luxuriate in useless concepts and propositions.

---

we particularize by means of combinations of C, an equivalent one obtained by replacing C with C'.

We therefore establish the following rule, valid for all domains of deduction: that among the elementary types of combination, equivalents are not to be tolerated. Out of any entirety of equivalent, elementary types of combination, a single one is to be selected; all the remaining ones are to be ignored.

In the derivation of this rule we have had no regard for the algorithmic methods which play so large a role in the deductive sciences; and yet we have stressed above that equivalences to which one would, in virtue of their triviality, otherwise pay no attention (as for example $a + b = b + a$), play an important role in this method, and <418> therefore they too must be formally expressed. Our later investigations will show that any algorithm begins by coordinating rigorously parallel algorithmic correlates to the basic concepts, basic judgments and basic inferences of the domain. Namely, the indeterminately represented objects of the domain are replaced by simple signs, the combinations by composite signs produced from the combination signs corresponding to the different combination concepts, and the relations by signs for relations. Further, the axioms are replaced by sign conventions that state which symbolic modifications are permitted (provided a correct judgment runs parallel to them) and which are not. These sign conventions at the same time confer conventional meanings upon the signs, and thus to the original concepts are correlated, in univocal correspondence, algorithmic concepts. Now since for equivalent concepts of combination equivalent axioms hold true, so the corresponding algorithmic concepts likewise must be equivalent. In the domain of algorithmic concepts there holds, however, what we have shown for arbitrary domains: combinations that are equivalent are useless. This also remains in force, as we will later recognize, for any "valid extension" of the original algorithm, although the full parallelism with the material domain falls away; for the "extended" concepts of combination must be equivalent if the original ones were.

We learn from the foregoing considerations of what type the equivalences are that have an essential function in the deductive disciplines. Disregarding the general logical ones, they are those which occur in the axioms or follow from them. The one type concern the general relationships which are possible between the objects of the domain as such. We mean, of course, the equivalences of the correlative relationships, as, let us say, between a = b and b = a, or a > b and b < a. The other type concern equivalent transformations of such relationships by means of combinations which join their terms to new objects. The equivalence between a = b and a + c = b + c may serve as an example. Apart from equivalences between judgments, equivalences between concepts also are found among the axioms. We think here of course not of equivalences between elementary concepts of combination (<419> as between union and adjoining) – for such are excluded by our rule – but rather of equivalences that relate to determinate particularizations of these concepts, though not precisely "determinations" in the sense of traditional logic. Thus, for example, a + b and b + a are two determinate particularizations of the additive combination, and the axiom a + b = b + a is valid. Conceptual equivalences open up the possibility of discerning, through a sequence of equivalent transformations, equivalences of complicated determinations that are extremely remote and not immediately intelligible at all. The equivalences between judgments either function as forms of inference or supply general major premises for inferences to be carried out, and serve, disregarding the equation inference, in the advance of knowledge beyond any domain of equivalence.

We likewise add here the determination of the concept of the basic combination. A type of combination is said to be "*reducible* to the combination types $C_1, C_2, \ldots, C_k$," if among them, or among the composite types formed from them, there is one which is equivalent to C, if not identical to it (this can happen only in the latter case). If a type of combination C is equivalent to certain others C', C", ..., then of course each determinate combination of the sphere [*Bereiches*] C is equivalent to a determinate combination of the sphere (C', C", ...), and conversely. Hereby, for the sake of brevity, we understand by the *sphere* of certain types of combinations the entirety of the determinate combinations which fall under those types or arise by composition out of combinations belonging to them. A group of combination types is said to be *irreducible* if none of them is reducible to the others. An irreducible group to which all combination types of a domain are reducible is said to be a *group of basic combinations*. Accordingly, each determinate combination of the domain belongs to the sphere of this group. Since every type of composite combination is reducible to the types of the combinations that make it up, the basic combinations can only be elementary combinations. An elementary type of combination can be determined by two members, but a composite type of combination can only be determined by three or more members. Hence the equivalence between an elementary and a composite type of combination is an impossibility. We can therefore say: a group of basic <420> combinations is a group of elementary and non-equivalent combinations to which all types of combination of the domain are reducible – in other words, in the sphere of which all combinations of the domain fall, whether immediately or through the mere instrumentality of equivalent transformations. That all groups of basic combinations are of the same number and can be coordinated univocally one-to-one, and that the ones so coordinated are equivalent, is easily comprehended.

⟨ II. Combinations (or Operations) ⟩

⟨ 1. *Division* ⟩

⟨ Division into: ⟩
1) Simple and composite,
2) Symmetrical and Non-symmetrical.
(For additional divisions in terms of number of members and types of combination see below. ⟨3⟩ ff⟩)
We then distinguish:
1) The Type [*Typus*] of a combination (of a combining thought). The Type of simple combinations is called the *combination kind*: a) the material [*sachliche*] or inner Type, b) the formal or exterior, pure Type.
2) The Mode [*Modus*] of a Type. Two combinations of equal Type, built up from the same members, we distinguish according to the position of the members within the Type. And therewith

again: a) Mode of the combination within a formal Type, b) Mode of combination within an objective Type.

3) We further distinguish (we classify) combinations of 2, 3, . . . members.

4) We classify them according to the kinds of combinations that are involved: e.g., combinations that are exclusively additive, multiplicative, etc.: a) Modes within one and the same operation, b) mixed Modes involving different operations.

Then we come to the mixed Modes. These lead to divisions according to Types (inventory of all possible Types for 2, 3, . . . members) and to sub-divisions according to the number of different Modes involved.

5) Reducible and non-reducible basic combinations.

One will begin with the simple combinations and their Modes <421> and then move forward in the degree of compositeness.

1) Here the first question to present itself will be: Does commutation apply or not? Is the simple combination in question a symmetrical one?

2) The next question will be: What effect have modifications of Mode, perhaps also modifications of Type, and in the first place for combinations composed out of the same kind? In case the simple combination is commutative, is the continued conjoining of members by means of the same combination also commutative, ((a o b) o c) . . . ?

Further: Does the associative law hold? All of this in the first place for combinations of three members.

3) We consider other composite combinations, mixed ones arising from combinations of various kinds. First, again, the three-membered ones, then those of several members.

4) We ask what results from the invoking of the same operations for two, three, . . . members. One can raise these questions for two-member combinations already, according to 1).

1) Laws of relation: If $a = b$, then $b = a$, etc.

2) Laws which proceed from the fact that for equality or other relations (relationships) both sides are subjected to the same operations.

3) Laws of equivalent transformations of combinations, e.g., laws of commutation, association.

4) Further laws for Modes of combination.

As with the forms of operations, the "combinations," so also with propositions we must introduce concepts of material [*materialer*] Type and of formal Type.

Every proposition expresses a relationship in the members of which these or those combinations, forms of operation, occur. Now, if two relationships are constructed in the same manner from operational forms of equivalent Type [*äquitypen*], then they themselves are said to be of equivalent Type: e.g., $a \cdot b = b \cdot a$ and $a + b = b + a$ have the same formal Type.

$$a + (b + c) = (a + b) + c$$
$$b + (a + c) = (b + a) + c$$

If the letters are signs for arbitrary objects of the domain, then we are dealing with two expressions for one and the same proposition. <422> But if the letters are signs for certain numbers, then we can say: The two propositions fall under the "form":

$$m + (n + q) = (m + n) + q.$$

In this way we generalize. We try to collect various special cases together in the one form. The generalization proceeds yet further when we consider the formal Types:

$$m \; o \; (n \; o \; q) = (m \; o \; n) \; o \; q.$$

Likewise for laws of combination from various domains. The general expression is: "Form of the proposition." The same can be said of groups of propositions, algorithms and systems of deduction.

All of this is closely connected with the method of expansion or generalization of propositions, deductions, algorithms.

⟨ 2. *On the Concept of Combination* [*Verknüpfung*] ⟩

Wherever a conceptual determination is present which determines one object by means of other objects, we speak of a *combination* (synthesis) of the latter objects with the former, the "result of the combination." Thus, for example, we call any kind of conceptual arrival at one number from two or more numbers a combination (additive, multiplicative, etc.) of those numbers.

The objects can be *given*, or they can be *determinate*, i.e., thought merely as objects of certain concepts. If the objects combined are all determined through the same concept, then we have the case where several objects of one class, as such, conceptually determine an object of the same class. If the objects are not all determined through the same concept, then it can happen that some object determined as a and some object determined as b, both as such, conceptually determine a new object as a c. Then not merely is the resulting object determined through the composing objects, but rather at the same time the conceptual determination of the first is determined through that of the latter. Of course the resulting determination can determine its object equivocally, perhaps infinitely equivocally. It is only in such cases that we speak of *the combination of conceptual determinations*. <423>

1) Simple and composite ⟨combinations⟩.

2) The combinations divide into *non-symmetrical* and *symmetrical*, depending on whether they combine their members with or without favoring any of them. In the former case certain members or all of them acquire differences of status in the context of the combination, such that through exchange of members the thought constituting this combination undergoes a modification, whereas with the symmetrical combinations this is not the case for any exchange of the members.

Differences of status arise in particular through composition, thus through the fact that results of combinations serve as members for new combinations, and so on. It is clear that, if we attend to the conceptual unity of the ultimate terms of the web of combinations, these are not collectively combined in a symmetrical manner. All composite combinations are therefore non-symmetrical: a proposition which, of course, cannot be reversed.

One could be tempted here to speak of the "combination" of concepts. If a, b, c are all different, then certainly one concept appears to be determined through the two others. However, a and b also can also be representative of the same concept, and one cannot speak of the "combination" of identical [*gleich*] concepts. Concepts are identical or they are different. On the other hand, two determinations can be equal.

If in a non-symmetrical combination we interchange the members occupying different positions, then the "mode of combination" of the members changes, and thus, in a certain manner, the concept of that combination as well. Nevertheless, the general concept of combination – the Type and the kind [*Art*] of the combination – remains unaltered. Let us consider, for example, the combination a + b. In b + a these same numbers are combined in a different way, but the Type is the same: the addition of one number to another. We retain, generally speaking, the *Type* of the combination when we, disregarding the special determinations of the members, consider each one only as an object [*Objekt*] in general of the respective domain. For each one specifically determinate as a, b, . . . we therefore posit in the number domain each time "a number," or in the domain of segments "a segment," etc., retaining the combining thought in all other respects. Accordingly, (a + b) + c and (a + c) + b <424> are of the same Type, and both are, further, of the same Type as (m + n) + p. Similarly for (a + b) + (c + d) and (m + n) + (p + q), etc. The same members can, as we said above, be combined within the same Type "in a different Mode [*Weise*]." We compare, for example, a + b and b + a, $\frac{a}{b}$ and $\frac{b}{a}$, a · (b · c) and c · (b · a), and so on. Here just as much attention is paid to the peculiar character of the members as is necessary in order to be able to distinguish the position of each in the context of combination and to confirm possible alterations in them. For that, it suffices to regard each particular member merely as "a certain one"; only provision must be made, by means of distinctive designations, for the identifiability of these "certain" members.

If we imagine the special characteristics of the particular members in some combination being suppressed to the point where each member is to be regarded only as "a certain one" – the one as a certain a, the other as a certain b, etc., and this in such a way that the same letter always indicates the same determination – then the conceptual representation thus originating is called the Mode of combination appertaining to this combination before us. Consequently if, for instance, a, b, c signify some numbers or other, then (a – b) – c is the common Mode for all of the specific combinations contained within this expression. In the Mode each member is thought merely with the determination that it is endowed with a "certain" determination, and indeed the one with a certain determination a, the other with a certain determination b, etc. With these determinations being designated by letters, and consequently being individually fixed before our thought, it is indicated that the contents of those determinations are to matter for our thought ⟨only⟩ insofar as they are different ones. We therefore will always, within the same thought process, have to designate the same determinations or same determinate objects by means of the same letters; whereas from the difference of the designations alone, just because of the possible equivalence of the determinations, the conceptual difference of objects certainly cannot be inferred. Also, since any letter *in concreto* can mean any determination, the possibility must remain open that different members of a Mode also are interpreted by means of the same determinations. Since a + b and b + a are different in Mode, this also holds true specifically of 7 + 5 and 5 + 7. Indeed it could seem <425> that both belong under the same Mode b + a: however, once we have understood under a, 7 and under b, 5, it is then necessary to write the second sum as b + a because of the significations of a and b, which are to be held constant.

All laws that apply to combinations in general concern modally determinate combinations, e.g., a + b = b + a, or (a + b) + c = a + (b + c).

We call the entirety of combinations that arise from the combination kinds $C_1 \ldots, C_n$ through specialization and composition the *sphere* of those combination kinds.

A combination kind is said to be *reducible* (in relation) to the totality of the combination kinds $C_1, C_2, \ldots, C_n$, or *reducible* to the combinations $C_1, \ldots, C_n$, if to each combination of its domain there is an equivalent property [*Merkmal*] in the domain of $C_1, \ldots, C_n$.

A group of combination kinds is said to be irreducible, if none of its combinations is reducible to the totality of the remaining ones. *An irreducible group* to which *all combination kinds of a domain* are *reducible* is said to be a *group of basic combinations* of this domain.

All groups of the domain are reducible to one another and all consist exclusively of kinds of elementary combinations; for each kind of composite combinations is *eo ipso* reducible to the kinds of the elementary combinations which it contains. The number of the members is the same in each equivalent group.

We still have, finally, to discuss the concept of the *combination kind*. If we direct our attention entirely away from the members and attend exclusively to the thought that brings about their synthesis, then we obtain the combination kind. We speak, for example, of addition, subtraction, etc., as different kinds of number combinations; and whenever in so doing we make clear to ourselves the concept of addition *in concreto*, we obviously pay no attention at all to the individual members of the particular addition that serves as our basis. Therefore, for the concept of the combination kind the number of the members is non-essential; whereas, in contrast to that, the Type is immediately modified with any modification of the number of the members. <426>

In every combination we distinguish the members of the combination and the combining thought. This latter remains unchanged under arbitrary variation of the members, and immediately stands out in its purity and universality when we abstract from the members to the highest degree possible.

We distinguish as many *essentially distinct combination kinds* as there are combination kinds with no equivalent. One recognizes the equivalence by the fact that to each determinate combination of the one kind there corresponds an equivalent combination of similar [*gleicher*] form for the other kind, and conversely.

We distinguish, further, kinds of elementary combinations and kinds of derived combinations. Every composite combination is a specialization of an elementary one; the members of the former are determined as results of combination. If the kinds of the elementary combinations are distinct, then the establishment of the derived combinations is a merely combinatory matter. The entire totality of elementary combination kinds of a domain that have no equivalent we call a "totality of basic combinations."

By the *formal Type* of a combination we understand the concept proceeding out of its Type through the following abstraction: With the members we also abstract from the fact that they are objects of domain D, thus retaining each one of them merely as an object [*Objekt*] in general. On the other hand, we also abstract from the specific nature [*Artung*] of the elementary combinations constituting the combination as a whole – from combinations of the same kind merely retaining the fact that they are, in general, combinations of the same kind, and from combinations of different kinds retaining that they are, in general, combinations of different kinds. In contrast to the *formal* Type we call the Type pure and simple the *material* [*sachlichen*] Type.

In the same way we place alongside the objective *Mode* the formal one. It proceeds from the former in virtue of our taking the letters as signs for "objects in general," but bringing to bear upon the combinations ⟨of signs⟩ the same abstractions as are requisite for the formation of the formal Type. The formal Mode thus proceeds from the formal Type through the same determinations as the objective Mode does from the objective Type. In order, for example, to determine the formal Mode of (7 + 5) − 8, we replace the number signs with Latin letters, as signs for ⟨427⟩ any objects whatever. The sign "+" is replaced by ρ, the sign of "a certain" combination, the sign "−" by ρ', which by the index indicates that a certain *other* combination is meant. And thus we obtain:

(a ρ b) ρ' c.

If we switch the letters for the members and form, perhaps, (b ρ a) ρ' c, then we obtain another Mode, but the formal Type is the same.

When one speaks of the *form* of a combination, one means sometimes the formal Type and sometimes the formal Mode, but usually the latter, as, on the whole, the consideration of the formal Modes is not less important than the consideration of the material [*sachlichen*] ones, whereas the Types can be of interest only for logical deliberations.

⟨ III. Addendum ⟩

On the Concept of Basic Operation

It is questionable whether at least two objects belong to the basic operation. In the logical calculus we have negation. Certainly one can also take another view of this matter. To every class there corresponds a negative: There is to every class a another class which is opposed to it: a_1; and therefore also to a_1 another, $(a_1)_1$, etc., and thus not the former one. We can interpret the situation as involving a relation: $a\phi b$. Then b is to be a_1; a_1 unambiguously determinate. $a\phi b\phi c$; $aNbNc$; $b = c_1$; $a = b_1$; $a = b_1 = c_{11}$. Segment a; $-a$; $-(-a)$: operation of inversion of a segment.

1) An operation is a way of deriving new numbers from one or several numbers; or, from one object: negation, inversion, coincidence.

2) An operation is either unambiguous or ambiguous. But it cannot mean just anything [*alldeutig sein*], for otherwise it determines nothing.

3) What distinguishes an operational formation of number by means of a relative determination, such as $x > b$? We could in fact also speak of an operation of magnifying and say: Every number can be magnified. We could even state propositions for it: A magnification of a magnification is, again, a magnification: $>> b => b$. b is enlarged if I assume a number x which is $> b$. But if $x > b$, then (or rather: that is to say nothing <428> other than) $x = b + u$. I enlarge b by adding some number or other to it. So if I have defined addition, I need no operation of enlargement. The distinction consists only in this: that with addition I have in

mind the formation of a number from two numbers, whereas with enlargement my only interest is that any number in general, it is all the same to me which, is added to b. In any case, I can derive all of what I would accomplish through the operation of enlargement by making use of the equation, or rather I must make use of it here. I thus obtain the laws for inequalities.

The operation of magnification is therefore the same as the operation of the adjoining of a number, to be arbitrarily selected, to the given number. But why is x ≠ a no operation? Can I not speak of the *operation* that leads me from a class to a subordinate class? That no doubt lies in indeterminacy. We therefore must distinguish the basic *relations* from the basic operations.

4) In the concept of the operation resides something of the production of an object. Some sort of activity directs itself upon the given object and produces a new object or directs itself upon given objects.

However, the representation of production is not so essential. The main thing is: Operation is a manner of conceptual transformation of the given whereby something novel originates, but something such that owing to the transformation I can also regard it as given. When can I do that? I must have the Evidence that it exists, either always or under certain conditions. In the latter case I operate precisely only when this requirement is fulfilled.

Is that perhaps necessary? If the objects are actually [*wirklich*] given to me, then the conceptual determination must also provide me with the novel object itself. I must then be able to produce it. But this also is not necessary for a deductive system. It suffices that my knowledge interest is grounded through such forms of construction, so that I am able to regard what is so determined as given.

Addition of numbers: I have the Evidence that the sum exists. An actual arriving at it *in concreto* I cannot require, for that surpasses our capabilities. Only this can I require: that the name of the respective natural number can be derived, that it thus can be defined in Idea by recursion. And that too is only possible within limits. But sufficiently for my theoretical needs. I therefore have at my disposal a method for "evaluating" the conceptual determination. (But the evaluation for systematic numbers rests already upon general arithmetic.)

One can perhaps say: Certain operations provide me with numbers which I can regard as given, because these conceptual determinations are required by my knowledge interests, e.g., because they describe to me a way of actually arriving *in concreto* at the corresponding objects from those given – or at least arriving at them within the range of practical <429> cases, whereas this is not the case with other operations. Consequently what is or is not a basic operation will depend upon the knowledge interest that is guiding us. But we here restrict the concept 1) to forms of determination that are non-equivalent, 2) to those which are not mere combinations of other forms of determination of the same kind.

The propositions $a \cdot b = b \cdot a$ are consequences of the additive laws. Nevertheless, $a \cdot b$ stands as a new operation. They are certainly not formal consequences in the narrower sense. They of course cannot be that because the sign "$a \cdot b$" does not occur among the formal laws of addition. The solution to an equation of second degree is a purely formal consequence of the laws of operation. This, as it seems, is not true for *Pell*'s equation. The general law of association is no purely formal consequence, inasmuch as, in general, the inference from $\underline{n}$ to $n + 1$ is employed, unless I adjoin this latter as a premise.

Why, then, consider multiplication as a new operation? The distinction lies in the fact that, in order to prove the laws of multiplication, we must go back to the natural number or cardinal as a sum or equality [*Gleichheit*] of units, whereas this is not necessary with the proof of the generalized law of association. All deductions which do not go back to the concept of the domain are formal. We will therefore distinguish as many basic operations as we have non-equivalent concepts of combination the laws of which cannot be derived purely formally from one another. Further, the following is to be considered: A determination is a formal consequence of certain presupposed ones if it can be formed from them without ever having to have recourse to the nature of the domain. In this sense $a + a$, $(a + a) + a$, etc., are formal, but not $a + a + \ldots$ $\underline{b}$-times. For this determination loses its sense if I do not think of the fact that b̲ is a number. The same holds for raising to a power. And because this is so, the propositions about the new operations also cannot be discovered from those about the old ones without recourse to the number concepts. Of course this does not mean that, where a determination rests upon mere composition, the laws which it falls under necessarily have to be formal consequences of the laws for determinations that are not composite. Consider $(a - b) + b = a$. But then such new laws, which are grounded in the concept of number, are conceived as laws for the combining of the two operations. We will therefore distinguish as many basic operations as we have forms of determination that are non-equivalent and formally independent from one another. Also formal is any determination by means of an equation, by means of a relation or a system of relations in general.

ESSAY III[1]

⟨ DOUBLE LECTURE:

ON THE TRANSITION THROUGH THE IMPOSSIBLE
("IMAGINARY") AND THE COMPLETENESS OF
AN AXIOM SYSTEM ⟩

⟨ I. For a Lecture before the Mathematical
Society of Göttingen 1901 ⟩

⟨ 1. Introduction ⟩

<K I 26/76>[2] The theme which I wish to deal with in this lecture concerns a fundamental question of mathematical method, and belongs as such to that difficult domain in which mathematicians and philosophers to the same degree, even if not entirely in the same sense, have an interest. The mathematician is the theoretician of deduction. Originally limited to the domain of {91}[3] number and quantity, mathematics has grown far beyond that domain. It has increasingly approximated to the goal that *Leibniz* had already clearly conceived, namely, the goal of being

---

[1] This largely corresponds to the "VIth Abhandlung" in "*Husserliana XII*," pp. 430-451, dated by the editor {LE} as from 1901. However, *Elizabeth Schuhmann* and *Karl Schuhmann* have recently re-edited much of the material related to the two lectures from the documents in the Husserl Archives, and have published a part of their results in *Husserl Studies* 17:87-123, 2001. Their edition includes much previously unpublished material and many revisions of what was published in *Husserliana XII*. They have graciously permitted me to utilize their new text, which runs up to p. 453 of this translation. I have also incorporated various important changes indicated to me by them for the remaining selections from *Husserliana XII* to the end of this present volume, most of which selections remain close in form and content to texts in *Husserliana XII*. {DW}

[2] Symbols inserted into the text in this manner refer to manuscripts as classified in the Husserl Archives.

[3] Numbers in braces refer to pages of the *Husserl Studies* edition.

a pure theory of theory, free from all special domains of knowledge and, insofar, formal. Mathematics in the highest and most inclusive sense is the science of theoretical systems in general, in abstraction from that which is theorized in the given theories of the various sciences. If for some given theory, for some given deductive system, we abstract from its matter, from the particular species of objects whose theoretical mastery it has in view, and if we substitute for the materially determinate representations of objects the merely formal ones – thus, the representation of objects in general – which are mastered through such a theory, through a theory of this form, then we have carried out a generalization that grasps the given theory as a mere singular case of a class of theories, or rather of a form of theories, <431> which we grasp in a unified way and in virtue of which we then can say that all these particular scientific domains have, in form, the same theory. A systematically elaborated theory in this sense is defined by a totality of formal axioms, i.e., by a limited number of purely formal basic propositions, mutually consistent and independent of one another. Systematic deduction supplies in a purely logical manner, i.e., purely according to the principle of contradiction, the dependent propositions, and therewith the entire totality of propositions that belong to the theory defined. But the object domain is defined through the axioms in the sense that it is delimited as a certain sphere of objects in general, irrespective of whether real or Ideal, for which basic propositions of such and such forms hold true. An object domain thus defined we call a determinate, but formally defined, manifold.

The theory forms defined by such abstraction can, then, be set into relation to one another; they can be systematically classified; one can broaden or narrow such forms; one can bring a certain previously given theory form into systematic interconnection with other forms of determinately defined classes and draw important conclusions concerning their interrelationship. A well known example makes this clear: *Euclid*ean geometry is a concrete theory which when formalized yields the form of theory that we designate as a theory of the three-dimensional *Euclid*ean manifold; and this in turn is only one particular case from the systematically interconnected class of manifolds of varying degrees of curvature.

Mathematics is thus, in terms of its highest-level Idea, a theory of theories, the most general science of possible deductive systems in general.

With this generalization, through which the spheres of the old bearers of mathematical investigation – the cardinal numbers, ordinal numbers, the scalar and vectoral magnitudes and the like – are entirely surpassed, there are now connected unsolved methodological problems. The mathematical theory of theories, which {92} we commonly designate as formal arithmetic or formal mathematics, has precisely *not* arisen out of a purely theoretical interest, out of interest in the lawlike order of the forms of theories, and its aim is also not merely to serve that interest. Formal mathematics <432> aims to be the instrument of concretely mathematical discoveries. As the old mathematics of quantity was the great instrument of natural scientific investigation – namely, the instrument of deductive theorizing for the various domains of physical knowledge – for which induction had supplied the appropriate first principles, so the new formal mathematics intends to accomplish not only the same, but rather very much more. It aims to create a method of incomparably greater generality and power that renders superfluous all methodological works of a substantive [*real*] mathematical variety.

But the difficulties lie precisely in the relationship between formal mathematics and its employment in substantive mathematics or in the particular domains of knowledge.

The problems associated with that relationship are of interest to the philosophers because they are problems upon whose solution depends the understanding of the general essence of the deductive sciences and of theories in general. Certainly these problems are of only limited interest to the mathematicians, namely, so far as the certitude of their methodological work depends on them. However, the development of the science shows again and again that obscurity on questions of principle finally one day takes revenge – that, once certain levels of progress are attained, further progress is obstructed by errors that arose from obscure methodological Ideas.

The problem of imaginary numbers arose within the form of pure mathematics that was historically first, within arithmetic, and particularly in the form of arithmetical algebra. The tendency toward formalization built into algebraic calculation led to forms of operation that were arithmetically meaningless, but which manifested the remarkable character that they nevertheless could be assimilated into calculation. It turned out, namely, that if the calculation was mechanically executed according to the rules of operation, as if everything were meaningful, then, at least in the broad range of cases, every result of calculation free of the imaginaries could be claimed as correct, as one could empirically establish by means of direct verification.

(Here I of course take the term "imaginary" in the widest possible sense, according to which also the negative, indeed even the fraction, the irrational <433> number, and so forth, can be regarded as imaginary. Historically, only the imaginary in the sense of the negative and the lateral number [*Lateralzahl*] has given offense.)

This situation, then, led to the habit – becoming more and more widespread – of freely operating with the imaginary. The errors and obscurities arising therefrom, like the question about the sense of the logic of negative numbers, gradually led to the construction [*Konstitution*] of rigorous rules of calculation for the imaginary forms, {93} rules which made it possible to operate with indubitable certainty and correctness. But the problem of the imaginary itself remained unsolved thereby. We can formulate it thus:

Problem: Suppose a domain of objects given in which, through the peculiar nature of the objects, forms of combination and relationship are determined that are expressed in a certain axiom system A. On the basis of this system, and thus on the basis of the particular nature of the objects, certain forms of combination have no signification for reality [*reale Bedeutung*], i.e., they are absurd forms of combination. With what justification can the absurd be assimilated into calculation – with what justification, therefore, can the absurd be utilized in deductive thinking – as if it were meaningful? How is it to be explained that one can operate with the absurd according to rules, and that, if the absurd is then eliminated from the propositions, the propositions obtained are correct?

It seems that in this case no clarification resolving the matter proceeds from general logical considerations. Logicians have repeatedly insisted upon the rule of operating with clear, rigorous, consistently defined concepts. They have emphasized clarity and rigor of concepts mainly because in that way unnoticed contradictions would be avoided. Contradiction would be banished. Only in one form, the logicians have said, can contradiction play a role in thinking: namely, where it is a matter of exhibiting the contradiction in order thereby to demonstrate that the concept is without an object. In indirect proof one operates with contradictory concepts: in full awareness. But in that case all is clear. From the self-contradictory concept we draw consequences that are self-contradictory, and we infer from this that the concept is, precisely, self-contradictory, that to it an object will not correspond. But in the mathematics of the imaginary this obviously is totally beside the point. <434>

***

⟨ 2. Theories Concerning the Imaginary ⟩

<K I 26/78> First theory: ⟨The imaginary is vindicated⟩ empirically, through induction (*Bain, Baumann*).

2) ⟨The imaginary is⟩ directly Evident apriori (*Boole*).

3) Third theory: Fractions, negative numbers etc., are indeed impossible numbers from the viewpoint of the cardinal numbers, originally the only ones defined. They are concepts to which no object can correspond. But who forces us to stay within the restricted number domain? Numbers, after all, are mere creations of our mind through the act of counting. If we create new numbers, we widen the domain of numbers, and indeed do so in such a way that then all inverse operations will be executable and that simultaneously the new numbers comply with the rules of the old to the furthest possible extent. "Precisely this narrow limitation on the executability of the indirect operations has become in each case the true cause of a novel act of creation; thus are the negative and fractional numbers created by the human

mind." (*Dedekind*)[4] If we now ask {94} in what way the human mind brings this creation about, one responds thus: by means of definition. The equation a + x = c cannot be satisfied by a cardinal number in case c ≺ a. The difficulty is simple to eliminate: We define a new number that does so, and we understand by c − a the number thus defined, for which we have only to convince ourselves that the laws of operation for this number, which are carried over from the numbers defined as primordially valid and possible, can yield no contradiction in the total system of the operations.

Refutation. This theory is, as here set forth, simply unintelligible. If we understand by *number* the answer to the question "How many?" then the number series is the closed manifold of particularizations that are possible in the sphere of the concept *how many*. Now I certainly can give various definitions on the basis of the operations which are grounded in the Idea of the cardinal number. But certain results of operations are contradictory to the Idea of <435> "how many"; and if I define these, then I have defined, precisely, contradictory numbers. The sphere of the concept of cardinal number I cannot, without absurdity, arbitrarily expand on the basis of creative definitions, for it is this concept, indeed, which imposes limits on me. But if I nevertheless so define, then I define absurd concepts, and if these are supposed to be admitted into calculating, if it is to be permitted to operate with them in thought in spite of their absurdity, then the justification of the operations must be established. It is incomprehensible how one can claim that the difficulty is in some way eliminated by means of arbitrary definition. The definition is an arbitrary stipulation of the signification of a word. In this we are certainly unrestricted. But once a word − e.g., the word "number" − is confined to a given domain of objects, one that clearly presents itself as possible, then I cannot decree through some sort of arbitrary stipulation that the domain shall admit of an expansion by means of new objects. It would be as if in geometry one would

---

[4] Compare R. Dedekind, *Stetigkeit und irrational Zahlen*, Braunschweig, 1872, p. 6. *Husserl* writes in the Ms. "inverse operations" instead of "indirect operations." {LE}

decree: There are to be round squares, if not in the plane, then in a higher dimension of space.

What lies back of this theory, and confers plausibility on it among mathematicians, is a certain conceptual displacement. That we cannot arbitrarily expand the concept of number is certain. But we no doubt can *abandon* the concept of number and, by means of the formal system of the definitions and operations that are valid for cardinal numbers, define a novel, purely formal concept, that of the positive whole numbers. And this formal concept of the positive numbers can, just as it itself is delimited by definition, be expanded by new definitions, and indeed in a manner free of contradiction.

***

<K I 26/79> Fourth theory: Arithmetic is constructed in the following manner: one begins with the explication of the concept of number and that of the series of {95} numbers, defines for this domain the operations of calculation, establishes the rules of calculation, and considers the well known restrictions. To explain away the latter one has recourse to the real applications of arithmetic. One says: These restrictions are inessential. In life and in science the units of enumeration are magnitudes: of time, of force, of line segments, etc. Now it lies in the nature of the various concepts of magnitude that the operations for them can be carried out with complete generality. <436> And thus in the domain of magnitudes divisible into equal parts the fractions acquire a real sense, in the domain of continuous magnitudes the irrational numbers do, and in the domain of line segments the negative and imaginary numbers do. Thus, one says, the impossibility of the forms of number in question is only an apparent one, and one is fully justified in operating with them.

Those who follow other justifying lines of reasoning that invoke the creative power of definition or the notions of formal arithmetic also like to invoke the real applications, as if thereby a decisive proof were supplied that the 'imaginariness' is only an appearance.

Refutation: This theory rests, in my opinion, upon a deficient analysis of the relationship of the various kinds of real numbers and arithmetics to each other, and indeed the formal arithmetics. If we define number as cardinal number, as for example *Weierstrass*, *Cantor* and others do, then in this sphere there are no fractions, no negative numbers, etc., irrespective of whether or not we take as the enumerated units magnitudes that are divisible or indivisible, continuous or discontinuous, imbued with direction and sense in the manner of sequences and segments. A closer examination would show that the fractions do not obtain a real sense owing to the fact that one retains the cardinal number concept and presumes the units to be divisible, but rather owing to the fact that one in general abandons the domain of cardinal numbers and lays as basis a novel concept, that of the divisible magnitude. This leads to an operation system which partially coincides with that of the cardinal numbers, but which in part is broader, i.e., includes additional basic elements and axioms.

And so one then changes, with each kind of magnitude, the arithmetic as well. The various arithmetics do not have parts in common; rather they have wholly different spheres, but an analogous structure; they have partially the same forms of operation, although different concepts of operation. This explains why it is absurd to say that the possibility of the Representation of the imaginary as real in geometry proves that the imaginary is not imaginary at all. In the first place, it is totally false that the imaginary is ever in any way Represented in cardinal arithmetic, nor is it Represented in geometry. The imaginary of arithmetic in the sense of the cardinal number means: a cardinal number, ⟨437⟩ e.g., which would be the ⟨square⟩ root of -1. But what one Represents is the concept of a line segment that has undergone such a rotation relative to a reference segment, that if I apply the same rotation to it I would arrive at the segment that has been rotated 180° relative to the originally rotated segment. {96}

The number 7 - 9 cannot be Represented. What is Represented is a certain operation, symbolized by the same signs in a simple sequence.

What is confusing here is the sameness of the designation for concepts that are intrinsically different, and this sameness of the designation rests in turn upon the sameness of the forms of operation, which alone matter in the deduction. But forms of operation that are meaningful in one conceptual domain are contradictory in another. And from the fact that they are meaningful in the one, one cannot conclude that the contradiction in the other is mere appearance.

The tendency to formalization is inherent in arithmetic, and that is what lies behind such obscure theories. If one rises to the level of the pure system of operations, if one abandons the original, real-object domain – whether it is one of line segments or cardinal numbers or forces and progressions [*Grössenlehren*] – and focuses in the most extreme generality upon *a domain in general*, which is defined through such forms of operation, <K I 26/80> then one can modify in various ways the Idea of such a domain, now in the sense of a broader, now in that of a narrower system of operations or system of axioms. And in this generality, which indeed is restricted by no limitations of given concepts, there is of course no absurdity, unless one has furnished the defining operations with contradictory conventions or axioms.

But certainly the problem still remains.[5] If we have defined the formal Idea of a manifold in general by means of a series of stipulations, $a + b = b + a$, etc, then it is obvious that every real domain where we, on the basis of the real peculiarities of its objects, come to laws of exactly this form is governed by the algorithm of the manifold. Every theory of the manifold obviously governs all theories of specific, determinate manifolds of the real having the same form. But how so, if a manifold of the real has only in part axioms of the same form? How, if <438> the general manifold has more objects and more axioms? It then is clear that the particular manifold is not a mere special case of the general one, the calculation system of which can then not be directly applicable. Signs and structures in the particular manifold either become, for the given case, totally senseless, or else they take on the character of impossible, i.e., contradictory, concepts.

With that we come to a fifth theory, which states: We rise,

---

[5] Parallelism: real – formal. But how, if not parallel? {HS}

"{HS}" indicates a comment by *Husserl* that is footnoted in the *Husserl Studies* edition.

according to the principle of permanence, above the particular domain, pass over into the sphere of the formal, and there can freely operate with $\sqrt{-1}$. Now the algorithm of the formal operation is indeed broader than the algorithm of the narrower operations, which alone are really presupposed in a given conceptual domain. But if the formal arithmetic is internally consistent, then the broader operating can exhibit no contradiction with the narrower. Therefore what I have formally deduced in such a way that it contains only signs of the narrower domain must also be true for the narrower domain. {97}

***

<K I 26/81> We take up a completely general line of thought. We conceive the matter in utter generality. — Assume that some given domain, whether apriori or real, leads us to certain fundamental propositions suited to govern the entire domain theoretically, thus through pure deduction. Then it can be, and will in general be the case, that – whether already in the basic propositions, or in the derived propositions – a restriction on the composite concepts, let us say on the forms of operation, is required in order that the propositions retain a sense, or in order that they not become absurd through the absurdity of the concepts themselves. Now one can raise the question of under what conditions one can freely deduce – i.e., under what conditions one, untroubled about the possibility or the absurdity of the concepts, can compose them and freely apply the axiomatic propositions to the composites without ever arriving at a contradiction. And then, further, under what conditions the propositions that are free of absurdity are also actually valid.

Imitating mathematical procedure, one could think as follows:

We take the basic propositions and the definitions of the concepts in a purely formal sense. Instead of starting out from the real domain, let us totally abstract from it and take the assertions <439> as formal definitions of a domain in general, which is defined precisely as such, for which assertions of such a form are proper. Let this domain be called D. Given this formalization, then to each proposition determined by the axioms of the real domain there corresponds a proposition in the formal domain, and conversely. The formal domain will have the same limitations as the real, limitations that are pre-formed [*präformiert*] in the axioms. Now let us conceive of the formal domain as expanded in such a way that, so far as is in general possible, it no longer has these limitations. It is inherent in the concept of the expansion, if we relate that concept to the axioms, that the new axioms include also the old ones and that the new in addition admit cases of operations which the old axioms rule out. Now one could say: An obvious presupposition of the expansion is that the new axiom system be internally consistent. For from what is inconsistent one can obviously prove everything possible – or rather: There is no proof at all there. But if the new system is a consistent one and includes the old one in itself, then in the entire range of deduction no inconsistency can occur. Thus, a proposition which is somehow derived in such a way that it contains none of the "impossible" forms of operation, cannot possibly include an inconsistency, and thus it is true.

However, this manner of reasoning is questionable.

First, it certainly is correct that no derived proposition which includes imaginary expressions can contain an inconsistency, that it can conflict neither with the expanded nor with the original and narrower axioms. But how do we know that what is free of contradiction also is true; {98} or, as it must be expressed here, how do we know of a proposition that exclusively contains concepts which occur in the narrower domain and are there defined, and which does not conflict with the axioms of the narrower domain, that such a proposition is valid for the narrower domain?

What does this mean: A proposition is valid for the narrower domain? That means: It holds true on the basis of the axioms of that domain, it is a purely logical consequence of those axioms. How do we know, thus, that a proposition not conflicting with the narrower axioms is a consequence of those axioms?

Let us consider the following: The narrower domain D has the axioms A_D, and the entirety of purely logical consequences C_D; the broader ⟨domain⟩ Δ, e.g., $A_D + A' = A_\Delta$; or $A_\Delta \supset A_D$, and thus the consequence (C = consequence): <440>

$$C_\Delta = C_D + C_{A'}$$
$$= C_{D+A'}$$

If some proposition or other does not contain the compounds of the broadened operations, it is surely not obvious that it belongs to the C_D.

***

<K I 26/43> *Transcript from the Lecture.*[6] <470>

Domain. Establishment of what is to be understood by the domain of an axiom system.

The axiom system "defines" a domain, a manifold, and this does not exclude that there are still further objects which satisfy the axiom system besides those defined. But these further objects are "not defined."

What does that mean? We indeed define: "A manifold has such and such axiomatic properties." The manifold, as a totality of objects, is therefore defined by means of the relational formal properties expressed in the axioms.

Response: In an indeterminate manner we can say of any consistent totality of formal conventions (or formal sentence forms) that it defines a domain, a manifold of objects. But now a special case is to be marked out; the concept of the domain of an axiom system, and the sense of the assertion "an axiom system has a domain," is to be fixed with a richer content:

Any axiom system must – in order that one can say of it at all, even in only an indeterminate manner, that it defines a manifold – include existence axioms. For example, in the manifold there is to be a combination "+" (which implies that there are to be determinate pairs of elements a b, which are combinable in the form a + b, and "combinable" means in turn: there is in it at least one new element, which = a + b), and for this combination such and such laws are valid.

But now these existence axioms can be univocal or equivocal,

---

[6] What follows, through page 422, was included in *Husserliana XII*, at pp. 470-472, line 3. {DW}

ESSAY III   421

and in the latter case again either determinately or indeterminately <471> equivocal. If we now totally exclude the case of the indeterminately equivocal, then the {99} determinate equivocality can once again be eliminated by the joint force of the axioms, so that we are enabled univocally to determine new and ever new elements from given elements (and here that can only mean elements assumed as given and, as it were, named by means of proper names) on the basis of the axioms, and consequently to regard them likewise as given. An axiom system which, in this manner, delimits a general sphere of univocally determinate existence, and thus contains forms – whether simple or more complicated – of univocal determination of objects, out of which forms, upon arbitrary substitution of given values for the indeterminate signs for givens in general, new and ever new givens of a derived kind result: Of such an axiom system we say that it has a domain.

In the domain belong all univocally determining object forms out of which constructively arise, through singularization, univocally defined totalities of objects.

The ground for this terminology resides in the fact that we can compare two axiom systems of this kind with each other with respect to domain, that we can perhaps prove that the domain of the one is contained in that of the other, and that we therefore can speak of the expansion or the contraction [*Verengerung*] of the domain.

If it can be shown, for example, that every object of the one system constructible from any totality of individually assumed objects also must occur in the other (which of course presupposes that the forms of operation of the one also have sense in the other), but not conversely, then we can say that the domain of the one is contained in the other, and so forth.

A more restricted case is the one where all existence assertions valid on the basis of the axioms, even if they are equivocal – indeed, infinitely equivocal – find their equivalents in the sphere of univocally constructive existence, and thus where all objects existing on the basis of the axioms occur among those which are to be produced constructively, and thus univocally, out of those that where previously given.

The most restricted case is the one where the domain coalesces to form one single field of operations: that is, where all existence is enclosed in that which is constructible from a finite number <472> of pre-given objects. We can also say: a distinctive case is the one where the entire domain is constructible from a finite number of objects of the domain.

***

<K I 26/93> *For the Second Lecture.*
Are mathematical manifold and definite manifold equivalent concepts?
a) A definite manifold is ruled out by the inessential closure axiom. {100}
b) Can a purely algebraic manifold, which defines [*definiert*] no individual of the domain whatever – can such a manifold have the character of a definite [*definiten*] manifold?
One can certainly say: If only one operation is defined, and if I know that in the deductive sphere in which we are moving only the most general of formulas hold true – thus, more precisely, those alone where the forms of combination have the greatest conceivable generality[7] – then the associative and commutative laws form a definite combination. Likewise it is clear: If I add further operations, and also with the expansion the same holds true, then these can change nothing with regard to commutation and association. That is, any sentence which contains only the "+," and regardless of how I have derived it, is decided as to truth and falsehood. Likewise, the well-known laws of addition and multiplication are definite in this sense, under presupposition of the said

---

[7] <K I 26/92> The most general relational formulas: Any formula that sets expressions into relations, in which each letter on the one side can be replaced by any arbitrary letter. Therefore, if on one side the same letters occur, then I can replace them with different letters without the formula ceasing to be valid. Thus: a relation law is "analytic" (or a specialization of an analytic one), if it contains on the one side no identical letters (or, if it contains such, it also remains valid for non-identity of the letters). The most general formulas admit of no further generalization, as long as the number of their terms is held constant. Each term has the greatest possible generality, it is variable without restriction, and indeed independently of every other term on the same side of the relation. Special formulas are those which arise through limitation, where therefore the same letters show up on one side of the relation, or also various specific values (constants). {HS}

supplementary axiom.

But it does require precisely the supplementary axiom, and without this that would not hold true.

c) We now take up operation systems which do not exclude the introduction of individuals which give rise to operative results.

α) Suppose certain operational results are established, but in such a way that not every generally defined and existing operational result belongs in the sphere of the operationally producible and distinguished individuals.

Then divide the individuals which exist according to the axioms, and thus the existing objects of the domain, into:

1) Indeterminate objects, i.e., objects which do not, by means of the axioms, undergo a characterization that turns them into given objects for us.

2) Determinate objects, i.e., into the individuals defined, with their operational axiomatic properties, and into the ones derived from them, i.e., characterized by operationally determined properties.

As to the former, they are not merely indeterminate *for us*: Such are also the many generally defined objects of the second domain, with respect to the fact that we cannot subjectively work out the general definition, cannot resolve the problematic objects into the determinate objects – and in that sense cannot reduce them to known and given objects.

But rather *apriori and objectively* ⟨indeterminate⟩: It is inherent in the nature of the axiom system. But then objective characteristics (new axioms) are conceivable which, when adjoined to the previous ones, objectively integrate all of the indeterminate objects into the new sphere of determinate and individually given objects. Why? Could the axiom system leave no determination open, by means of which some object would be determined? And indeed some object still indeterminate?

In a formal system "objects" can, in general, be determined only through operations and operationally characterizing concepts. Namely, I certainly can, as discussed above, define: Let ω be a determinate object. But then I can assign to ω still many further kinds of determinations, a proof that it is not formally determined in an operational manner. If, then, already determinate objects

{101} are determined in an operationally univocal manner (in the sense of the axioms) then it is a case to be considered as apriori possible that no more operational specifications [*Festsetzungen*] (individually fixable) are possible, that henceforth no kind of operational results whatever are individually fixable. <K I 26/94> If this is the case, then the sphere of individuals is to that extent closed, as henceforth no further individual can be added that still admits of an operational specification. The question is, whether there can still be individuals apart from these – thus, individuals which can undergo no operational determination and which are absent from the individuals demarcated. It appears quite certain that that is possible. It would certainly be so if the ordering axioms were absent from arithmetic.

The further question: Would such a system be definite?

It would be definite if, for the demarcated sphere of existence, for the given individuals, and for the individuals not given, no further new axiom were possible.

How is that to be proven? One would have to set up combinatorially all axiomatic specifications (we perhaps restrict ourselves to binary operations – to more than two terms for the associative) for the operations in question, and then prove that all specifications are accounted for by the given ones: either as consequent or as contradiction. Then nothing further is to be said. I do not see what else needs to be said here.

β) Let the system be a mathematical one. That is, any individual existing on the basis of the axioms admits of an operational determination and must belong within the sphere of specific operational results (which are obtained on the basis of a certain finite number of objects, whether originally assumed as given in the definition of the manifold, or whether to be arbitrarily selected and given). Can such a system be indefinite?

All that exists ⟨in the domain⟩ is univocally determined operationally and has, accordingly, only the properties which arise out of the operational determinations. Therefore any sentence which holds true of the manifold in general is either a consequence of those determinations, is to be satisfied by means of them, and thus is conformable to the axioms, or else it is false. If the existent objects have only the properties which the axioms confer upon them, and if all objects that come into consideration are objectively of determinate, univocal properties, then for those objects, neither in particular nor in general, can anything remain to be settled. In particular not: because each object is determinate and can receive no properties to be arbitrarily assigned to it. In general not: for any general assertion about objects that are determinate in themselves must prove true in each single case, i.e., must be valid as a singular assertion. But I cannot express singular assertions about given objects. Thus, any mathematical manifold is definite.[8]

I can therefore say:

An axiomatically defined manifold can have the property that any of its objects is operationally determinable, and indeed univocally. I.e., any {102} object which is defined for it as existing (belongs in the sphere of existence which the axioms circumscribe) is to be directly or indirectly determined – and indeed univocally – by means of the determinate existents founded ⟨by the definitions⟩ or by a finite number of arbitrarily assumed ones. Such a manifold is a mathematical one and is definite (i.e., its axiom system is definite).

NB: We call any partial manifold which is extracted by an operational stipulation – or a complex of operational delimitations – from the previously given manifold, a structure [*Gebilde*] of the mathematical manifold. Such a "structure" does not have the character of a mathematical manifold, provided no axiom system is given here that refers exclusively to the partial manifold. Only when such an axiom system comes to light is it no longer a structure, but rather an independent manifold. And of course if the axiom system actually and exclusively deals with this manifold, and indeed genuinely operationally, then it also is definite. Thus any spatial structure which has an independent geometry is determined by means of a definite axiom system; and, conversely, any spatial structure so determinable has an independent geometry.

⟨K I 26/95⟩ With reference to *Hilbert*'s requirement that consistency and completeness be proven for *Euclid*ean geometry – and so, of course, for any meta-geometry proceeding synthetically – and thus that it be proven in the latter respect, alone of

---

[8] No, that is an unclear way of operating with "determinateness." {HS}

interest to us here, that the system of axioms suffices for the demonstration of all geometrical propositions, I observe: As soon as the axiom system is so formed and filled out that each point of the space (if we reduce everything to relationships between points) can be grasped as a determinate operational result with respect to the originally introduced individual operation arguments, then completeness also is already proven; it is proven that the axioms are capable of proving every geometrical proposition. (Likewise for any "element" instead of "point.")

It results, moreover, that there are not merely continuous, but rather also discrete definite "structures" (groups of points and the like) in space.

On my view, completeness is never an axiom – but rather a theorem, for definite axiom systems and manifolds.

Finally, I further distinguish *relatively* and *absolutely* definite axiom systems. An axiom system is *relatively definite* if, for its domain of existence it indeed admits of no additional axioms, but it does admit that for a broader domain the same, and then of course also new, axioms are valid. New axioms, since the old axioms alone in fact determine only the old domain. Relatively definite is the sphere of the whole and the fractional numbers, of the rational numbers, likewise of the discrete sequence of ordered pairs of numbers (complex numbers). I call a manifold absolutely definite if there is no other manifold which has the same axioms (all together) as it has. Continuous number sequence, continuous sequence of ordered pairs of numbers. {103}

In this way it can always be capable of further expansion. Namely, in such a way that it indeed introduces new axioms, and in return allows old ones to drop away, but that the new axiom system defines a domain which includes the old domain and, consequently, in a certain way, also has all the old axioms in itself – but now no longer as mere parts, but rather as deductive consequences. Through restriction of the manifold the new axioms pass over into the old.

Every manifold is capable of expansion, but certainly in general under modification of the axioms.

Every consistent expansion of a mathematical system, even if that expanded system is no mathematical one, yields a permissible system of calculation. (We call any system which can yield only correct results for a manifold a system of calculation for that manifold.)

If a manifold is relatively definite, then for its objects there is no further axiom which can be added to the axioms defined.

If a manifold is absolutely definite, then there is, in general, no further axiom which could be added to the axioms.[9]

***

⟨ 3. The Transition through the Imaginary ⟩

<K I 26/82> Definite under restriction or relatively definite = definite in the sense given up to now.

Absolutely definite:

(1) An axiom system is relatively definite if every proposition meaningful according to it is decided under restriction to its domain. An axiom system is absolutely definite if every proposition meaningful according to it is decided in general.[10]

Therefore, absolutely definite = complete, in *Hilbert*'s sense.

(2) If it is not only "for the objects of the domain" (which gets its sense through the axioms already given) that no axiom can be added, but rather if no axiom can be added at all.

(3) But this means that the manifold (the domain) cannot be broadened in such a way that the same axiom system is valid for the broadened manifold as was valid for the old one. For in the broadened domain it cannot be that merely the old axioms are valid. Otherwise the domain would not be broadened. Therefore in the broader domain, in addition to the old axioms, yet further propositions must be valid – and, indeed, propositions that are not mere consequences of the old axioms. But every possible proposition is already decided in advance.

***

---

[9] At this point the text of *Husserliana XII* resumes at p. 440. {DW}

[10] Here a question mark in the margin. {HS}

<K I 26/83> Clearly there is a peculiar property of the deduction system that comes into question here. In order to bring this character to light, let us consider the special case of the *arithmetica universalis*. What is familiar to us there? 1) That no proposition which is deduced in the broadened arithmetic but falls within the domain of the narrower one can conflict with the axioms of the latter. 2) But we know still more. Such a proposition can be reduced to {104} an equation, no matter whether it is an equation with all elements determinate or with partially indeterminate ones. Just as we then understand $a > b$ as the equation $b + u = a$. But now it is <441> clear that the axiom system of the arithmetic, however narrowly it may be defined, has the property that each equation falling within that arithmetic either is valid on the basis of the axioms or is invalid on the basis of the axioms. I.e., either the proposition is a consequence of the axioms or it contradicts the axioms. Accordingly we will state that for the arithmetic the problem resolves itself in this way:

Every proposition falling within the narrower, but deduced on the basis of the broader arithmetic, is an equation. Now every equation falling within the narrower arithmetic is either true in it or contradictory in it. An equation deduced within the broader domain cannot be in contradiction with the axioms of the narrower domain. Otherwise the entire broader domain would be inconsistent. Therefore it is true.

According to this the following general law seems to result: A transition through the imaginary is permitted 1) if the imaginary can be *formally* defined in a consistent and comprehensive system of deduction, and 2) if the original domain of deduction when formalized has the property that every proposition falling within that domain is either true on the basis of the axioms of that domain or else is false on the same basis (i.e., is contradictory to the axioms).

However, it is easily seen that this formulation does not suffice, although it already brings to expression the most essential part of the truth.

Namely: The difficulty lies in the expression, "proposition falling within the domain." What does this mean? How do I know that a proposition falls within the domain? A proposition which can be proven true on the basis of the axioms – i.e., which follows from them – or can be proven contradictory, violating them, falls within the domain. But there is still the question whether the derived propositions of the broader domain fall in this sense within the narrower domain. If that is not determined in advance, we can say absolutely nothing about it.

But it can be determined in advance only if I can from the outset read off of the proposition whether it falls in this sense under the axioms. And this is possible in such a way that the axioms completely delimit the domain, inasmuch as for this domain no other axioms are to be possible. An object domain can be delimited by axioms completely or incompletely. Namely, if I have all of the first principles, from which all possible propositions of the domain are deducible, then I have with them the entire theory of the domain. If we formalize <442> these first principles then we obtain a formal system of deduction in which to each proposition of the domain there corresponds a formal sentence. But thereby to the formal axioms, through which the formal domain is in fact to be defined, there belongs an axiom of closure, which states: Through such and such axioms the domain is determined, and no others are valid for it. Where this axiom of closure is not added, there the domain is not closed, inasmuch as perhaps still {105} further axioms can be admitted and, consequently, the objects of the domain can be formally defined through new determinations.

This concept of completeness is to be characterized as "spurious" completeness. <K I 26/84> Such "completeness" is, of course, not something peculiarly characteristic of axiom systems. For we can make any axiom system, any system of first principles for possible deductions, quasi-complete by means of such a negative closure axiom, and therefore this "completeness" can be of no use at all to us here. In the moment we broaden an axiom system we obviously abandon the axiom of closure. An axiom system with the closure axiom cannot be broadened: The concept of broadening presupposes that the negative closure axiom is not simultaneously intended.

On the other hand it is certainly true that such an axiom system, closed in an exterior and spurious manner, already has the property which we have in mind: namely, it can be read off of each proposition whether it is or is not a consequence of the axiom system. The proposition need only contain the relational forms, the forms of combination, or in short the concepts that are formally defined by means of the axiom system. If it has sense at all in terms of those definitions, then it is either true or contradictory.

E.g., $a + b = b + a$ and the axioms of equality and the closure axiom. Accordingly, I know that the associative law is either valid as a consequence or is in contradiction with the axioms. If it is, namely, no consequence of the axioms (excluding the axiom of closure), then it can be valid only on the basis of a new axiom; but that is in fact excluded.

We thus now raise the question of whether there are axiom systems which contain no axiom of closure and yet – namely, on the basis of their peculiar nature – permit it to be read off of each proposition whether that proposition belongs within the sphere of its deductions as to truth and falsity.

Now that is, in fact, precisely the case with arithmetic. Every <443> arithmetic, regardless of how constricted – whether it has reference to the whole positive numbers, or to the whole real numbers, or to the positive rational numbers, or to rational numbers in general, etc. – every arithmetic is defined through an axiom system such that on the basis of it we can prove: every proposition in general that is constructed exclusively of concepts which are established as valid through the axioms (or are axiomatically admitted), every such proposition falls in the domain, i.e., it is either a consequence of the axioms or is contradictory to them. The proof of this assertion lies of course in the fact that every defined operational formation is a natural number and that each natural number stands to every natural number in a relation of magnitude determinable on the basis of the axioms. If therefore a proposition is meaningful at all in virtue of the axioms, then it is either valid as consequence of the axioms, or it is invalid as a contradiction to the axioms. For a numerical equation to obtain, that of course means that given the execution of the operations in the sense of the axioms the identity $a = a$ is produced. Every {106} numerical equation is true if it can be transformed into an identity, and otherwise false. Every algebraic formula is, then, also decided, for it is decided for each numerical case. This can be formulated in general thus:

A formal axiom system which contains no extra-essential closure axiom is said to be a definite one if each proposition that has a sense at all through the axiom system *eo ipso* falls under the axiom system, be it as consequence or be it as contradiction. And this will prove true in all cases where it can be shown on the basis of the axioms that every object of the domain reduces to the group of the "numerical objects" for which in every relation the true is fulfilled in an identity and every other one is therefore false. Wherever, for example, each defined proposition is reducible to an equation or to the ≻/≺ between numerical objects, there the axiom system is definite.

One could have drawn still further distinctions here. If, for example, only propositions of a certain form have this property with certainty, then one could designate the axiom system as definite with respect to the propositions of that form.

The transition through the imaginary is therefore linked with the condition of definiteness, and in fact this partially definite axiom system already suffices.

<K I 26/85> Thus every proposition is integrated into the domain through its definite sense, and the broadening must therefore consist in the fact that if the same forms of operation occur in both domains <444>, the results of the operation either belong to both domains or are so defined in the narrower domain that the definitions remain in force for the broader one – except that possible but not defined cases of operation are added by definitions in the wider domain.

We therefore have the following situation:

1) Certain axioms delimit a domain perfectly inasmuch as, by hypothesis, they are all of the axioms of the domain.

2) The system of operations is capable of broadening to the extent that for the same forms of operation and relation still further axioms are possible: a domain is conceivable which has formally the same axioms and yet others besides.

3) The old axioms give a determinate sense to the general operations and to certain special forms of operation. What is not defined, that is excluded in this narrower domain. The new axioms retain all of these axioms, but give a sense to special

operational formations which previously were not defined and were previously left open. That requires supplementary axioms.

The broader domain possesses expanded axioms, which can be divided up in this way:

a) One formulates the axioms so that they contain only the narrower concepts, and, consequently, one also separates off the special axioms with reference to the broadened concepts.

b) One adds on the supplementary axioms. {107}

Every proposition which contains concepts defined in the first domain then belongs in the narrower domain. We then know apriori that in this domain it is true or false, thus in virtue of the correlative axioms.

Therefore any axiom system is said to be complete in a limited way if, according to its definitions, it leaves open no possible operational result.

An axiom system is said to be complete absolutely if it is so far reaching in the definitions that no possible operational result whatsoever remains open.[11]

***

<K I 26/33> An operation system is assuredly definite if it is related to a determinate (existing, given) operational domain in such a way that every defined operation has its validation or its rejection (its contradiction) in that domain.

(When is an axiom system definite? Certainly it is so if it refers all operations back to a rigorously determinate domain of specific results of operations in such a way that that domain is the total domain of axiomatically enclosed existence – if, therefore, anything at all that exists in the domain remains not merely more or less indeterminate, but rather is completely determinate. All propositions refer, then, to a given field of operation, to the "domain," and express relations which appertain to the determinate magnitudes: Univocally determinate magnitudes have only univocally determinate relations.)

---

[11] Pp. 444-451 of *Husserliana XII* are omitted here and reinserted as Appendix III below. What immediately follows here corresponds to the first "Study" of the "VII Abhandlung" in *Husserliana XII*, pp. 452-457. No date assigned by the editor. {DW}

That is not sufficiently precise.

An axiom system defines a mathematical or "constructible" manifold if the sphere of formal objects which it defines has the property that any object existing in this sphere on the basis of the axioms is univocally determined through its operational relations to "determinately given" objects – i.e., objects univocally established and not further discriminable as to kind. If, therefore, all objects which exist in the domain at all, and are thought through arbitrary concepts defined by means of the axiomatically defined concepts or objects, belong to those objects which proceed from those "determinately given" objects through defined conceptual formation, through defined combinations and relations.

Every object concept which <453> is defined through the axioms is a concept either of objects of the domain of the determinate objects, or is "objectless" (in the domain) – i.e., among the objects the existence of which the axiom system establishes (can demonstrate), there is none which corresponds to the concept.

***

<K I 26/34> *Axiom system* A, expanded *axiom system* A_w.

How do I know that a proposition deduced from A_w follows from A alone? {108}

Object domain of A (defined by means of A). Object domain of A_w (defined by means of A_w).

Imaginary objects = objects which do not occur in A, are not defined there, are not established by means of the axioms and existential definitions of A, so that, therefore, if we regard A as the axiom system of a domain which has no other axioms – and thus also no other objects – those objects are in fact "impossible."

The situation now is that we derive from A_w an assertion which refers purely to the objects of A. That presupposes: We see on the face of an assertion, or can prove at any time, that it has a "sense" purely for the objects of the narrower domain, i.e., that it, if it is true, presupposes the validity of no concept (the being of no object) which owes its validity only to the supplementary axioms.

The assertion is then also true for the narrower domain in and for itself – that is, it is a consequence of the A – if I know apriori either that this assertion belongs to the class of assertions which must be decided apriori by means of the A, or if I know quite in general that every assertion which has a sense for the narrower domain (which thus contains nothing undefined, presupposes the existence of no objects that are imaginary, if only the axioms of the narrower domain hold true) must be a consequence of the A.

It resides in the concept of the expanded axiom system, of course, that the newly added axioms are not inherently self-contradictory and are not simple consequences of the old axioms, and thus that they formulate assertions left open by the narrower axioms, but on the other hand are also consistent with the old ones. Now if the new axioms refer to the same operations (if not, there would be no problem), it is then clear that the old axiom system does not prejudicate everything with regard to those operations, but, instead, precisely leaves many things open, which appear added in the form of axioms. <454>

An *irreducible* axiom system is definite which delimits (or grounds as existing) a formal domain of objects in such a way that for this domain – that is, if one preserves the identity of the axiom system, and if one presupposes that no new objects are defined and thereby assumed as existing – no independent axiom can be added which is constructed purely from the concepts already defined (of course, also, none can be withdrawn, since otherwise the axiom system would not be irreducible). But I can also say: An axiom system is definite which formally defines an object domain in such a way that every meaningful question for this object domain would find its answer by means of the axiom system; or that every proposition meaningful by means of the axioms, if we limit it exclusively to the objects grounded as existing through the axioms, either follows from the axioms or contradicts them. If P is a proposition which says: For the manifold proven as existent by A, P holds true – then for this manifold this proposition is either true on the basis of A or false on the basis of A. {109}

The manifold thereby includes all objects proven as existent by A, which does not rule out that the same axioms are, in addition, valid for a more inclusive manifold, but in such a manner that the surplus of objects is not defined as existent or proven as existent by the axioms (without reinforcement by new axioms).

***

<K I 26/35> The general forms of operation, and the axioms and laws of relation belonging to them, leave various specific operational forms open to determination. In a formal manner there is here a generally finite number of possibilities which can be established by combinatory techniques, e.g., the particularization of the axioms by means of equality ⟨involving⟩ $a + a$, $a - a$, $a : a$, and the like. If we determine such operational formations and if we establish the special laws of operation concerning them, then we must prove that they are compatible with the general laws, but are not included in them. If this is done, two cases are then possible:

Either we have set up a series of formulations in such a way that, in fact, possible operational forms are still left open, but that, on the other hand, so far as we have defined it, the axiom system is definite; or else there remain open no further operational results whatsoever – *nota bene*, none that are possible compatibly with the general basic laws and with the <455> specific laws already defined, and thus preserving the identity of the entire axiom system for the whole domain. This is the case with the axiom systems that are absolute or essentially complete. The essentially complete axiom systems – and we could also say: the essentially closed operational systems – form therefore the outermost sphere within which the expansion of extra-essential axioms can move, given retention of the original axiom system. An essentially complete operational system is, thus, one such that, with regard to the forms of relation and combination which it in general founds, no possibility remains open; i.e., it defines by means of laws of a general and specific type in such a way that any formulation that is in general formally possible is already accounted for, or that – better still – no further arbitrary formal formulation with regard to some operational formation or another can be set up without violating the prevailing axioms (in their general applicability for the entire existential domain).

Such an operational system is also definite in the sense that one can now see on each proposition whether it falls in the domain or not. In general it need only contain the operation signs and relational signs, and combine them meaningfully. If a proposition has sense according to the axioms, then it is decided as to truth or falsity by the axioms. *NB*: But if the axiom system is not to be expanded, perhaps the domain can be, by means of a new axiom system that deductively includes the old one in itself. In this way {110} every domain can certainly be expanded, e.g., a two-dimensional *Euclid*ean manifold to an n-dimensional one.

From this point we then easily arrive at axiom systems that are "complete" in Mr. *Hilbert*'s sense. I have designated as "definite" axiom systems in the sense employed up to now. I will now designate them as "extra-essentially complete," in contrast to those that are complete in *Hilbert*'s sense, which I will designate as "essentially complete." This latter concept remained concealed from me, since for my purposes all was accomplished by means of extra-essential completeness.

It is, I said, an intrinsic property of an axiom system in case it is definite. But the system will then in general <456> still be capable of expansion. Inasmuch as it is definite it determines a sphere of objects of operation so complete that those objects of operation permit no new axioms deductively distinct from the ones already given. The object domain of the arithmetic of the whole numbers is completely determined by its operations: If I restrict myself to those numbers, i.e., to the sphere of existence delimited by means of the axioms of the whole numbers, which constitute the points of reference for the general propositions of the domain, then I can no longer bring those numbers under any new axioms, unless they concern totally novel operations, unrelated to the arithmetical ones. For example, nothing stands in the way of duplicating each operation and each relation, etc. A "definite" axiom system leaves for its operational substrate absolutely nothing open with respect to the operations defined. If it left something open, then there would in fact be relations which are not true or false on the basis of the axioms.

And yet something remains open. Namely, a restricted arithmetic, and a domain of deduction restricted in an analogous sense, is restricted in that not all specific operational formations remaining free on the basis of the general laws of operation are defined, and then of course the relevant specific laws of operation also are not introduced. So far as the definitions reach, so far all is definite.

***

<K I 26/36> The problem is fundamental for applied mathematics and for pure mathematics.
The problem in applied mathematics.
The problem in pure mathematics.
Axiom System. Expansion: A larger system of axioms or a system of axioms that deductively includes within itself the old one = deductively, logically more inclusive.
System of objects, manifold. Extended manifold. Its axiom system either larger or logically more inclusive.
In general, a proposition which has a sense according to the definitions of the narrower axiom system and which follows from the extended one will not immediately {111} be a proposition of the narrower axiom system. If the narrower system alone is valid, or if it alone is to be taken into account, then every proposition that follows from it belongs to it.
The utilization of a broader system in order to bring forth propositions of the narrower one can only be permitted if we possess some characterizing mark by which we recognize that every proposition that has a sense in the narrower <457> domain also is decided in the broader one, thus must be its consequence or its contradictory.
Distinctions: The inference from the imaginary is permitted in the singular case or for a class, if we can know in advance and can see that for this case or for this class the inference is decided by the narrower system.
But if the inference is to be permitted in general, starting from principles, then we must know in general that every meaningful proposition is decided.
Further distinctions:
An axiom system can delimit a sphere of existence and leave open a vague, broader sphere. Existence by definition/existence dependent on deduction. Domain of the axiom system. We restrict ourselves to axiom systems that have a domain. (Why not directly:

to totalities of objects which satisfy the axiom system?).

We define:

An axiom system that delimits a domain is said to be "definite" if every proposition intelligible on the basis of the axiom system, understood as a proposition of the domain, is either true on the basis of the axioms or false on the basis of them. Or, put otherwise: If only two things are possible, either the proposition follows from the axioms or contradicts them.[12]

Equivalent to this is the following statement:

An axiom system with a domain is definite if it leaves open or undecided no question related to the domain and meaningful in terms of this system of axioms.

Equivalent to this, once again, is the following crucial statement: An axiom system is definite if it delimits an object domain as existing, and indeed in such a way that for that domain no new axiom (deductively independent of the axiom system) is possible. For that domain: If I hold the axiom system fixed, and I do not, perhaps, add new specifications of existence, do not expand the domain, then no deductively new axiom can be added without contradiction. But obviously that concerns only axioms constructed purely from concepts already defined, and which thus are intelligible propositions. <458>[13]

***

<K I 26/92> *"Every proposition having a sense for the axiom system."* — *What does that mean?*

Let us consider, for example, the axiom system of the whole numbers, positive {112} and negative. Then $x^2 = -a$, $x = \pm\sqrt{-a}$ certainly has a sense. For square is defined, and $-a$, and $=$ also. But "in the domain" there exists no $\sqrt{-a}$. The equation is false in

---

[12] I believe that here the model of arithmetic is unnecessarily taken in a stern and rigorous manner. All remains in order if we simply take "domain" in the natural sense of the term: objects which satisfy the axioms. {HS}

[13] This concludes "I. Studie" of "VII. Abhandlung" in *Husserliana XII*. What now follows up to "Appendix I" below is from the *Husserl Studies* edition and not previously published. {DW}

the domain, inasmuch as such an equation cannot hold at all in the domain. Therefore I cannot pose the problem: "A certain magnitude ⟨x satisfies⟩ $x^2 = a$. Which is that magnitude?"

If I formulate the axiom system of the positive whole numbers, I do not, perhaps, state the axiom that no other axiom is to hold true, and also not the axiom (which does not say the same thing) that no axiom is to hold true which posits new magnitudes besides the magnitudes grounded as existing. Rather, I define: These and those axioms hold (whether still others, I do not say). And I consider, then, the sphere of the magnitudes posited as existing thereby. This then is the domain of the axiom system, which, in our present case, is a completely specified domain. But it can also be a partially indeterminate and partially given one, or a completely indeterminate one.

I then say: If I suppose some meaningful sentence constructed, then I can ask whether it is valid if I take it to be a sentence about the objects of the domain, in the previously defined sense. The domain is definite if the truth and falsity of any such sentence is decided for the domain on the basis of the axioms.

***

⟨K I 26/97⟩ System of objects, $O(\alpha, \beta \ldots)$, defined by means of an axiom system. Impossible operations $U(\alpha, \beta \ldots)$

Can one not say: If an expansion of the system of operations succeeds in such a way that the objects newly introduced are defined solely by means of operational results ⟨objects⟩ which, although formed from the operations defined in the narrower manifold, are imaginary in it, then one can say: The expansion must be definable in the manner of number groups? If not, when is that the case?

How can one make all of this suitably precise?

1) A number group is an arrangement [*Verhältnis*] of a certain number of indeterminately general objects of the domain O, which is defined in terms of $=, \succ, \prec$, by means of certain genuine [*reelle*] occurrences $(=, \succ, \prec)$ in the domain O, without the arrangement itself, however, being for that reason an object of the domain. Thus, two arrangements V_1, V_2 are equal to each other, if between

their members a certain equation, $\phi(a, b...) = \psi(a, b...)$ holds etc.[14]

2) A special case of the arrangement "supplies" from time to time determinate objects; namely, where the specialization of the arrangement runs exactly parallel to the objects, to the determinately defined objects. We have reduced the original domain to a determinately given manifold of mutually exclusive species (all univocally derived from a few given ones). Each determinate object, each determinate number, corresponds to a determinate specific arrangement {113} (thus the same holds true of the arrangements) in such a way that two specific arrangements are equal to each other, >, or <, whenever two determinate numbers are equal to each other, >, or <.[15] Likewise, each axiom for the determinate numbers is satisfied by specific arrangements, i.e., there is a sentence for them. To each simple number sentence there corresponds a sentence for determinate arrangements, and conversely.

If I expand the old system of operations in such a way that the new one is, precisely, an expansion, and therefore wholly includes the old one, then any operation of the old domain is a specialization of an operation of the new, whereas on the other hand the new domain contains only operations which are composed of objects of the old domain, and contains new objects which are operational combinations of the old: Is not then *eo ipso* the new domain a domain of "arrangements" among objects of the old domain?

To maintain that in general would certainly be bold, but for which exact arrangements?

a) The expanded manifold would also be a numerical one ("mathematical"? I still do not have a name for it.). That is certain if all that is newly introduced is an "imaginary" operational formation of the old, or if the axiom system is modified in such a way that every object of the manifold occurs among the real or imaginary operational formations of the old . . . .

Perhaps better conversely: The numerical manifold would be so constituted that all determinate numbers consist of the old

---

[14] $a : b$; $ab' = ba'$; $a + b' = b + a'$. {HS}
[15] $a + bi = a' + b'i$, if $a = a'$, $b = b'$. {HS}

numbers and operational combinations of the old numbers.

The operations with the expanded numbers must pass over into operations with the old numbers if we restrict the present numbers in such a way that the old numbers result.

An operation with new numbers will then reduce to operations with old numbers where old numbers enter into combination, as well as the primitive new units or newly introduced complexes of old numbers. But inasmuch as these latter are "impossible" old numbers, and as such have a determinate definition, e.g., $\sqrt{-1}$ is defined by $x^2 = -1$, or $a:b$ is defined by $bx = a$, then every new number will be defined by means of a relationship to old numbers. But then the equality of new numbers is defined by means of relationships [*Verhältnisse*] between old numbers. Thus each new number will certainly be determinable in the manner of relationships.[16]

This would have to be rigorously thought out.

***

<K I 26/87> *Axiom System*

General signs $\underline{a}$, $\underline{b}$: for objects of the domain to be defined.

Defining axioms: $a + b = b + a$.

Particular cases of operation: $a - a$, $a : a$, and axioms related to them, which define "determinate" objects of the domain, and to be sure univocally define them. {114}

What is "determinateness"?

Any arbitrary object of the domain, any arbitrary general or specific operational combination, is equivalent to a univocal operational result in the class of the specific.[17] Determinate objects could also first be defined, e.g., let $\pi, \rho \ldots \mu$ be given objects. Then they have operational meaning if they have determinate operational properties. And then for those objects these or those properties expressed by operations must hold true. It certainly is

---

[16] In fact that is certain, but yet not always in such a way that all operations with the new numbers reduce to groups of mere operations with the old and entirely in conformity with the principle of permanence. {HS}

[17] Object of the domain = Object which is provable as existing by means of the existence axioms. {HS}

not necessary that they be univocal results of determinate operations. I could in fact define a - a = π or ρ and the like.

But it is a peculiar property of the arithmetical operational system that every ⟨indeterminately⟩ general operation which is grounded or defined by axioms as univocally executable, and every existing a̱, is equivalent to a determinate object of the "number series." That surely does not suffice. Also, in the case of such disjunctive stipulations of them the same could hold true. It could hold true that any a, a + b, etc. = "one" of π, ρ, or any of the objects to be constructed therefrom by means of operations, while it were impossible to pick it out in any given case on the basis of the axioms. How do matters stand in arithmetic?

I have 0 and 1, and from them is to be derived, through operation, 2, 3 . . . . That is: From 0, by means of the univocal operation "+1," in a finite number of steps, any other number of the series can be produced. Therefore every number whatever is equivalent to one which is univocally derivable from 0 through a finite number of steps. Therefore every indeterminately general number a̱ is in itself determinate, or the concept "object of the domain," a̱, in general, specifies itself into a series of species, the determinate numbers, which all proceed from one number through a univocal operation, "increase by 1." Each number is accordingly either 0 or 1 or 2 . . . .[18] 0 is defined by means of a - a = b - b, and 1 by means of a : a = b : b, a : 0 excluded. Can 0 receive a still further determination by means of the axioms?

0 is univocally determinate and can undergo no further determination, unless it be one which, grounded in that determination and in the axioms, is its consequence. That lies in the sense of determinateness, and it is also because in arithmetic we study exclusively operational properties.[19]

In any case, 0 is univocally defined, and to be sure it is a

---

[18] Each number of the domain. That means that I assume as existing 1) every generally defined operational result; 2) every result defined under limitation, within the limits; 3) what is specifically stipulated as existing: 0, 1. {HS}

[19] But in fact, were some proposition about 0 left open, then I could not add this on. It would be compatible with the "definition" of 0 and yet would not be its consequence. That depends upon the nature of the remaining axioms. Therefore the univocality in itself tells us nothing. {HS}

univocal operational result, on the basis of general numbers but independent of the generality of those numbers. Likewise for any number of the series of natural numbers.

If the arithmetic is an arithmetic of fractional numbers (rational numbers), then every fraction a:b is equivalent to a fraction of the natural numbers, and every such fraction is again univocally determined. Likewise then every irrational number. Each has its equivalent in a natural irrational number.

The series of natural numbers (the whole, fractional, real, imaginary) is the field of operation of arithmetic, i.e., any number whatever has its equivalent in a unique number of this field. A number is said to be given if {115} the natural number equivalent to it is given. Every natural number is different from every other one, and each equation, $>$, or $<$ between natural numbers is determinate in itself, because each is axiomatically provable or refutable.

***

<K I 26/88> [. . .] or a number which is the sum of two other numbers. Any number can be that. But I cannot bring it about that a determinate number is the sum of two arbitrary numbers. The natural numbers are what they are only through the definitions. Since the definitions univocally determine the numbers (in virtue of the axioms), then a number or group of numbers can indeed have infinitely many properties, but none which is not grounded in the definitions and axioms and determined through them. It would be a contradiction against the determinateness of the natural numbers if one wished to reckon among their properties those which are not covered by the definitions.[20] In fact, it is the peculiar property of the natural numbers that they are "determined" in this sense. Not only are they in general univocally determinate objects of the domain, but rather they are determinate in such a way that they can undergo no other determination, i.e., that for them, fixed as they are by the axioms, no additional property can be newly adjoined axiomatically. But that must be proven. There may be new kinds of numbers which

---

[20] That is unclear and probably also false. {HS}

satisfy the same rules of operation, the same axioms; but the "defined numbers" are completely determined by the rules of operation, and all properties that they have and which include no "imaginary" number concepts, no relations to the imaginary, are determined by means of the axioms and definitions of the narrower domain. New properties cannot be ascribed to them axiomatically.

The "defined numbers" – that does not mean defined in the logical sense, but rather the numbers presupposed as existing. If we restrict ourselves to these, assume as existing only that which is established as existing by the axioms, then all existing numbers reduce to the "number series."

What constitutes the determinateness of the natural numbers? That it is a series of logically singular ("individual," and, as individual, determinately defined) numbers, and therefore not further specifiable. If I say: a + b is univocally determinate for any pair a, b, that it yields a certain number c, then that is indeterminate. Indeed, it merely expresses the fact that there is a certain number c (which I, however, do not know, and whose determinate being depends upon the determinate character of the a and the b).

0 and 1 are determinate numbers, and subsequently 2, 3, 4 . . . are too, i.e., they are species of objects of the axiom system which (through certain specific cases of operation) undergo a univocal stipulation as lowest species, and consequently, as such, can have no other properties than those that are determined through their operational definition and the other axioms consistent therewith.[21]

{116} All that 0 is, it is in virtue of the definition: $0 = a - a$. What 0 has otherwise in the way of properties, that must proceed from this property, which is a univocally determining one, and from the other axioms.[22]

If I were to say, there is an object, a determinate "0" in the domain, then this certainly would be understood as determinate, but not differently than if I were to say, a is a determinate object. The distinct determinate character of the 0 resides in the fact that

---

[21] But that they are lowest species, that they are not further specifiable by means of new axioms, that must be proven. {HS}

[22] Indeed, this must be proven. {HS}

I designate it not merely as a determinate particular object, but rather that I associate with it an instance of operation that is independent of the generality of the terms of the operation: a − a, for any arbitrary a, always yields the same value, named "0."

Now generally characterized instances of operations lead me to 0, which is defined as the identical value of these general operational occurrences. What occurs in the formal domain is, since it is a domain for operations, to be determined through operations according to the axioms. Therefore all properties of 0 are to be determined through the operations that are valid for it. 0 would here not be operationally distinguished from any other a if I only knew that it is determinate. Its basic operational property is 0 = a − a, and this, with the other axioms (in which 0 also still occurs), then determines the totality of the properties that 0 has.[23]

* * *

<K I 26/89> That was not entirely clear:

The essential point is the following: In the axiom system I define not only sentences which hold true for all members of the manifold in general. I therefore operate not only with general, indeterminate concepts of objects, but rather I also introduce individually designating concepts of objects − as it were proper names for objects (or species of objects) − and I axiomatically establish their existence.

Were I now merely to say that 0, 1, are determinant (given) individuals of the manifold, then these would be determined only through the fact that I can apply the general axioms to them, and the fact that I can operate with them just as I do with a, b . . . in general, or with objects of the manifold in general. Only because I introduce them as particular results of operations do I first introduce characteristic formal properties of the axiom system, do I first characterize the manifold more closely in a determining manner − and all of this in an extremely important way: for only through these determinate definitions does the domain first become altogether definite.

---

[23] Indeed, why can there not be, nevertheless, yet further properties that hold of 0? That is not immediately obvious. {HS}

Only in this manner – namely, in terms of the axioms – does the sphere of the natural, whole, fractional, rational, or irrational numbers become a determinate domain of operations, a domain individually given, although containing infinitely many individuals.[24]

The relationship between the general axioms and the axioms pertaining to individuals is thereby such that the domain of the manifold {117} in general coincides with this domain of individuals. The domain of the manifold = the totality of the objects provable as existing by means of the axioms, the objects which belong in the extension of the axiomatic generality. But that the natural numbers are the lowest specific differences given of numbers in general (= object of the axiomatic domain in general) – that is proven.

***

<K I 26/90> $a + b$; $a - b$; $a : b$ — A sphere of two-member operations.

Requirement:

Every operation is executable in the numerical domain, and any relationship in the numerical domain is determined.

1) Axioms of combination: Thus there must be expressed for each operation a general law of existence, a law of univocal determination of magnitudes by means of some given magnitudes, whether it is an axiomatic law or one which proceeds from others as a consequence.

2) Axioms of calculation: The equality or inequality of two operational formations assembled purely out of elementary operations (each of which therefore must be univocal) is always determined numerically. Therefore, conversely, there must be general laws which algebraically establish the equivalence of algebraic operational formations, and to be sure in such a way that the totality of the algebraic rules of operation is a definite whole. It indeed must, before all determination by means of particular values, be decided which is an analytic formula and which is

---

[24] Individually given = contains only given lowest specific differences. {HS}

not. Thus one will no doubt be allowed to say: The "axioms of calculation," as a totality of calculations of equality, must yield a definite whole under abstraction from all particular value determination.[25]

3) Axioms of "incorporation": Besides equality there are other relations. For these there hold peculiar laws of relation and laws of combination of them with the relation of equality, and with respect to these and those special operations.

Let $>$ be such a relation, and the contradictory case be $<$ or $=$: For any pair of numerical formations it is decided whether they stand in the relationship $n_1 > n_2$ or not. Every numerical formation is operationally derivable from the primitive elements given. Therefore the relation $>$ must be determined for the primitive elements, and for every elementary operational combination there must further be determined a law which in general makes it possible to set up the $> <$ relation.

E.g.:

If $u > 0$, $a > 0$, then $a + u > 0$.
If $a > b$, then $a + x > b + x$.

If we consider all univocal formations which are derivable from an arbitrary, determinately conceived $\underline{a}$, then it can be decided for any such $\underline{a}$ in relation to any of the formations of this group how it stands with respect to magnitude, {118} presupposing that the formation is not 0 (that it changes nothing in the respective operation) and that I know how it stands to 0, whether it is $>$ or < 0. If $a > 0 : a; a + a; a \cdot a$.

Consider all the axioms.

Suppose it proceeds from the axioms that the domain is a "mathematical" one. Further, that for each of the numerical objects it can be decided whether any primitive $a > b$, etc., occurring in the axioms identically holds true or does not hold true, and that any primitive combination is executable (namely, univocally belongs in the sphere of the numerical). Can one then say: that is sufficient for the definiteness of the entire domain? Indeed.

---

[25] Why? {HS}

Executability means: If N is the totality of evaluated [*numerische*] numbers, then every a + b, a · b, etc., belongs to this totality, and, to be sure, it is a unique member of this totality. If not only every relation >, <, = between two evaluated numbers, but also every relation between two formations of the form a + b etc., made up solely from evaluated numbers, is securely determined, then that suffices.

It is not exactly necessary that a *series* of numbers be there, but rather it suffices that there are particular evaluated numbers and formations made up of such, and that for this totality of simple and complex evaluated numbers every relationship is established. One then can designate the entire totality as a totality of numerical elements.

***

<K I 26/91> Of what type, therefore, is the structure of levels in arithmetic?

All levels have in common: certain general axioms. The distinction between the levels resides in the existence axioms, in which the existence of particular forms of operation is stipulated under narrower or broader conditions.

1) By means of the existents stipulated in the given case, one domain is then marked out, i.e., it is indicated that all magnitudes of the domain which result operationally – if in each step the conditions (restrictions) on existence are adhered to – reduce to the "number series" in question, i.e., to an ordered totality of given species. Any operation that adheres to these conditions is "executable," i.e., yields a number of the number series in question. Any operation that does not adhere to these conditions is not executable, yields no number of the number series in question.

2) Every number domain of lower level is completely contained in every number domain of higher level.

Every operational domain of lower level is completely contained in every operational domain of higher level.

Every axiom system of lower level is completely contained in every axiom system of higher level, either as an actual part or as a logical part (deductively).

We said: The distinction between axiom systems resides in the {119} existence axioms, which are either broader or narrower (within the sphere of real numbers).

Any such existence axiom we can then split: There is for any pair $\underline{a}$, $\underline{b}$, an a − b. For any pair $\underline{a}$, $\underline{b}$, there are two possibilities: Either a ≻, = b, or a ≺ b.

The axiom is therefore equivalent to two axioms: There is for each a ≻, = b, for any a ≺ b, an a − b. Thus only one new axiom is added, which defines what was left open.

3) Every level is definite. How can I now more closely determine that?

Axioms of calculation relate to the number series, i.e., connect to special axioms which define the number series. No other objects are to be considered than those which are there defined. The calculations are executable subject to conditions.

The axiom system is definite if for the domain which it delimits it leaves no question open. The entirety of objects of the domain is determined by means of the definitions.

If a ≻ b, then there exists an a − b, that is, a univocally determinate number which . . . . If a = mb, then a : b is the univocally determinate number which . . . .

If a ≺ b, and a ≠ mb, then there are no such numbers as a − b, a : b. There are none: That means, any number bound to the conditions a ≺ b and a = mb occurs in the series of positive whole numbers. That closes the domain. And there is in this domain no number which I could designate as a − b if b ≻ a. Outside there may be such. That means, a − b is "not defined," remains open; concerning it nothing is determined by means of the axioms. If I limit myself in operations in such a way that only such differences and quotients occur in calculation, then I remain bound to the domain of the positive whole numbers, and each number is then a positive whole number. And then I can show that the "not definite" differences are not among the numbers.

***

<K I 26/75>[26] In the axioms of an operational domain certain basic relations are formally established. Certain laws are formally fixed which characterize the operations as such, and therefore determine the concepts of +, -, etc. Laws are fixed which formally determine the kinds of relation.

Conversely, laws are established which characterize certain kinds of relation that hold true in exclusive disjunction between any two objects of a domain. If we speak of relations, now only these are to be meant.

Laws are fixed which characterize certain operations by means of the relations between certain types of operation, and indeed formal laws and existence laws.

What kinds of questions can then be raised in this axiom system?

1) What general relations obtain between types of operation (formulas) other than the {120} ones originally defined? Entirety of all formulas.

2) What existence laws obtain . . . .

3) Equations of determinations and relations of determinations. To the existence axioms belong some that say: For a determination of the forms $F(x) > a$ or $F(x) < a$, and the like, there is an x which satisfies it.

But a further question: For which operational combination is this a formula (this combination posited for x)?

4) Properties of "determined" objects which are assumed in the axioms with peculiar properties, and properties of classes of such objects, which are delimited by means of operational determinations that are direct or indirect.

***

<K I 26/90> I have not overlooked from the outset the issue of whether an essentially closed axiom system is possible in only one or in several ways; whether, therefore, the sequence of stipulations that concern specific operational formations do not play a role.

---

[26] This MS page was previously published on p. 551 of the "Textkritische Anmerkungen" for *Husserliana XII*.

That is in reality to be expected from the outset. The situation of arithmetic and of all mathematical systems of deduction is not obvious, according to which, namely, within the sphere of an essentially closed axiom system, extra-essentially closed ones can be marked out in stages. But precisely this relationship between axiom systems – according to which a narrower is contained within a broader, an essential one within a still more essential one – is the presupposition for the possibility of the transition through the imaginary.

1) An axiom system is definite if it leaves no question (no operational relation or combination) open, leaves no relation undetermined, for the formal objects of its domain, or in accordance with its definition.

2) An axiom system is definite if the addition of a new axiom restricted to those objects signifies a contradiction with the axioms defining the objects.

3) An axiom system is definite if every proposition that is meaningful (intelligible) in virtue of the axioms is also true or false on the basis of the axioms. All of these determinations are equivalent.

The peculiar character of the *Hilbert*ian closed axiom system is this, that it leaves open no question whatsoever that the operation system offers, therefore also leaves no operational formations undefined and unregulated, and consequently admits of no expansion of the operational domain by new objects brought under the same prevailing operations. And, second, that it admits of the addition of axioms in general which belong to the same forms of operation as those already defined, without contradiction. And, third, that no proposition meaningful in virtue of the axioms becomes false because of the fact that it has recourse to operational formations which are not defined, and thus are not accounted for in the domain. {121}

But there is still required here a more detailed discussion.

If we start out from a subject domain and formulate certain basic laws for the constitution of a deductive system, then, in general, various sorts of propositions will have restricted generality to the extent that they do not yield a sense for all possible combinations of variables. The combinations will not always provide results which have a sense in the domain.

We can therefore raise the question of when it is permitted to not burden oneself with such boundary conditions, thus when to deduce freely, untroubled about impossibility of combinations.

One can then say again: We abandon the subject domain and set forth the basic laws formally; we give them a merely formal sense. We expand the formal domain of deduction so obtained in such a way that the excluded operation combinations in question obtain a defined sense through certain new axioms. Always presupposing consistency, which of course can never be violated. Our intention can be only upon such "imaginaries" as do not violate consistency.

Then all expositions remain good. Any proposition there derived cannot stand in contradiction with the special axioms.

Each of the propositions falling within the original formal domain is then true; and true also from the pure standpoint of the narrower domain, *if* I know that each proposition falling in that domain is either true or false on the basis of the narrower domain. But one cannot conclude that. For if I expand the axiom system, then the old axioms may remain in force also, but the new axioms will in general also contain assertions about the old operations; and the question then becomes, whether through the new axioms determinations are not also introduced which were not contained in the old laws of operation. Only if I am certain that not only do the old laws remain in force, but rather that the originally defined operations – and, formally considered, the formal objects of the domain – have undergone no new determination that could become deductively efficacious, could we so conclude.

***

APPENDIX I [27]

If the elements of a manifold are left undetermined by the axioms in such a way that from no totality of elements that can be determinantly given all the remaining elements can be derived, then one also cannot speak of expansion of the manifold. For in that case how should it be decided that a new manifold included in itself all of the identical elements of another, the earlier one? One would then have to add an empty axiom which stated that.

If the old axioms are A and the new A + B, I might then say that any object which satisfies A + B also satisfies A, but that the objects of the latter manifold must be contained in one of the former, or conversely. — That, of course, cannot be said.

Only if the axiom system defines a manifold in such a way that by means of that axiom system its domain is given, and indeed a domain which includes in itself all objects of the manifold, can one speak seriously of "an expansion."

An axiom system can either include in its definitions determinate existence of objects, so that through the univocal forms [*Gebilde*] of operation with these determinately given elements ever new elements are determined, which can then be regarded as given. And if these are all elements of the domain, then the manifold is given by means of the results of operation upon that group. Or it can also be that only through the arbitrary assumption of a finite number of determinate elements all others are univocally determinable, as a totality of the possible operational forms, from those determinate elements. This is the case which I have dealt with under the heading "mathematical manifold." But, properly speaking, I would have to say "constructible manifold." "An axiom system defines a constructible manifold."

Is it then already obvious that the axiom system of the expanded manifold can assign no new characteristics to the old objects? One could perhaps say: It cannot do so, for otherwise the old

---

[27] Christmas 1901/1902. {LE} This is the "II. Studie" of *Husserliana XII*, pp. 458-463. <K I 26/37a — K I 26/39b> The remainder of this present volume incorporates numerous crucial corrections of the *Husserliana XII* text by the Schuhmanns, to whom I am greatly indebted. But, since their corrections are still unpublished, I have stayed as close to the form of the *Husserliana XII* text as possible, while incorporating their corrections. {DW}

objects would not be univocally determinate. Only extrinsic characteristics could be assigned, <459> those through relations to new objects, and further new operations could again be formulated. But what would hold true for the old objects with respect to newly assigned operations, and what is inferred for them from the expanded system, is quite certainly settled by the fact that we of course only consider propositions that have a determinate sense in the old axiom system. But this conclusion is not sufficient. From whence do I know that every operation is executable for the determinate (numerical) objects, and from whence that every previously given relation between numerical objects, perhaps $a_0 > b_0$, is decided as true or false on the basis of the axioms? That could, in spite of the univocal determinateness, certainly be left open.

Consider a manifold M. It consists of a two-fold Euclidean manifold and a totality of other elements. Then a special case is where the manifold ⟨call it M_0⟩ consists of a two-fold Euclidean manifold and one isolated straight line. We then could set up the expansion in such a way that, in addition to them, a two-fold manifold belongs to it, and one that includes in itself these "other elements" – that is to say, that straight line – in such a way, however, that the expanding determinations make the elements excluded from the plane appear as belonging to a straight line ($M_0 \notin M_E$).

Then in the expansion the proposition holds true: There is no point of the manifold excluded which does not lie on a segment of a straight line. But in M that proposition does not hold true. There it is indeterminate whether the excluded elements lie on a straight line or in general on a figure that is already defined in the domain.

M_E is to be an expansion of M_0. Thus M_E consists of the elements of M_0 plus other elements. But that does not suffice. The M_0 must be a *part* of M_E. M_E has a part that falls under the concept M_0. But that too is not sufficient. The expansion to M_E must not disturb M_0 as that which it is, and above all must not specialize it. I.e., the definitional determinations for M_E must be a mere expansion of those for M_0. Thus, if we consider the laws for M_0, and these are L_0, and the newly added elements are e_N, then for these latter, in their exclusive interrelationship to one another,

L_N must hold true. And if we then consider the elements of the two kinds in their relationships to one another, <460> the laws would be L_{0N}. Then it must be the case that $L_E = L_0 + L_N + L_{0N}$.

The laws must be representable in this form. By that we mean: the assumption that the manifold reduces to the partial manifold M_0, or the limitation to the elements of the new manifold belonging to M_0, should have the consequence that the L_E reduce to the L_0, without a further determination for M_0 thereby resulting. For example, if I set out from a manifold consisting of a two-fold plane manifold and a totality of further elements, and I expand it by saying, "this manifold is to be part of a three-fold plane manifold," then there holds true for the expanded manifold that system of basic propositions through which a three-fold manifold is defined in general.

For the partial manifold there then follows not merely that which proceeds from the concept of a two-fold manifold + a totality of elements; but rather, for the partial manifold we obtain the determination, richer in content: A two-fold manifold plus a totality of elements which belong to a three-fold Euclidean space just as does the plane one itself. The two-fold manifold is lowest species in the three-fold. All that is left is for it to individualize itself. Were the original manifold to consist merely of a two-fold plane manifold, then the limitation to the elements of the same would lead to the basic propositions for such manifolds. In virtue of the indeterminateness of this manifold and in virtue of the required abstraction from the external relations – those, namely, to the newly added elements – nothing would result which did not follow from the concept of a two-fold manifold in general.

It is otherwise with the assumed modification. The abstraction from the elements that are the surplus in the expanded manifold leaves over a determination for the elements of the original manifold that was earlier not expressed. Namely, that the elements are elements of a Euclidean manifold; that, therefore, of the elements which earlier were wholly indeterminate, any two have a certain distance between them; that two such distances are comparable to each other; that such distances also exist between the elements of the original two-fold manifold <461> and the "further elements"; similarly for angles and so forth.

Therefore the L_E must be characterized in such a way that the limitation to the partial manifold L_0 results, and nothing more. In other words: The basic principles which result for the interior relations of M_0 as a structure [*Gebilde*] within M_E through simple application of L_E to M_0, are precisely the laws that from the start defined M_0 in an absolute manner. Thus the definition of the expansion is such that its immediate and, simultaneously, complete consequence for the limited domain – given that we abstract from all external elements – is the original definition of precisely this domain. The simplest case is where formally the same laws hold in the expanded domain as in the narrower. L_0, L_N, L_{0N} then have the same form. Through limitation to the Domain O, L_0 then proceeds immediately from L_E, and nothing more does. Just as the domains come together to form one domain, so the laws do to form one law. In the same way, as the whole restricts itself . . . . The situation is simply this: Every structure of a manifold that is given in terms of its concept, I can regard as a constriction of that manifold. Conversely: If a manifold is given to me as an M_0, then M is an expansion of M_0 if M_0 undergoes no further "specialization" within M, that is, no closer determination out of which new propositions for M and the elements and formations contained therein could then proceed.

If I expand an M_0 to M, then the M_0 remains in M, thus as structure still an M_0. It is not thereby modified in species.

One could say: If I demonstrate for the structure M_0 that *some* A's are B's, that will certainly also be valid for the autonomous M_0. Likewise if I demonstrate that there are A's. But if I demonstrate that *all* A's are B's, then I can only say: For the M_0's which are sub-structures [*Gebilde*] of such an expanded manifold, it is valid that all A's are B's. In general I can therefore only say: In M_0 for itself it holds true that some A's are B's. Likewise for "There is no A." That is only proven for M_0's which are sub-structures. I can therefore assert no proposition at all for autonomous M_0's.

But right here lies a harmful confusion.

We have, namely, considered M_0 for itself, defined some manifold by certain L_0. Then we considered M_0 as a sub-structure: i.e., we considered a manifold that includes in itself an M_0 as substructure, let us say M_0'. There then will be propositions for M_0' which proceed exclusively <462> from the definitional determinations; which therefore hold true for any M_0 in general.

Now if I demonstrate in some other way that only some A_0' are B_0', then of course all of the A in M_0 cannot in general be B. But it does not follow from this that in M_0 it then holds true that only some A are B. For there can be kinds of M_0's where the one, and kinds where the other holds true. Likewise, if I demonstrate: there are A_0. So it must in general also be true that there are A's. How, then, if I demonstrate that all A_0 are B_0 = there are none among the A_0 which are not B_0, does that leave open the possibility that in M there are A's which are not B? It does not. For from that it will indeed follow that there are A_0 which are not B_0, etc. But that presupposes that through the expansion no differentiation results, or that M_0 in general admits of absolutely no distinction in kind.

("A circle which is the intersection of two spheres," that is not possible in the plane. If I say in the plane: there is no circle which is the intersection of two spheres, then I do not mean that there is no such thing at all. In the plane there are no spheres, thus no intersection of spheres. A circle which is acute angled: The definition of the circle excludes acute angles, regardless of whether in the plane or in another manifold.)

The series of the positive whole numbers is a part of the series of numbers that is infinite at both ends. This in turn is part of the two-fold manifold of the complex numbers. The system of the positive whole numbers is defined by certain elementary relations. In these latter nothing is modified through expansion of the number series. No new elementary relations are added, but rather only new elements and relations between the new and the old. The laws of the expanded domain include those of the narrower one, but in such a way, however, that for the old domain no new laws are established.

In the old domain are defined at one stroke the elements, the numbers, and then the relations and the laws for the relations and combinations of the numbers. In the new domain new relations as well as new elements may be defined. In the new domain there then will be such conceivable relations as include the old elements and old relations, as well as those which . . . .

The situation is then this: That every relationship between any existing numbers which possibly inheres in the concept of number <463> is also determinate on the basis of the concept of number and not, perhaps, left open by it. If I represent to myself some proposition about numbers – $f(a, b...) = 0$, or $M(\alpha, \beta...)$ is prime relative to α, and the like – these are possible propositions, since being equal to 0, and being prime relative to a number α, are possible predicates. There are number formations equal to 0, etc. Now either the propositions are correct in general, or they are correct for the special cases, or they are correct in no case.

Are there no propositions concerning any characteristics of, or any relations between, numbers which remain undetermined by the basic propositions of arithmetic? Which remain open? (That question can of course only refer to formal arithmetic. For cardinal arithmetic it is obvious that every relation between numbers is either true or false.) Let us take another example. If I define a manifold as rectilinear, then on the basis of that concept it is indeterminate whether the sum of the angles of a triangle is equal to two right angles or not. Yes, but that is not how matters stand here. The definition contains nothing about angles. But as soon as I define a manifold in such a way that the existence of triangles with angles is included, then for any triangle we have the disjunction: either the sum of the angles is two right angles or not. And that suffices. Whether the definition provides the means for actually deciding that for every triangle, or perhaps for distinguishing the species of triangle for which it holds true and for which it does not, is a matter of indifference.

If I then demonstrate from the expanded domain that a subject existing in the narrower domain does or does not have a predicate existing in the narrower domain, is that then a theorem of the narrower domain itself? No. That does not suffice. Why should the definition of the manifold not leave that proposition open?

***

APPENDIX II[28]

Manifolds can be more or less completely determinate. If I define a manifold by saying that it is a totality of elements in which there is a series that links up to form a continuous sequence, then <464> this manifold is less completely determinate than when I define another by saying that any two of its elements belong to a continuous sequence.

But this determination also is not complete. All elements can belong to one and the same sequence or to several sequences. In the latter case the sequences can coincide or not coincide. In the latter case again there are required certain more detailed determinations having to do with the conceptual distinction of these many sequences. For example, can all sequences link up to form a continuous or a discrete two-fold sequence? Consider the example of the number series. Here every element of the manifold is completely determinate in its relation to the lowest member. By means of the concept of the 1 and the relation increase-by-1, every element of the manifold obtains a univocal determinateness, such that further specification is impossible. Or rather, since the elements here are species, we can say that the element is individually determinate. We therefore have conceptual determinations that determine each individual also in its individual distinctiveness from every other.

Consider a straight line. We can arbitrarily select a point in it. Then every point can be fixed conceptually in its individual determinateness. We then have concepts making it possible to fix each point relative to any arbitrarily selected one. Likewise in the plane in space.

Thus the concept of the number, the concept of the straight line (or of the point on a straight line), of a plane, of a *Euclid*ean manifold in general, is so constituted that on the basis thereof every element of the respective manifold can be fixed relative to a determinate or an arbitrary element individually, in its distinctiveness from every other. Of course the same is true for the

---

[28] This is "Study III" from *Husserliana XII*, pp. 463-469. <K I 26/42a – K I 26/40b> {DW}

manifold of relations belonging to a straight line, etc. The relations between relations are completely determinate. For otherwise the relationships of the elements could not be completely determinate. (E.g., in the cyclical manifolds of relations with respect to a point.) That manifold indeed serves for the determination of the points themselves.

All laws that obtain for the manifolds mentioned, or for their elements and sub-structures [*Gebilde*] in general, can be reduced to these elementary relations, and since these <465> are determined through the concept of the manifold, then the indirect relations between elements and element-compounds are too.

(*Nota bene*: The unity of the manifold resides in the interlinking of the elements. Any two elements are linked on the basis of the concept of the manifold in a manner that is given completely conceptually. It would, then, be possible for there to be different, non-equivalent determinations that determined one point in relation to another. So it is if two sentences coincide. — We will explicitly make the presupposition that when two determinations are given which determine e_2 in relation to e_1, they are equivalent to each other. We do that for the sake of simplicity.)[29]

We now assume that no other characteristics of the manifold lie in the definition than those which are grounded in these elementary objects and relations.

There must then be formal equivalence between the definition of the manifold and the concept of a manifold in which the elements are linked by these elementary relations (of course formally; thus, for example – "From the concept of number, a + b, etc., follows"; conversely, "From the concept of a manifold in which 1 + 1 = 2 holds true, it follows that a + b can be defined thus," etc.)[30]

Or better thus: Let the manifold be defined in such a way that all of its objects in general are included in the determinate numerical objects, the determinateness of which is exhausted in their univocal operational definitions. What therefore holds true of the numerical objects is exhausted by the axioms and their definition.

---

[29] *Husserl*'s marginal note: This is quite superfluous. {LE}

[30] A question mark in pencil over this section. {LE}

Nothing else can be true of them. But what holds true for the objects of the manifold in general is also exhausted thereby, for all objects of the manifold are to be found among these numerical ones.

It is of course clear that all that is demonstrated from the concept of a concrete manifold of the form M, provided we infer purely formally from the definitional determinations, also is valid in general and would be in general provable from the concept M, and conversely. All that is valid of the number series is valid for ⟨466⟩ any concept that has the same formal laws, and (obviously) conversely.

We consider, now, an axiomatically defined, completely determinate manifold as a sub-structure within a more encompassing manifold. Then in this latter there will be new elements and new relations, perhaps even new relations for the old elements. But the new relations cannot, in whatever manner, disturb the completely determinate old ones, and consequently also cannot disturb any of the concepts and truths that receive their determinateness precisely through that determinateness. A completely determinate manifold cannot leave open the obtaining of propositions, *nota bene* – of propositions which have to be based exclusively in the elementary relations. Therefore, as soon as I come upon a proposition which is, in general, a subjectively possible one on the basis of the concept of the restricted manifold – that is, a proposition which in the last analysis asserts a certain relationship between the elements or their relations – then that proposition must also be one that is true for the restricted manifold, even if that manifold is not a sub-structure.

Perhaps one could say: If a manifold is so qualified that it puts each element into a determinate conceptual relation to every element – if, thus, each element, with reference to one or a number of elements, receives conceptual properties firmly fixed and well-distinguished, which then can serve as surrogates for it – then the manifold must be determinate in the above sense also with reference to all determinations deducible from those determinations ⟨conceptual properties⟩.

If, therefore, we can prove that this is the case, or if from the outset we define a manifold of elements in this manner, and if we then inquire only into all the propositions which are based purely logically in the elementary relations, then the total domain of propositions is a closed one.

It is then also clear that every part of such a manifold is definite, and every sub-structure likewise.

But from what do I know that such a partial manifold can be axiomatically defined for itself by means of a finite number of axioms? The following should be correct: So long as the primitive relations between any two elements are not completely determined (formally, of course) with regard to the original relations that we attribute to them in general, the manifold also is not defined. If there is in the manifold a certain combination of elements <467> or a relation, and it is not completely determinate for any two elements whether they stand in that relation or not, I then immediately have a proposition that is left open. That of course holds true only for elements that are determinately defined.

A concept falls within the domain of a concept if it sets objects of this domain into relations or combinations in such a way that the existence of the corresponding objects is certain. We can also say: There *explicitly* falls within the domain of a concept B any concept which belongs to the formal definition of B, but, *implicitly*, all concepts that are derivable from B in a logically general way. A state of affairs must, accordingly, be explained as belonging to the domain if it could be recognized as true purely on the basis of the formal definition of the concept: — In a broader sense, any state of affairs that is constructed exclusively from concepts that belong within the domain.

Consider a plane in space. In it there are various kinds of sub-structures, and we can imagine a geometry set up for this plane. Can some proposition or other now be demonstrated from the geometry of space which concerns the geometry for this plane and yet is not deducible within that geometry itself?

Can I, in order to construct the geometry for a manifold, hold myself to a more encompassing manifold within which the previous manifold itself is a mere sub-structure [*Gebilde*] or kind of sub-structure?

If we speak of a geometry for the plane, then we mean the total complex of propositions which exclusively concern the substructures of the plane in their reciprocal relationships; whereas all relations to exterior sub-structures remain out of consideration. Likewise for characteristics of the sub-structures which do not presuppose relations to elements or sub-structures outside the plane.

Can the laws for the intrinsic relations of a partial manifold be influenced by the relations of its elements to external elements belonging to an encompassing manifold? And if not, why not? Let us define a manifold exclusively by means of relations between the elements. A manifold is a whole composed of diverse elements. We distinguish as many different manifolds *in genere* as there are different ways which certain elements have of uniting to form a whole. Should the elements have qualities, we do not take these into consideration. Thus, qualitatively <468> heterogeneous elements can link up to form a manifold of the same genus, provided, precisely, that the manner of combination is the same. But the combining factors are the relations. But it also is not a question of the specific quality of the relations, but rather of the form in which they interweave. The definition of a partial manifold therefore consists exclusively in it being stated: In the total manifold there are elements that, through such and such relations, are bound together to form a whole. And the definition will then either exclusively take the intrinsic relations into consideration, or the exterior ones too. There plainly must be a definition of the first kind. For as a manifold of a determinate genus it must be definable by a mere glimpse of its inner constitution or form. The determination of the form can be a more or less complete one. It is a complete one if a further specification of it is not possible, and therefore a differentiation can be based only in the *quality* of the elements and of their relations, or in the connection of the manifold to other elements foreign to it.

The elementary relations between any two elements must be completely determined. Each relational network leads to certain elemental relations. On their combination as well as on their laws then rest all other relations. The relational framework can then be part of an encompassing one. Thereby new elements and new relations are to obtain, but in such a way that for the old elements no new elemental relations are introduced. Were, now, a new proposition for the old elements in their interrelationship to one another to result, then it would have to lead to elemental relations which were not defined in the original domain, and which thus, from the standpoint of the latter, would be new.

Let us nonetheless consider the manifold for itself, given through its definition. The definition emphasizes certain ones from among the manifold of relations that obtain between the elements or sub-structures. From these all the remaining ones follow by deductive consequence. If the manifold is one of lowest species, then it is incapable of a further specification. Let us now consider the manifold as a part of a more encompassing manifold. Can this circumstance, that it is a part, add to (of course it could not take away from) its relations some new one, and indeed an intrinsic one? <469> No. For otherwise there would have to occur in it elements or sub-structures that grounded the possibility of a relation without a corresponding relation obtaining. Thus a further specification would be conceivable, for every modification of relational form conditions a new species of the form of overall combination, and the modification here would take place within a genus because the original form included a certain indeterminateness.

***

APPENDIX III [31]

⟨ Notes on a Lecture by *Hilbert*. ⟩

· *Hilbert*, Mathematical Society. November 5, 1901 (reproduced from memory): <445>

1) An axiom system is closed (unfortunately I no longer recall the term *Hilbert* used) if it so determines the domain of objects of thought which it governs that no new (new kind of) object can be added to the domain in such a way that that axiom system then also governs the expanded domain (cannot be added – that is, without a contradiction resulting).

---

[31] This is the "Beilagen" of *Husserliana XII*, pp. 444-451, containing notes on the views of Hilbert and Frege. <K I 26/3a – K I 26/14a>

2) One can, then, attempt with a given axiom system to close it by adding to it the closure axiom: There is to be no new kind of object ε which (besides the other objects) satisfies the axiom system. But in general that is impossible. In particular, the axiom system of arithmetic (on *Hilbert*'s account) is not closed, given the exclusion of the *Archimede*an axiom (and without the axioms pertaining to the irrationals).

3) If one adds the *Archimede*an axiom, then the system of arithmetic is closed. One can prove that any addition of a new object ε would give rise to a contradiction.

And then, says *Hilbert*, one can prove the existence of the irrational numbers in the domain. One therefore need not presuppose infinite processes from the outset.

Hilbert's objection: — Am I justified in saying that every proposition containing only the whole positive numbers is true or false on the basis of the axioms for whole positive numbers? Here the following would have to be added: If we assert that a proposition is decided on the basis of the axioms of a domain, what may we thereby use besides those axioms? All that is logical. What is that? All propositions which are free of all particularity of a knowledge domain, or that which is valid independently of all "special axioms," of all matter of knowledge.

But here one falls into a fine dilemma: In the domain of algorithmic logic, in the domain of cardinal number, in the domain of combinatorics, in the domain of the general theory of sequences and ordinal numbers. And finally, is not the most general theory of manifolds itself purely logical?

If the axioms are formal statements, then they determine for me the indeterminate domain of a manifold through the form of their combination and relation. In them the terms stand for the material elements of such a domain. The axioms are no purely logical laws, nor are the theorems of the manifold. Only the connection between the axioms and the theorems, or the validity of the theorems on the basis of the axioms, is purely logical.

Therefore I can very well say: In order to evaluate such a theory I can utilize the whole of pure logic, the sphere of all propositions which are not materially determined. Of course to avoid circular reasoning I must not employ the relational propositions which refer to the relation between the axioms and theorems of the theory form before us. All others besides these.

Now, to be sure, a certain order must be established so <446> that the easier will not be proven by means of something more difficult. If we had a chart of all axioms in the genuine sense, as the basic principles of all propositions valid independently of all matter, that would be the sphere which one can have at one's disposal unconditionally.

The Idea of an ordering of the theories of a purely logical type, according to which a "first" theory is developed as far as it can be without bringing in other propositions from other theories, or in the worst case the basic principles. Then a second theory, which is partly developed and which at most utilizes the propositions of the first theory, to the extent it was developed. Then the first theory can, in turn, be developed further, utilizing perhaps something from the propositions which were just deduced, etc.

Definite? I cannot add to the "axioms," i.e., to the forms of basic principles hypothetically taken for a basis, any new "axiom," any new statement of substance, without evoking a contradiction. Or: Every proposition which utilizes the definite concepts of the domain, and, besides them, only absolute concepts (i.e., those which are purely logical, which have validity in every domain because they contain nothing of the matter of knowledge, but on the other hand are not concepts that define the domain formally) – thus, every proposition composed from purely logical concepts and the concepts defining the domain – is true or false in the domain.

How do I know that?

⟨1⟩⟩ Every direct operational combination, however often it may contain each operation, is equal to a number. That will be proven. Therefore every proposition which asserts two algebraically general, closed expressions to be equal – and likewise every mixed equation built up from algebraic and number signs – will of course have to be necessarily true or false on the basis of the axioms. For: Whichever group of numbers I may substitute for the a, b, c, ... p in a formula, there is always one determinate number for each side of the equation. And, indeed, on the basis of the axioms. If it is satisfied for all possible combinations of numbers, then the formula is valid. If not, it is not. How I may derive the formal symbolism is of no significance. It suffices that I can demonstrate from the axioms that every expression is a number, and consequently it is Evident that two expressions either always present the same number or different numbers.

2) Deciding equality. Either an equation is satisfied by an expression, by some number of the domain, or it is not. If we imagine any number being substituted, then the proposition in fact holds true that for each one the value of the left side of the equation, as well as of the right side, is determinate; and thus it is objectively determined whether or not the equation has a solution in the domain. It cannot remain open whether it has one or not.

3) Equations with unknowns. Here there is nothing new.

4) Problems which arise from the fact that the numbers are objects of enumerations, that one asks how many of these or those values show up in this or that domain, etc. For these problems the axioms of the domain and, on the other hand, the propositions of <447> number theory come into consideration. In any case it may in fact be, and certainly is so, that here not the mere axioms are utilized, but rather propositions about numbers as well, and perhaps also propositions of combinatorics, etc. But to begin with it also can be that such propositions are utilized in the prior domain and that they are not excluded. And 2) these propositions are objectively true independently of the given domain. But to us it is not a matter of whether everything can be proven solely from the axioms of the given domain, but rather of whether I can justifiably say that, if these axioms are established as conventions which refer to material elements that are left indeterminate, could it then be said: Each "proposition of the domain" is true or false on the basis of the axioms.

Of course every actual proposition is either true or false. But here the issue is: If M is defined by such and such axioms, then is proposition P within M, i.e., the proposition P falling under M, true or false? If the M are valid, then either P is valid or non-P is, provided it does not remain undetermined by M, but always with the support of the absolutely valid propositions of pure logic.

1) The problem recapitulated:
   a) That of the free combination of imaginary expressions in real domains.
   b) The formal principle.
2) The conception of the solution:
   a) Singular and particular solution. – I know: Certain groups of propositions are decided apriori by the axioms.
   b) Universal solution. – Necessary and sufficient condition: The axiom system must be definite, i.e., it defines a manifold in such a way that for each proposition utilizing the defined concepts – at the most also utilizing the purely logical concepts – it is objectively determined that it is valid or not valid for this manifold, rather than that it remains indeterminate.

***

Husserl's Excerpts from an Exchange of Letters between *Hilbert* and *Frege*.[32]

Frege, in a letter to *Hilbert* of December 27, 1899: — Mathematical propositions divide into 1) definitions, 2) all the remaining propositions.

"Every definition contains a sign (an expression, a word) which previously had no signification and which, through the definition, is given a signification for the first time. After this is done one can make of the definition a self-evident proposition that is to be used as an axiom. But with this it must be kept in mind that in the definition nothing is asserted, but something is stipulated. Thus something must never be set forth as a definition that requires a proof or some other type of grounding for its truth. I utilize the = sign as a sign of identity."

If +, 1, 3 are already explained, then $3 + 1 = 4$ is a definition of 4. Then the equation $3 + 1 = 4$ is "true of itself and needs no further proof." <448>

---

[32] In these excerpts there is some unclarity as to what is quotation and what is paraphrase. I have made some adjustments of quotation marks in the light of the letters. {DW}

Frege distinguishes between definitions and "explanatory propositions," which he, however, "would not locate *within* mathematics proper, but rather would prefer to assign to the 'vestibule,' to a propaedeutic."

"Also in their case it is a matter of a stipulation of the signification of a sign (word). They too therefore contain something the signification of which cannot be presupposed as completely and indubitably known, because it is perhaps used in a vacillating and ambiguous manner in the language of ordinary life. If in such ⟨a⟩ case the signification to be attributed is logically simple, then one can give no genuine definition ⟨of it⟩, but rather must limit oneself to rejecting those significations occurring in linguistic usage but which are not wanted, and to indicating the one wanted, in the course of which one certainly must always count on a cooperative and ingenious understanding."

"Such elucidatory propositions cannot be used in proof equally with definitions, because in them the necessary exactitude for that is lacking, which is why" *Frege* "prefers to assign them to the 'vestibule'."

Here *Frege* perhaps has in mind the distinctive character of such explanations-in-use as are given for the point and other "elements" in geometry, and for the 1 in arithmetic, in contrast to definitions of a kind with those for the remaining cardinal numbers, or for $a:b$, a^n and the like.

The elementary concepts, by means of which we define, and the definitions themselves.

Axioms: "Propositions which are true, but are not proven because knowledge of them flows from an entirely different knowledge source than the logical, which one can call spatial intuition(!).

From the truth of the axioms it follows that they do not contradict one another. That therefore requires no further proof.... If they ⟨the definitions⟩ do contradict, then they are defective."

The basic principles of definition must be constituted in such a way that in adhering to them a contradiction cannot appear.

(I observe here:

Frege does not understand the sense of *Hilbert*'s "axiomatic" grounding of geometry. Namely, that it is a matter of a purely formal system of conventions, which coincides, as to the form of the theory, with the *Euclid*ean.)

When are we certain we have not introduced the matter of a knowledge domain into the course of the deductions and in actuality have drawn our conclusions by purely logical means? When are we certain of having captured, in the basic principles laid down, all of the bits of knowledge flowing from the material substance of the domain that are sufficient for the system of the whole theory and its theorems? Only when we express the matter symbolically and raise ourselves to the level of a formal system, to a theory form, which is defined through the sentence forms of the basic material principles of the domain. Only so can we resolve the questions of dependence and independence, etc. <449>

In a formal deductive system (an algorithm) there are no "explanations" in *Frege*'s sense. In the definitional foundations, the spaces for possible "explanation" correspond to the elementary signs. To the general sentence forms there correspond the axioms. To the sentence forms by which determinate numerical values are operationally defined there correspond definitions – and indeed, real definitions, which simultaneously have the character of existence statements. *Mere* definitions do not occur in the foundations.

Therefore the axiomatic foundations define the formal domain.

However, we also distinguish within them between axioms or basic principles (although, there, absolutely nothing axiomatic in the logical sense obtains) and definitions. More precisely, we distinguish between basic laws (laws of relations), axiomatic existence propositions of a more general and of a more particular type, and definitions.

Likewise we distinguish in the formal domain, in the form of the theories, between basic principles and theorems – although the basic principles are no propositions at all and the theorems likewise are no genuine propositions, but rather are only forms of hypothetical consequents.

But "axiom" obviously now means: A proposition which would have to be formulated as an axiom in a knowledge domain whose theory form would be the present one. Likewise "basic principle": A proposition which would be valid as a starting point of the theoretical deductions. And "theorem": A proposition demonstrated within the theory, etc.

From *Hilbert*'s answer:

The sentence, "From the truth of the axioms it follows that they do not contradict one another," *Hilbert* found very interesting, because, of course, as long as he has thought, written and lectured on such things, he has been saying precisely the opposite: "If the arbitrarily assumed axioms do not contradict one another, . . . then they are true, and then the things defined by means of the axioms exist.

That is for me the criterion of truth and existence.

The proposition, 'Every equation has a root' is true," and therein lies the proof that the root exists, once that proposition can be added to the ⟨other⟩ axioms of arithmetic without ever leading to a contradiction.

In this sense we speak "of the existence of the real numbers," of the "non-existence" of the system of all ⟨*Cantor*ian⟩ cardinalities.

Point in geometry. Its complete definition in the full system of all axioms.

"Every axiom in fact contributes something to the definition. Every new axiom therefore modifies the concept *point*. In the *Euclid*ean, non-*Euclid*ean, *Archimede*an, non-*Archimede*an geometry the point is each time something different. After the perfectly complete and univocal establishment of the concept, the addition of some axiom or another . . . is something absolutely unpermitted and illogical – an error often committed by physicists, in that they are always making up these and those new axioms as they go along, and totally failing to bring them into confrontation with the earlier ones . . . resulting . . . in sheer nonsense." <450>

"Precisely the procedure of producing an axiom, invoking the truth (?) of the axiom (the question mark is *Hilbert*'s), and inferring therefrom that the axiom is consistent with the concepts defined, is the unending source of errors and misunderstandings. That is precisely what I wanted to avoid in the *Festschrift*."[33]

---

[33] D. Hilbert, *Grundlagen der Geometrie* (1st edition), in *Festschrift zur Feier der Enthüllung des Gauss-Weber-Denkmals in Göttingen*, Leipzig 1899. {LE}

"In fact it is surely obvious that any theory is only a framework (schema) of concepts together with necessary interrelations, whereby the basic elements can be conceived of in an arbitrary manner, e.g., posited as point, love, law, chimney sweep, and then all the axioms are satisfied, and then for these things the *Pythagor*ean theorem holds true."

"Any theory can be applied to infinitely many systems of basic elements. One only need apply a transformation that is one-to-one and invertible, and to stipulate that the axioms must be correspondingly the same for the transformed things (as happens for example in the principle of duality and in my proof of independence)."

"For the application of a theory to the world of appearance a measure of tact and good will is requisite. Point as smallest possible body, straight line a body as long as possible, e.g., rays. Also, the testing of propositions need not be too pedantic."

"Moreover, the further developed a theory and the more ramified it is, the more obvious becomes its manner of application to the world of appearances. No doubt a certain measure of ill will is involved if one would, e.g., apply the subtler propositions of the theory of surfaces or the propositions of *Maxwell*'s theory of electricity to other appearances than those for which they are intended."

Thus *Hilbert*.

Now *Frege*, January 6, 1900:

Thanks for sending the Munich lecture ("Concerning the Axioms of Arithmetic").

"It seems that you wish to completely separate geometry from spatial intuition and to make of it a purely logical science like arithmetic(!). The axioms which otherwise, as secured by means of intuition of space, are usually made the foundation of the entire structure, are, if I understand you correctly, to be carried along as conditions in every theorem – indeed not expressed in full wording, but included in the words 'point,' 'straight line,' etc. You wish to prove the reciprocal independence and absence of contradiction between certain presuppositions (axioms), and the unprovability of certain propositions from certain presuppositions (axioms).

"From the general logical point of view this is always the same case: The non-contradictory character of certain determinations is to be shown. 'D is no consequence of A, B and C' says the same thing as 'the occurrence of A, B and <451> C does not stand in contradiction with the non-occurrence of D.' 'A B C are independent of each other' = 'B is no consequence of A, C; C is no consequence of A, B; A is no consequence of B, C.'

"But what means do we have of proving that certain properties, certain requirements (or whatsoever one wants to call them) do not stand in contradiction with each other? The only means known to me is this: to pick out an object which has all those properties, to specify a case in which all those requirements are fulfilled. To show the absence of contradiction in another way would not be possible."

ESSAY IV [1]

⟨ THE DOMAIN OF AN AXIOM SYSTEM/ AXIOM SYSTEM – OPERATION SYSTEM ⟩

If the definition of a manifold does not unambiguously determine its objects in relation to each other, it then expresses formal relationships which can belong to manifolds that are not only individually, but specifically and indeed *formally* different.

The defining axiom system constitutes a formal Idea of a manifold of such and such forms of relations and corresponding laws of relations; or it defines a manifold in general as a manifold that is to be characterized by basic propositions of such and such form. In the basic propositions are established the kinds of object combinations and object relations and the laws they are subject to. If the definition is such that the kinds of objects and kinds of combinations and relations established do not formally determine, in an unambiguous manner, the kinds of objects possible at all, in virtue of the total definition, then possible determinations of kinds are left open, and the formal Idea of the manifold is further specifiable. A formally defined manifold is completely determined as to its form if nothing more remains open formally. The manifold is in that case only materially determinable, its concept is an ultimate specific difference.

However, it is here necessary to consider the following: In the formal definition of a manifold possibilities remain open under all circumstances, if no closure axioms of any kind are added. There always remains open the possibility of novel kinds of

---

[1] This corresponds to "VIII Abhandlung" in "*Husserliana XII*" (pp. 470-488), dated by the editor as from 1901. However, the first two pages of that "Abhandlung" are now re-located to pp. 420-422, line 10, above. <K I 26/44a – K I 26/51b> Numerous changes and additions have been made to the texts of *Husserliana XII*, following the re-editings by the *Schuhmanns*, yet unpublished. {DW}

combinations and relations that have nothing to do with those already defined, but also the possibility of novel kinds of relations that, through distributive laws, are posited in combination with those already defined – although the nature of these distributive laws cannot be arbitrary. Thus, expansion of the axioms is always possible if no closure axiom is added that states: No further kinds of relations and combinations are to be valid or come into consideration.

If we therefore introduce such a closure axiom, or if we only consider specifications with respect to the kinds of relations and combinations defined, then there is the possibility <473> of an ultimate specific differentiation with respect to the form. If we set out from certain kinds of operations and relations, and if we have already prescribed certain laws for them – certain existence-laws and laws of calculation – then the starting point will have some arbitrary aspects to it; but once we have begun, we are restricted in the addition of further axioms, and different possibilities of internal closure result. Internal closure is present as soon as the formal Idea of the manifold no longer admits of a further formal determination (if we disregard expansion by new kinds of operations and relations), so that the manifold is only still individualizable, but is not further determinable (differentiable) as to its form. The manifold forms (and, with them, the theory forms) thus constitute a realm [*Reich*] that is graduated in terms of genera and species. The lowest species are, so to speak, the individuals of this realm. These "individuals" are the complete [*perfekten*] forms of manifolds.

A manifold is *completely* [*perfekt*] *defined* if it is defined, not by a mere manifold genus, but rather by a lowest specific difference. An axiom system is a *complete* one if, with reference to the presupposed forms of operation and relation (and for its domain), no new axiom is any longer possible.

Question: But must an axiom system that is complete in this sense permit the unambiguous determination of every object on the basis of determinately given objects? Further: An existential domain can be so delimited by fundamental existence specifications that no new axiom is possible any longer for that domain. Such a manifold still admits of new specifications of existence and of corresponding rules of operation, but of no new general law of operation, and above all of no new axiom for the old domain. If we therefore disregard the expansion of the domain, then the manifold is not further specifiable. (But we will not call an expansion of the domain a specification in the true sense.)

We therefore must distinguish two things:

1). Expansion by means of operations and forms of relation. This is ruled out. <474>

2). Expansion by means of existence axioms, and thus expansion of the domain (within the sphere of the same operations):

α). Defined completely [*perfekt*] while preserving the domain closed in virtue of the existents already defined.

β). Defined completely without qualification, even if expansion of the domain is permitted. Expansion is, precisely, no longer possible.

Again, two cases:

a). The axiom system is to be identically retained.

b). The axiom system is preserved only for the old domain. But new objects are defined and an axiom system so constructed that when restricted to the old domain it becomes the old axiom system. But completeness [*Perfektion*] in the sense that such an expansion must not be possible is not to be required.

⟨ System of Numbers ⟩

Starting from these considerations, the definition of the general concept (the formal concept) of number should succeed. We distinguish a definite operation system from a definite manifold. The distinction consists merely in the fact that in the one case "operations" are defined for a domain of species, in the other case relations and relational networks are defined for a domain of elements.

We call a domain of species that is defined in a purely formal manner by means of laws of operations (axioms of operation in general) an "operational system."

I further believe I can rigorously prove that every definite operational system is a mathematical one, and that, therefore, every operational substrate and every operation is reducible by means of equivalences to a systematically producible totality of "elementary" forms of operation.

It can no doubt be said: Where every operational species, thought in indeterminate generality, is, on the basis of the axioms, equal (equivalent in the logical sense) to a certain form of operation that is derived in a lawlike manner from the primitive operational elements (operational invariants); and again, where every operational complex that is constructed in unambiguous steps from simple operational elements is reducible – namely <475>, it can be traced back through equivalences in a finite number of steps – to an unambiguously determinate operational complex made up of primitive elements: There we speak of number systems.

But this must admit of being formulated in a still better way: To an operational system there belongs the possibility that every operation can be equivalently replaced in various ways by another operation. Now where all operation forms are equivalently reducible to a finite totality of systematically derivable operation forms (systematically derivable starting from certain primitive operational elements), in which no two equivalent forms are present, but rather every one is dissimilar from every other, there we call these forms of operations "numbers," and the primitive elements "units."

It belongs to the concept of operation to be a transformation that is reversible – has its inverse. Is, therefore, operation nothing other than some sort of relational determination, and operation form not to be separated from form of relational determination? "Transformation" is only a simile. If aRb, and thus ...Rb, is a relational determination and bR'a the inverse, then aRbR'a returns back to $\underline{a}$. If $\underline{a}$ is grandfather of $\underline{b}$, $\underline{b}$ grandchild of $\underline{a}$, then $\underline{a}$ is also grandfather of anyone who is the grandchild of $\underline{a}$. These compositions, if they are unambiguous, lead back to $\underline{a}$.

Numbers are the standards of operation in a definite operational domain. They are the members of a complete collective totality, one that can neither be augmented nor diminished, and is made up of unambiguous operational characters not equivalent to one another, which are lowest specific differences in this sphere of operations, and which have the property that every real operation within this domain must have its demonstrable equivalent in a characterization [*Charakteristik*] from that totality.

If one grasps the concept of number that broadly, then any definite manifold can be interpreted in terms of numbers, and to any definite manifold a number system can be coordinated that governs all of its relations. And this is possible by means of a number system and the determination that the manifold has the property that it can be arithmetically constructed, defined, etc., starting from any point or from some group of elements, only subject to certain conditions.

⟨ Arithmetizability of a Manifold ⟩

If a system of axioms is to define its objects through a network of relations (or through the form of one) in such a way that they are susceptible to only a single additional kind of determination, the material, then every object must be unambiguously determined by means of its interconnectedness. Each object, formally considered, is the mere locus in the network of relations, i.e., in the relational form, where objects can be situated; and the relational form must be so rigid – it must be formally differentiated to the last degree – that it unambiguously determines each locus in relation to the other locii. If indeterminacies still remain here, then there would also be the possibility of formally characterizing the network of relations further. <476>

The objects will have to be so linked to one another by unambiguous combinations, and these combinations will have to be governed by such laws, that, starting from any point, with a certain number of points given in advance, the entire manifold must be capable of being generated unambiguously by advancing under the guidance of unambiguous modes of determination.

The question arises whether this conceptualization is fully and rigorously correct, and how it can be elevated to complete rigor. It is certainly unassailable from the start that a *formal* axiom system can define the objects it does there define (points, straight lines, planes), whether of one or of several kinds, only by means of the forms of relations. What is meant by "relations"? Through the forms of combination – whereby, say, two objects of such and such classes or having these and those specific determinations (which, of course, are designated only indeterminately) determine one object of a correlative class or of a correlative specifically peculiar nature – and, further, through the forms of relations that span objects formally determined in one way or another, we exclude the material determinations, in such a way that the relational terms of the combinations and relations can be designated only algebraically.

Material determinations ⟨are excluded⟩ also in the case of the combinations, so that the signs for combinations and relations are to have no material interpretation. Then the domain can be completely determined in a formal manner – thus so rigidly that only material determinations remain open – so long as we do not define new forms of combination (+, -) that are independent of the prevailing forms, likewise for independent forms of relation (>, =); and, finally, so long as we do not want to determine the objects that are delimited by the axioms already established through augmentation by new objects and through relation to those new objects; thus no expansion by new operations, no expansion by new relations, no expansion by new objects and consequently by exterior relations.

Therefore we have not genuinely determined the objects, but rather have determined the system of relations, the system of combinations and relations, and this indeed as a system which establishes univocality for every term of a relation. Were a class of relations to entail ambiguity, and were the ambiguity not a determinate one, which within <477> the relational system is transformed into univocality through addition of available relative determinations, then the values of the pertinent class of relative determinations would remain open, and therefore stipulations could be made that would remain compatible with the axioms. Every "operation" must therefore, in virtue of its laws, be an unambiguous one, and must permit the possibility of an unambiguous determination for every operational composite. All that is ambiguous within the manifold must only be ambiguous through incompleteness of determination, and must be transformable into something unambiguous. In the manifold or in the axioms nothing can remain that is ambiguous "in principle."

But the question now is: Must one be able to produce the entire manifold from a restricted number of given elements, whether they are singled out by the special nature of the axioms or arbitrarily selected?

If we assume certain objects given in advance, then with these we could undertake all operations, or establish all propositions determined by them, specify all of the correlative formations [*Gebilde*] that they determine, and so forth. If in the sphere of these objects so determined not all objects of the domain were included (however many previously given objects I may have assumed), then we would be able to form innumerable such spheres, and it would not be decidable on the basis of the axioms, given quite different objects to start from, how those spheres are related to one another. If each sphere must have some connection with every other, and so much so that, in fact, every object and object formation in the manifold is completely determined through its pure form (admitting only of individualization), then the manifold cannot decompose into spheres the members of which lack determinateness in relation to one another: Otherwise there would always still remain the possibility of new axioms for the same objects and relations already defined.

Only if the entire manifold is capable of being systematically generated on the basis of a finite number of given fragmentary determinations, and we thus know on the basis of the axioms that all other fragments of determination occur among those so defined, etc., then nothing remains formally indeterminate. We have no possibility of adding a determination that would not already be predetermined in the axioms. <478>

Naturally I can never say how an a relates to a b, where a and b are given "indeterminately." But if I know that for every object an unambiguous determination is constructible – a construction consisting of nothing but unambiguous steps, the determinate form of which is defined in the axioms – then I can construct a system of formal determinations that extend to any object in general, and then in a given case I can construct the given a and given b by means of the pure form out of the materially given starting points.

In each member of the <479> construction sequence (number sequence) I have determinations for each possible element of a manifold so structured.

Accordingly, a form of manifold that no longer admits an essential, formal differentiation is *eo ipso* mathematical. Since a manifold defines an unlimited number of objects, the determinability of each element in relation to n̲ given elements is only realizable in a law-governed manner. The elementary relations and the laws of these relations must therefore be so ordered that general forms of determination result which differentiate themselves in a lawfully determinate manner, so that one can successively generate every element by complication of relations to arrive at the given element.

One will then certainly be able to say that every such manifold is arithmetizable. One will always be able to define the relation to the previously given elements as an operation. Those given will of course result from the forms of determination, since the pregiven elements always recur, since they are the identical reference points for all determinations. One will therefore consider only those relational predicates which have their obvious subject in those points. One will, for each generative operation, be able to define an opposed one that restores the starting point, and define this latter as the inverse operation. One will define combinations that link outcome with outcome in such a way that one unified outcome comes about. Equality will have the signification of the equality of two operational formations, and thus the signification that two formations unambiguously yield the same element (always in relation to the initial elements), etc. One will therefore obtain an operation system that necessarily must be a definite one because it traces all forms of operation that can be generated back to certain primitive forms, and always permits one to decide equality and inequality and the possible cases of inequality.

⟨ On the Concept of an Operation System ⟩

What is the distinction between an operation system and a manifold? A manifold is a totality of objects that are "individually" different. An operation system does not define objects, but rather operations, and operations are differentiable only down to the lowest kinds. The same operation can occur arbitrarily often, and therefore the same signs in one and the same context of combination. One can speak of the manifold of operational species, but that manifold is not defined in the way the manifold is, but rather it is precisely forms of operation that are defined, since indeed each "object" here can occur arbitrarily often. In the operation system, if it is definite, all is traced back to certain ultimate invariants of operation. The ultimately determining factor, out of which every form of operation is constructed and with reference to which every form of operation is unambiguously determinable, is here something well distinguished. In the manifold I can select a number of elements arbitrarily as point of reference. But not here. Here the ultimate points of reference for the determinations of all forms of operation are certain specifically distinguished forms: $0, 1, \sqrt{-1}$, etc.[2]

In the domain of operations, i.e., in the domain of forms of relation, we consider the determination of forms of relation from forms of relation; and in the concrete domain of operations we have to do with the determination of materially filled out forms of the genus of material relations concerned.

In the domain of elements we speak of elements that are the foundations of the relations and of determinate kinds of formations [*Gebilde*] and their properties. But in the mathematical treatment we are interested in the elements only as reference points of relations, and thus as possible bearers of the operational characterization [*Charakteristik*].

---

[2] The "determinate" forms of operation (the "numbers") admit only one kind of material fulfillment, namely, that the kind of the relations, and consequently the genus of the objects in which the relations are grounded, are materially determinate.

The determinate elements of a manifold (the points and basic formations) admit of materialization in the sense that the lowest specific differences of a material genus, whose lowest species display themselves in such a manifold, are specified. E.g., the determinate place, the determinate time, etc., the here and now, the determinate color nuance, the determinate intensity. Elements of that sort then ground such and such forms of relation and laws of relation. The relations are then likewise "materialized" and to them the laws refer. But that is in fact exactly the same on both sides. {HS}

If we define a manifold, then we define a domain of elements by means of their relations. If we define a <480> "number system," then we define a sphere of forms of operation and investigate the laws of their reciprocal determinations. To every mathematical manifold there belongs a number system.

1). Elements, points as ultimate points of reference of the relations, those defined in the axioms, constitute a definite relational network. In this the points can very well be species, only they then must have the character of coordinate lowest species within a genus. The same holds true for points in manifolds of the geometrical kind.

2). Numbers, the lowest species of operations in a system of operations. (This does not fit the 0, 1 of the logical calculus.) Numbers are the operation values determined to the lowest specific differentiation, to which all operational formations can be traced back by equivalence. (Which also fits the 0 and 1 of the logical calculus.) But these species are not determined relationally in the manner of points, which would be an entirely different "manifold." What constitutes the difference?

In an operation system we define and we determine "operations," which are predicates of relations and forms of combinations of predicates of relations. Two such predicates are "equal" to one another, i.e., they are of the same value, equivalent – of course with respect to the possible subjects. In their conceptual content they can however still be different. They are *identically* the same operations if the content of the concept (the Ideal content of the predicate) is the same: e.g., a + a and a + a. On the other hand, a + b = b + a, but the concepts are not identical. The combination sign expresses the combination of predicates to form new predicates, but of course does not express the mere determinative combinations – e.g., 2 + 3, i.e., the third after the second; or 2 · 3, i.e., the second third member – but rather combinations that come about by means of combinations of relations. The relation signs express either the equivalence of the relational predicates – i.e., that in general something that is a̲ is, as such, something that is b̲, and conversely – or relations between objects, insofar as they are characterized by means of such relations, e.g., a ≻ b, the ath is farther than the bth.

What, therefore, do we study in an arithmetic? The properties of operations in an operation system that is defined through certain <481> axioms. An operation, I said, is a relational predicate. Of course it is not one that is determinate in relation to an individual, but rather to something determined as being an a̱. If I generate a line segment from a unit, then I commit myself to the measurement of lengths by means of the unit of length. I have therefore characterized the concept of a segment as unit, and there is then an operation of combining segments with segments, and further operations founded thereon. In arithmetic the a̱, ḇ, etc., are themselves operations, and 1 is the identity element of multiplication and 0 the identity element of addition. Therefore, all the elements of operation themselves have there an operational significance. I have chosen a species of line segments as "unit." Then I can, in virtue of the relations that obtain between segments, refer all segments back to that species; and, with the help of the relations, I can unambiguously determine all segments. It is these forms of determination that submit to the arithmetical treatment, while the substrate of the unit and the matter of the relations remain undetermined.

We have a sphere of species. These species determine new species in virtue of certain combinations existing between them, and between these combinations relations again obtain, and these relations can then in turn serve to determine species – whether through equivalence, or indirectly, by means of inexplicit [*aenigmatische*] conceptual determinations.

Species can function as "points" if we specify manifolds of species, so that in the formal treatment it is all the same whether we have to do with species or non-species. But species function operationally if we do not specify manifolds of species, but rather we put the objects [*Gegenstände*] of species as such in relation and combination, and thereby obtain objects [*Objekte*] which, in virtue of their origination (of their conceptual character as object of this species and as bearer of those relations and combinations), are determined as objects of novel species, in such a way that thereby a connection is consequently instituted between the species. We therefore determine species operationally by means of species, and determine the possible forms of such determination, determine such forms of determination indirectly, etc., and therefore conceptual objects as such. And ever new generation of conceptual objects on the basis of the <482> concepts. But the modes of generation are the object of our concern. In geometry: a totality of objects characterized by means of axioms. And then the possible forms of relation between those objects are studied. In arithmetic, the totality of the modes of generation, of the relational characterizations [*Charakteristiken*] and the law-governed relationships between the relational characterizations, is studied. Therefore in geometry an arithmetic can be established, and the arithmetic can serve to characterize the forms of relation and the modes of determination of the elements. On the basis of the axioms the forms of determination are defined, and then the arithmetic of the modes of determination is developed – that arithmetic then serving to determine every element by reference to ones previously given, and then obtaining the forms of relation, the relational networks, which we call formations [*Gebilde*], on the basis of those modes of determination: equations in geometry.

Operations: That literally means "to generate." These are modes of (relational) determination of an object by means of given objects, and indeed such modes as can be newly executed again and again on the objects already determined, reinforced by new objects. An operation type can be newly executed over and over again on the result of the operation, e.g., $a + b$, $(a + b) + c$, etc., or $(a \cdot b)$, $(a \cdot b) \cdot c$, ... In arithmetic all modes of generating are associative.³ If each letter is itself an indicator of an operation, in such a way, namely, that it represents an object as produced by means of a certain operation type, then $a + a$ says: There is an operation which combines two things generated by the same operation, and there then originates once again an object that is generated by means of the entire complex operation. $a = b$ says: The objects generated by means of the operation $\underline{a}$ and the operation $\underline{b}$ are the same (in case of univocality: the object is the same). (Of course that cannot mean that the forms of operation are the same.) $a \succ b$ says: The object generated through the operation $\underline{a}$ stands to the one generated through the operation $\underline{b}$ in the relation-

---

³ Not in the case of subtraction. {LE}

ship ">". Starting from the conditions stipulated in the axioms to which the operational determinations of objects are subject, or the operations are subject, the general <483> rules of operation and the rules deriving from them are then investigated. Further, unclear forms of operation are determined, e.g., in equations: ax = b. What is sought here? A certain form of operation which, substituted for x, satisfies the equation.

In arithmetic we have to do with a sphere of unambiguous operations. These operations determine under different forms (those which arithmeticians call operations of addition) new operations of such a kind that the results can again and again be combined with other operational results according to the same forms of operation. Laws of equality and ≻ ≺ obtain between types of operational formations (which themselves in turn are operations). Derived laws are to be obtained in this way. Operational forms are to be determined from given relations between operational formations and unknown operations, etc.

In the logical calculus of predicates the individual a, b, etc., do not have the signification of "operations." There it is simply a matter of predicates and their determinative combination to form new predicates; however, thus:

a	a is a predicate.		
a · b	a and b are two predicates which determine:		
	the predicate "a and b"	a · a = a	
a + b	the predicate "a or b"	a + a = a.	

a ∉ b. If something is a, then it is also b. If something has the predicate a then that thing also has the predicate b. The relation here is a relation of predicates to one and the same subject.

In arithmetic a, b, ... are not mere predicates. They are, we might say, forms of relation of the same kind as sequences of segments. If a is a type of ordering (*Cantor*'s order type [*Ordnungstypus*]) and b is a type of ordering (the same or conceptually different), then they determine a novel type of ordering, but thus: Any ordering of type a can (in the concrete domain of sequences concerned) be combined with a sequence of type b, in such a way, namely, that the end point of the one is the initial point of the other. And thereby, then, a sequence is always determined between the initial point of the first and the end point of the second. This sequence <484> has a type c: a + b = c. Every sequence is producible from the simple relations out of which it is composed (in the case of discrete sequences) or out of the relations of units (which perhaps again are "divisible").

The forms of relation (species), the types of order, determine novel types of order, novel forms of relation, by the fact that any particular relational complex of the form a concerned, as a particular of that form, can be combined with a particular relational complex of the form b, and that then the compound arising through combination is determined in its form c by means of the forms a and b. Thereupon the novel relational forms are in turn possible members of novel combinations, and so on *in infinitum*.

Relational forms are certain species (order types, ordinal numbers, segment numbers). Let us attempt to conceptualize this for species in a general manner: In a sphere of species of lowest difference, encompassed by a genus concept G, it is to hold true: Any two individuals (conceptual objects) of the species a and b (arbitrarily selected from the sphere, whereby a and b may be of different species or of the same) determine, as individual objects of those species, a new object, which in turn falls under a species of the genus G, and indeed the species of the object thus determined must be unambiguously determinate: a + b = c. This is to hold true regardless of whatever species we may pick out. It holds also for c in combination with some species or other d, where d is either totally new or is the same as b or a, or even c itself, and so on *in infinitum*.

A special case of this is the one where a + a = a, where therefore the objects of a combined in pairs always yield again an object of a, and indeed unambiguously. That is the case of the logical calculus. Therewith the concept of operation is defined in the most general manner.

Under "operation" we understand a mode of determination of species of a certain genus G, through which those species are determined (and indeed unambiguously, as we can say by way of restriction) through modes of combination possible in general between objects of species of precisely that genus, *as* objects of those species, and indeed by modes of combination which are in

ESSAY IV 489

general possible for any object of any species with any <485> object of any species (the same or different) of precisely that genus.

Operational laws are laws for such modes of determination, namely, laws that put different modalities and types of such modes of determination into relationship, and that, as laws, are satisfied under unrestricted variation of the species. Different operations +, -, ×, etc.; pure and primitive types, mixed types; different modalities of one and the same operation, and forms of compounding within one operation. In an arithmetic, species of a genus Z are determined through operational laws. On the basis of the lawfully operational relations between species, new relations between species are derived, unknown species are determined by means of their operational relations to given species, etc. If species are determined only in virtue of the form of those relations, and are therefore themselves characterized only by means of such forms, then we stand within a formal arithmetic (form of an arithmetic).

It belongs to the concept of the operation that the result of the operation can itself in turn become the substratum of operation for the same operation type, and so on. The results of the operation are species of the same genus as the bases of the operation.

In a "geometry" we determine a totality of elements of a certain genus G (or even species of a genus) by means of relations. If these relations determine operations and we restrict ourselves to those operations and the operational laws and problems resulting therefrom, then we are engaged in arithmetic. But where we characterize relations that do not have the nature of operations (e.g., two points determine a straight line, etc.), there we also have no arithmetic.

Objects of a domain D are, as such, characterized through their relations. Those relations belong to certain genera corresponding to D. On the basis of the relations (which will, in general, be more or less complex relational formations) between some previously given elements or other, new elements or compounds of elements (which, as such, will be defined by means of the relational unity) will be determined, and so on. If now we have a totality of basic propositions, which define the domain by their relations, then we can study the modes of determination of elements in virtue of their relations to <486> other elements; we can compose the relations, and compose them ever further, and study the modes of determination resulting therefrom as to their equivalences or inequivalences, and as to their interrelationships [*Verhältnissen*] characterized in such and such ways.

Instead of considering elements in relation to elements – therefore, determinations of interrelationships – we can also consider complexes of elements, structures of the domain, i.e., groups of elements that stand out in virtue of relational networks that can be determinately characterized. We can study the properties of the structures and interrelationships between them, the simpler structures occurring within them and their interrelationships to one another. We can study the modes of generating structures out of other structures. And so forth.

Those are the tasks of a geometry as a discipline that deals with the elements and structures of a manifold; i.e., a science of the relations (of the forms of unity and the interrelationship determinations grounded in them) that are grounded in a genus of elements. If we think of the elements and their relations as formally defined through their laws, and of the relational points as undetermined in other respects, then we have a formal geometry, i.e., the form of a geometry.

A geometry, therefore, determines elements and structures of elements. An arithmetic determines numbers, which are operational species, on the basis of the laws of operations.

1) In arithmetic: the "objects" ["*Objekte*"] are numbers, and thus all of the objects of the domain are species that function operationally and are logically given through ultimately differentiating concepts: all numbers belong to a manifold or are "=" to those contained in a certain manifold, namely, the one which can be generated by construction starting from 0 and 1, etc.; and, indeed, in such a way that all properties of the natural numbers are deductively inherent in the properties given by definitions. (For no determinate natural number can I arbitrarily fix a property: what holds true of it is fixed by its concept on the basis of the axioms.)

2) In geometry: the elements (points, lines) are not determined. They cannot come forth logically. They are defined as relational

terms of the relations characterized in the axioms. On the other hand, if we regard a certain number of <487> elements as given, then, for all the remaining ones, ultimately differentiating concepts (numbers) can be established, which determine all formations (objects) of the manifold in relation to the ones presupposed as given, and do so unambiguously and exhaustively (in the manner of ultimate differentiation).

In the logical calculus we consider characteristics and propositions with respect to objectivity [*Gegenständlichkeit*] or truth, objectlessness or falsehood. There it is not a question of generating out of predicates further predicates, as if the calculus determined predicates from predicates, but rather it aims to derive true from true, and so in general to draw conclusions concerning truth and falsehood. Likewise, propositions are not to be derived from propositions (thus not purely grammatically), but rather, from propositions which are assumed to be true or false, inferences are drawn concerning propositions which are, again, true or false.

Therefore we are concerned with values of truth and falsehood of predicates and propositions. All truth-values are considered equal to one another, as are all values of falsehood in general. Truth-values only distinguish themselves through the predicates or propositions to which they relate. The laws of truth-values are conditioned through the pure forms that permit, purely grammatically, the combination of predicates with predicates, etc.

Thus, as the sphere of the calculus we have the truth-values of propositions (or of predicates) and the possible combinations which such truth-values permit, whereby the well known reductions come into play.

The logical calculus is also definite. For every letter symbol is either $= 0$ or $= 1$, and consequently it is apriori determined, for every relation presenting itself as a formula, whether it is satisfied or not. It is, in general, satisfied if it, in general, yields either $0 = 0$ or $1 = 1$, and otherwise it is false. Likewise for every "equation": either there is a truth-value which satisfies it, etc. An arithmetic, too, is definite. For, all indeterminates of the domain are only determinable as truth or falsehood, as 1 or 0. (The substitution of given concepts or propositions is not arithmetico-logical determination, but rather materialization.) The concept of arithmetic must therefore be so broadly conceived that it also includes this case. <488>

Truth of A, truth of B – combination: truth of A and B, of A or B. Relation: consequence, equivalence.

The case of the logical calculus shows that "equality" does not have to be immediately taken as identity. Equality is equivalence in the respect here under consideration. The truth-values are certainly all = 0 or 1. But they are given to us as truth-values of these or those propositions, and that is what really matters. We assess everything in terms of 0 and 1, but the domain does not genuinely consist of 0 and 1.

ESSAY V [1]

⟨ THE QUESTION ABOUT THE CLARIFICATION OF THE CONCEPT OF THE "NATURAL" NUMBERS AS "GIVEN," AS "INDIVIDUALLY DETERMINANT" ⟩

Determinateness has been very confusing to me. It does not consist in the fact that the numbers concerned are unambiguously defined, nor in the fact that one states that a_0 is a determinate number, nor yet in the fact that one defines unambiguous cases of operation, that is, establishes axiomatically that $a - a$ has, for any $\underline{a}$, identically the same "value." Rather, it lies in the nature of the entire system of axioms that any number derivable from 0 through augmentation by 1 (and any operational formation [*Gebilde*] directly derivable from such natural numbers) has the character of a lowest level species; and this implies that the concepts of the natural numbers are a sequence of kinds [*Arten*] of the concept *number* (object of the domain), delimited by means of the axiom system, kinds which admit of no further differentiation. No property can be assigned to these numbers that is not already settled for them in the axioms. But that must be demonstrated.

The demonstration lies in the fact that every proposition valid for numbers must follow naturally from the axioms, and again, that every proposition about natural numbers is decided on the basis of the axioms, since for every one of its given natural numbers every equality or $\succ \prec$ of those numbers, and of the operational formations that can be constructed from them, can be decided. Consequently, every general proposition valid for them is also decided. Thus, no proposition about natural numbers can be put forth as a new axiom.

---

[1] This corresponds to "IX. Abhandlung" in "*Husserliana XII*," (pp. 489-492), undated. K I 26/60a – K I 26/61b. {DW}

It is apriori possible for an axiom system to include a definite [*definites*] domain in the form of a domain of "natural" numbers while it is not defined [*definiert*] beyond that. In arithmetic there is the particular case where:

1) it is provable on the basis of the axioms that the numbers derivable from 0 and the 1 are lowest species;

2) that every number in general is to be found among <490> the natural numbers, and thus the definite domain exhausts the whole domain. The axiom system therefore defines the domain in such a way that every object can be characterized by means of a lowest species, and thus the concept of the manifold itself is not further differentiable. The manifold is itself a lowest species.

Better: The definition of the sequence of natural numbers is a sequence of definitions whereby the numbers concerned are represented in the manner of what is differentiated to the last degree. The numbers are given to us logically through definition of the sequence of natural numbers. All that is valid for a naturally defined number is an analytical consequence of the definition. Of course the numbers are not, but the definitions in question are distinctly marked concepts of numbers.

Important observation: Every genus concept A represents under the attributive form a lowest specific difference, e.g., color, a color: this or that red (shade of red) is *a* color. The genus [color] itself differentiates itself into red. And the concept red is the concept of a lowest difference. Consider, now, a concept like cardinal number, ordinal number, or formal number of this or that axiom system. Then a cardinal number, an ordinal number, a formal number always represents a lowest specific difference. And likewise every concept that is formed within the respective arithmetic.

If I form the Idea of a manifold by means of a formal axiom system, the element will not, in general, be unambiguously represented as lowest specific difference. But that is certainly the case in an arithmetic where I take species as objects of the domain and where I submit them to conditions whereby they must appear as lowest specific differences of a genus. If I define combinations where something determined as a̲ combines with something determined as b̲, then of course a̲ and b̲ are certain lowest specific differences, but not, in general, of one and the same genus. But in arithmetic it turns out that a and b . . . are natural numbers, and natural numbers are objects that exhaust the concept of number ("object of the respective axiomatic manifold") in the manner of a complete disjunction of species, and thereby they are all operational formations [*Bildungen*] in conformity with the axioms. <491> All possible operational formations lead through equivalents to a certain fundamental sequence of such formations, and the Ideas of these latter formations we can regard as the lowest differences of the Idea of operational unit in general.

Perhaps the best way of saying it is: An arithmetic defines species. The natural numbers belong among those species. But the "natural numbers" are so named in virtue of the *nature* of the concepts defining them. These concepts have the character that they differentiate their objects to the lowest level possible, i.e., that not only do they in general represent their objects, which are lowest differences, as lowest differences (that they could do indirectly and by means of the genus), but rather that they provide representations with which the lowest differences are given to us *logically*. I.e., again: Every determination of this species is contained logically (on the basis of the axioms) in the defining determinations, is equivalent to them or is contained in them as consequence. The unambiguous concepts (delimited by means of the axioms) of objects in general that satisfy the axioms, are either identical with or equivalent to the concepts that are constructed in the sequence of definitions, the sequence of numbers. And the natural concepts of numbers are so formed that every further determination that their objects can be subject to is deductively contained in the defining concepts.

We could say: The concept, "object that is determined by means of this axiom system," undergoes a lowest specific differentiation in the definitions of the sequence of natural numbers.

The *formal* Idea *number* undergoes its lowest specific differentiation by means of the sequence of natural numbers. The formal Idea *number* is, in the first place, the concept "concept of an object of the axiom system" (not the object itself!). Every determinate concept of such an object is a number, and every definition of the natural number is a natural number. Thus the concept "member of the sequence of concepts 1 + 1 . . ." is equivalent with the concept "unambiguous concept of objects of the axiom system."

Every unambiguous concept that defines some object or other of the axiom system purely from the conceptual material imprinted in that system is, therefore, a number; but thereby we designate two equivalent concepts as the same number – for example, 2 + 2 = 4. Therefore the objects themselves are actually the numbers. <492>

So we will finally have to say: The concept of number (= object of the axiom system) differentiates itself in the sequences of natural numbers; i.e., we have in it a sequence of concepts of numbers that represent the numbers in the manner of something differentiated to the last degree, in such a way that those concepts yield the numbers *logically*, insofar as, namely, all properties of the numbers concerned are logically contained in the natural concepts. And with that, enough! Of course one can also define number by means of the content of this description.

ESSAY VI[1]

⟨ ON THE FORMAL DETERMINATION OF MANIFOLD ⟩

A formal definition of a manifold is obviously a definition which, abstracting from the "particular nature" of the objects, defines certain objects in general by the form of their relations. A formal definition of a manifold, a formal axiom system, therefore defines, not an individually determinate manifold, but rather one determinately conceived of by means of a concept of a manifold, a genus concept.

The definition of the manifold delimits a concept of a manifold, a genus of a manifold, and represents a manifold in general as object of that concept. Probably better: a class concept for manifolds, or a species. The genus of manifolds divides into kinds, and ultimately into lowest kinds. But the lowest kind is a class of individuals.

The correlate of the manifolds and of the properties of its objects is the form of theory arising out of the concept of the manifold, out of its axiom system. On the one hand we have the axiom system and the systematic deduction of the theorems and theories, on the other hand the objects and their basic properties, as well as their derived properties expressed in the theorems. Just as the objects and their properties are only formal, so the axioms and theorems are only formal; that is, they are not genuine axioms and genuine propositions, but rather they are forms of axioms and propositions.

The formal genus concept that the "axioms" delimit is specifiable in various ways, and – to speak more generally – modifiable in various ways, while preserving the axioms. <494> We have: "a

---

[1] This corresponds to "X Abhandlung" in "*Husserliana XII*," (pp. 493-500), undated. K I 26/62a – K I 26/66a. {DW}

manifold of objects corresponding to axioms A." If we form the concept: "manifold which satisfies axioms A and B," then this is a concept of a kind [*Artbegriff*]. Every manifold that satisfies axioms A and B also in fact satisfies axioms A. And among the manifolds that are defined by the latter axioms are also to be found those which in addition satisfy yet other axioms.

We distinguish:

I. 1) Extra-essential (formal) specification, and we capture under this heading cases where to the original defining axioms there are to be added only those in which completely new combinations and relations are defined, and indeed such as are without relation to the old ones; e.g., when besides the operations +, −, etc., yet a (+) is introduced, with laws of operation that contain nothing of +, −, etc.

2) Essential specification.

a) Extrinsic specification.

New operations are introduced, but by means of laws of operation that combine new and old operations.

b) Intrinsic specification.

No new operations are introduced, but the old operations are determined more precisely, insofar as cases of operations or general laws for relations between operations that were left open are now axiomatically established. Here we must distinguish:

α) Intrinsic specification by specialization of the operations. An operation is defined by means of its laws of operation. It is specialized through new laws of operation. A law is a general rule of operation.

β) Intrinsic specification through securing cases of operation that were left open, by means of existential axioms and supplementary rules of operation for these new magnitudes.

The most complete *intrinsic* specification: Every object is unambiguously determined by relations, therefore through the operations; and no operation and no law of operation or relation is dispensable for this unambiguous determination.

The most complete specification: α). No new formulas are any longer possible, no new laws of operation can be added; β). No particular rules of operation can be prescribed for <495> magnitudes of the primitive or derived kind that were fixed by means of existential axioms. The univocality is already implied therein: for if essential ambiguity existed, univocality could be procured by prescription.

II. An axiom system can be modified not only through specification. The important possibilities here are the following:

α) Extra-essential alteration: equivalents replaced by equivalents.

β) The axiom system is replaced by another from which it follows, but with which it is not equivalent.

γ) The axiom system is modified as in β), but in such a way that the "domain" is an expanded one; or, the other way around, that the domain is a narrower one. A "complete" axiom system (*Hilbert*) can result if the expansion of the domain is possible only through an alteration of the axiom system; that is, it cannot result through a mere adjoining of new axioms.

The species of the genus *manifold* (that is, of the formal Idea *manifold*) must be a lowest species, not admitting a further specific differentiation, thus only admitting "individualization."

The relations and combinations must link every member of the manifold with every other. The manifold cannot divide up into unrelated fragments.

The Idea of a whole is determined by the "nature" of the elements (thus by their species) and by the species of combinations and relations belonging to the elements of the species. (Spatial points: right, left; combinations: linked by a line segment.) The form of the whole: the different kinds of elementary combinations and relations.

Within a whole belongs whatever is combined in a unitary manner. Therefore even if we can assemble a sequence of axioms that simultaneously define several wholes, we will be able to exclude this case if they are not wholes that themselves come together to form a whole. We will speak of a manifold only where a lawful order actually establishes, directly or indirectly, relations and combinations between all elements.

That, therefore, we presuppose. And constitutive for the respective Ideas of manifolds are the so and so determined combinations <496> and relations. Thus there are so and so many combinations specially named and assumed as given, +, -, ×, . . .; likewise, relations determinately assumed and specially named: =, >, <.

We therefore construct the Idea of the manifold by means of a − closed and presupposed as given − totality of combinations and relations. The Idea of the manifold is therefore modified if we:

a) change the number of these combinations and relations;

b) modify the laws pertaining to, and formally defining, the given combinations and relations.

α) Thereby we can modify the particular laws, or adopt others of another form in place of laws that we eliminate.

β) Or we can retain laws already adopted and add new laws to them.

We secure the case which obviously is the only important one: that a manifold be characterized by means of a fixed number of certain combinations and relations, with certain axioms defining them. This manifold Idea is then to be further elaborated, further determined. When and in what sense is that in general possible? We exclude modifications of this Idea that partially annul its content. Thus the laws are to remain in force. We permit only addition of new laws. New propositions express new properties. To that extent the Idea of the manifold becomes specifically differentiated. It was more general, it now becomes more specific.

Then we still have to distinguish: a) The expansion. The laws already defined delimit a domain. New objects are defined by means of new propositions, objects which, in virtue of the new propositions, are different from those already defined; b) The genuine specialization, interior specialization without expansion.

Interior specialization, we now require, is not to be possible: for the same objects, for those of the domain delimited, it must not be possible to add any new axioms. If we exclude expansion, then there shall only be one ⟨type of⟩ determination left to be given, and that is one which does not alter the Idea of the manifold, but rather only fills in matter [*materialisiert*].

I have not yet mentioned one kind of determination of the Idea of a manifold <497>: determination by power [*Mächtigkeit*]. Thus, for example, a manifold for which $a \succ b$, $b \prec a$, and $a \succ b \succ c$ hold true, and for which the number of its members is 5, is completely determined formally: $\alpha_1 \succ \alpha_2 \succ \alpha_3 \succ \alpha_4 \succ \alpha_5$. But of course from the nature of the axiom system the power of the manifold can also be determinable: it is then not left open. (However, will this not then quite often be based upon the tacit presupposition that the power is an unbounded one? No, that is not at all necessarily the case, and no tacit presuppositions are to be permitted.)

The question then is: Must every object of the manifold be unambiguously determined in terms of its relations to all other objects? Every combination must be unambiguous. For if it were left undetermined by the axiom system as a whole – indeterminately ambiguous in general, or indeterminate between 2, 3 . . . meanings – then further axioms could be added that would establish univocity. The "values" would in fact be left open. Every ambiguity, every indeterminateness, must be resolvable on the basis of the axioms into a complete disjunction of unambiguously determinable cases. Between any two elements it must be possible to bridge an unambiguous path of relations, for if any relation between them were ambiguous or indeterminate, then I could introduce axioms that would bring determination: every indeterminate relation must admit of being transformed into a determinate one on the basis of the axioms. If, now, the relations and the forms of relation and combination are restricted in number, then, with the given relations and their possible compositions, every element must determine itself with respect to every other one, and indeed either unambiguously or in a determinately ambiguous manner. In the latter case, by means of a finite number of previously given elements and relations bearing upon them, every element will have to become unambiguously determinable. That is the point of unclarity!

Suppose I start out from one element and I consider the unambiguous relations that emanate from it. Every term arrived at can be, in turn, the beginning term of new unambiguous determinations, etc. So, beginning from one point, I have characterized a manifold. Can there be infinitely many unambiguous determinations that emanate from one point which are independent of each other? No. For the axioms must formally include all modes of determination, and they can only contain a <498> finite number of independent ones. All the rest must, therefore, be unambiguous implication. If the modes of determination were not formally determinate, then we could arbitrarily make them so. If there are two modes of determination that yield the same result, then there must be an axiom which unites them. If, starting from 0, I determine α, and, through another determination, β, then γ . . ., then these determinations must reduce to a finite number of basic determinations. For there cannot be infinitely many determinate forms of determination, namely, as independent from one another. They must be determined by their relation to one another. All determinations emanating from one point must encompass all determinations in general. For no class of objects can be left over. That class would in fact be without determinate relation to 0, and therefore new determinations would be possible.

How would it work if we distinguish: operational systems and geometrical systems? When are operations of an operation system defined in such a way that they can undergo no further determination? The operations must be unambiguous. Every species of operation must be unambiguously determinable on the basis of the laws of operation in relation to a finite number of given species – must be constructible by purely axiomatically determined and given steps. Were a species or a class of species not constructible from the ones given, then it would not stand to them in operational relations that would make possible a determinability of the ones by means of the others. But now the combinations and relations obtain between all species. Therefore they would certainly be sufficient for unambiguous determination, and therefore I could not adopt new axioms for these combinations and relations.

The Idea of the manifold that the axioms define is the Idea of a lowest species of manifold. But the axioms are not supposed to merely describe this Idea indirectly, or generate and require it in part constructively and in part indirectly; but rather the axioms are to give this Idea itself. The manifold is to be, according to its form, a given one. Therefore the formal species must be constructed for us by the axioms. That is the main point, and upon it the demonstration must be grounded if it is at all to be shown with complete rigor that every constructible Idea <499> of a lowest specific difference of a manifold genus must be, precisely, a constructible manifold, a "mathematical" one.

Only if, starting from some element or other, every element can be constructively generated (thus on a conceptually formal level), and consequently if the entire manifold is constructed, will the

Idea of the manifold be in fact given and not merely indirectly characterized.

I have taken the concept of the operation, and therewith that of the arithmetic, in such a broad sense that it includes the logical calculus. The following concept is a narrower one: In some domain of objects (a manifold) a genus of relations, lowest species of which we designate by $\underline{a}$, $\underline{b}$, $\underline{c}$, . . ., may be grounded. Now if in each particular case, but purely on the basis of the specific nature of the relations, any pair of relations of the species $\underline{a}$, $\underline{b}$, . . . determine, under a determinate form, a third that belongs to the same genus of relation (thus also to the $\underline{a}$, $\underline{b}$, $\underline{c}$, . . .); if, therefore, in this mode of determination the species of relation concerned, $\underline{c}$, is determined by the species of the members of the combination – then we say: There exists an operation according to which, in general, for any two species $\underline{a}$, $\underline{b}$ a third is determined by means of a + b. We of course distinguish as many operations as there are essentially distinct forms of such combination.

Every relation that is composite contains other relations in itself, and so an operation which is in fact the concept of a composition of relations can be resolvable into simpler operations. Now a domain of relations can be homogeneous, so that all relations in it can be established from certain primitive relations of one and the same genus of relations, and are reducible to them by means of composition or division. Likewise, it can be that the relations are operationally determined as to their specific values; thus, that the possible species of operation and relation can all be reduced to certain elementary species, finite in number, and that then, on the basis of the laws of relation or operation, all possible combinations of relations can be determined in a manner purely conceptual, starting from relations; and therefore that all the possible determinations of species of relation can be realized starting from species of relation. That is the case of an arithmetic in the narrower sense, and the elements of operation are the units. <500>

It is certain that precisely the cardinal number and its 1 are compellingly adaptable to this conception, which regards operations as pure combinations of relations to form new relations of the same genus. But the forms of relation are by nature domains for possible arithmetics.

Index

abstraction, xvii, xx, xxiii, 16, 17, 19, 20, 22, 26, 30, 36, 40, 43, 46, 48, 50, 52, 79, 83, 85, 87, 88, 90, 91, 98, 114, 125, 135, 148, 150, 151, 152, 153, 154, 157, 159, 162, 171, 174, 221, 255, 275, 291, 314, 315, 316, 318, 330, 332, 333, 336, 344, 352, 354, 355, 363, 404, 410, 447, 455

acts of higher order, xxiii, xxxi, 72f., 77f., 97, 192f., 324

aggregate, 15, 146, 171, 246, 281, 314

analysis, xiv, xv, xvi, xvii, xx, xxv, xxviii, xxix, xxxi, xxxii, xxxv, xxxix, xl, l, li, lxi, 6, 7, 8, 13, 15, 16, 17, 23, 33, 40, 43, 58, 59, 64, 65, 70, 75, 78, 81, 83, 86, 89, 90, 95, 99, 110, 113, 115, 119, 124, 125, 128, 131, 136, 137, 141, 149, 160, 169, 170, 173, 176, 179, 180, 184, 191, 199, 202, 203, 207, 208, 216, 217, 218, 221, 223, 225, 227, 242, 246, 293, 301, 305, 307, 309, 310, 311, 312, 319, 327, 328, 342, 343, 346, 349, 352, 386, 391, 416, 461

analytic, 309, 422, 446, 494

ARISTOTLE, 33, 89, 327

arithmetica universalis, lix, 7, 310, 428

arithmetic, xxxviii, 13, 271, 310f., 485f., 489f., 495, 503

authentic, xv, xvii, xxix, xxx, xxxi, xxxii, xxxv, xliv, xlv, xlvi, xlvii, xlix, li, lii, liii, lx, lxiv, 7, 13, 15, 16, 17, 24, 29, 31, 55, 74, 97, 115, 122, 130, 137, 149, 164, 186, 198, 200, 201, 202, 203, 205, 206, 208, 209, 210, 211, 212, 226, 228, 229, 231, 233, 235, 236, 237, 239, 242, 248, 250, 251, 253, 254, 261, 271, 273, 275, 276, 277, 278, 279, 280, 284, 285, 339, 357, 359, 375, 376, 377, 378

axiom system, lx, 412, 419, 420, 421, 423, 425, 426, 427, 428, 429, 430, 431, 432, 433, 434, 435, 436, 437, 438, 439, 440, 444, 445, 448, 449, 450, 451, 452, 453, 464, 465, 468, 475, 476, 477, 479, 493, 494, 495, 496, 497, 499, 500, 501

BAIN, 34, 90, 413

BAUMANN, 35, 45, 46, 89, 136, 147, 158, 159, 162, 313, 329, 333, 335, 413
BERKELEY, 134, 161, 162, 179, 186
BRADLEY, F. H., xvii, xli
BRENTANO, xxi, xxii, xxiv, xxv, xxviii, xli, xliv, xlv, xlvi, lxiii, 20, 67, 71, 73, 89, 205, 317, 344, 347

calculation, xxix, xxxix, xlviii, liv, lv, lvii, lix, 7, 45, 139, 140, 141, 156, 160, 181, 194, 201, 247, 250, 253, 254, 257, 271, 273, 274, 282, 287, 288, 289, 290, 291, 292, 293, 295, 296, 309, 412, 415, 417, 427, 446, 449, 476
CANTOR, 120, 121, 140, 374, 416, 444, 452, 468, 471, 487
category, xix, xxviii, lx, 33, 34, 44, 86, 89, 327, 328, 332
collective combination, xvii, xx, xxi, xxiii, xxiv, xxv, xxviii, xxix, xxxiv, xxxv, xxxix, xlviii, 21, 36, 48, 57, 67, 68, 69, 73, 74, 75, 76, 77, 78, 81, 83, 84, 86, 93, 111, 112, 151, 152, 193, 194, 209, 248, 317, 331, 344, 345, 346, 349, 350, 351, 352, 353, 354, 355, 360, 361, 364
comparing, 44, 56, 64, 99, 108, 109, 112, 115, 129, 150, 153, 341

completeness, lix, 162, 226, 227, 228, 258, 425ff., 429, 432-436, 451, 477, 499
consistency, 55, 181, 288, 425, 452
constitution, liii, 14, 193, 259, 451, 463
content relation, xxi, xxii, 76, 348, 349, 350
continuous, xix, 20, 36, 68, 71, 75, 120, 307, 316, 329, 345, 347, 349, 415, 416, 426, 459
continuum, xxiv, 17, 20, 24, 36, 57, 71, 160, 219, 231, 316, 319, 329, 341, 347, 357

DEDEKIND, 131, 186, 202, 414
definite axiom system, 425, 426f., 431, 436, 438f., 449, 466, 468, 481f., 491
DELBOEUF, 147, 156
descriptive psychology, xxiv, xlvi, 69, 346
determinateness, 37, 233, 425, 441, 442, 443, 444, 454, 459, 460, 461, 481, 493
DEWEY, xvii
difference, 50, 56-58
distinctness, xxii, 56-58, 60-66, 72, 76, 99, 146, 148, 149, 153, 155, 348, 351, 355
distinguishing, 25, 30, 49, 50, 53, 54, 56, 58, 60, 62, 63, 64, 65, 68, 73, 90, 99, 148, 149, 153, 320, 336, 338, 339, 341, 342, 343, 360, 458
DÜHRING, 201

EHRENFELS, l, 223
eigentlich, lx, 7, 16, 55, 203
equality, xxxii, xxxiii, xxxiv,
    xxxv, 99, 101, 102, 103, 104,
    105, 106, 107, 108, 109, 110,
    111, 112, 115, 121, 125, 126,
    127, 129, 130, 156, 219, 365,
    366, 367, 368, 369, 388, 398,
    407, 430, 435, 441, 446, 447,
    467, 482, 487, 492, 493
equivalence, xxxiv, xxxv,
    xxxvi, xxxvii, xlvii, 41, 117,
    118, 119, 121, 122, 124, 126,
    129, 130, 131, 206, 298, 377,
    378, 388, 389, 390, 391, 396,
    397, 402, 403, 446, 460, 484,
    485, 492
essence, xv, xvi, xvii, xx, xxv,
    xxvii, xxxiii, xxxvii, xxxix,
    xlii, xliii, lvii, 16, 44, 45, 51,
    73, 104, 114, 121, 131, 136,
    157, 185, 200, 231, 252, 256,
    301, 352, 411
EUCLID, 15, 101, 105, 137,
    307, 313
Evidence, xix, xxiii, xl, lx, 27,
    30, 55, 158, 202, 216, 252,
    350, 367, 373, 388, 390, 406
evident, xxii, lx, 50, 56, 57, 59,
    72, 74, 137, 148, 202, 340,
    341, 343, 345, 348, 360, 361,
    363, 365, 376, 379, 388, 389,
    390, 394, 413, 467
Evidenz, lx, 108, 158
existence, xxi, xxvii, xlix, 52,
    121, 214, 215, 222, 223, 337,
    346, 351, 420, 421, 422, 424,
    425, 426, 432, 433, 434, 436,
    437, 438, 441, 445, 446, 448,
    449, 450, 453, 458, 462, 465,
    470, 471, 476, 477

FARBER, xxvi
FIELD, H., xxix
figural moment, l-li, 215ff.
form concepts, xxviii, 89
formal, xxvi, xxviii, xxxii,
    xxxvii, xlviii, lvii, lviii, lix,
    34, 41, 62, 101, 124, 125,
    132, 139, 193, 252, 298, 327,
    357, 368, 397, 398, 399, 404,
    405, 407, 410, 411, 415, 416,
    417, 418, 420, 423, 429, 431,
    433, 434, 435, 445, 450, 451,
    452, 458, 460, 461, 462, 465,
    467, 468, 470, 475, 476, 477,
    479, 480, 481, 482, 485, 489,
    490, 494, 495, 497, 498, 499,
    502
FREGE, xxv, xxvii, xxxvi,
    xxxvii, xxxix, xl, xlii, lxiv,
    18, 102, 111, 112, 123, 124,
    125, 126, 127, 128, 136, 141,
    148, 155, 158, 160, 163, 169,
    171, 174, 175, 176, 464, 468-
    472

GERBERT, 292
gestalt, l, 126, 217
GIAQUINTO, M., xxix
GRASSMANN, 101, 103, 312

HADDOCK, lxiv
HAMILTON, W. R., 12, 35,
    306, 312, 328

HANKEL, 35, 49, 248, 264, 291, 328, 357
HELMHOLTZ, xliii, 12, 35, 103, 115, 161, 179, 180, 181, 182, 183, 184, 185, 186, 309, 312, 328, 357
HERBART, xli, xlii, 32, 99, 137, 142, 147, 149, 158, 160, 165, 166, 169, 171, 174, 325
higher order, xviii, xxi, xxii, xxiii, xxvii, xxxi, xlix, 60, 97, 214, 229, 264, 324
HILBERT, lix, 425, 427, 436, 451, 464, 465, 468, 469, 471, 472, 499
HILL, CLAIRE ORTIZ, lxiv
HOBBES, 136, 147, 313
HUERTAS-JOURDA, JOSE, lxiv
HUME, xvi, 158

Idea, xvi, xxvi, xxvii, xxx, xxxi, xxxvii, xxxviii, xlv, lii, lviii, lx, 5, 11, 18, 49, 51, 97, 109, 114, 117, 127, 130, 131, 133, 136, 147, 156, 163, 179, 211, 232, 233, 247, 306, 313, 336, 339, 375, 406, 411, 414, 417, 466, 475, 476, 494, 495, 499, 500, 502, 503
Ideal, xxxvi, xliv, lx, 12, 124, 162, 202, 236, 241, 247, 258, 276, 288, 410, 484
idealization, 231, 236, 248, 255
identity, xxxii, xxxiii, 20, 50, 51, 52, 59, 68, 70, 71, 72, 74, 75, 78, 82, 89, 99, 100, 102, 104, 155, 238, 316, 335, 336,
341, 342, 343, 344, 346, 347, 348, 350, 351, 361, 362, 368, 370, 430, 431, 434, 435, 468, 485, 492
imaginary, lvii, lix, 7, 12, 92, 120, 160, 181, 233, 298, 307, 310, 357, 359, 411, 412, 413, 415, 416, 419, 427, 428, 431, 433, 434, 437, 439, 440, 443, 444, 451, 468
inauthentic, xviii, xxviii, xxxii, xliv-xlviii, lii, lvii, lviii, 97, 130, 205-208, 210, 214, 224, 225, 239, 248, 359, 378, 385
indefinable, xix, xxi, xxx-xxxii, xli, 51, 96, 101, 125
infinite, xiii, xv, xxxv, lviii, 24, 103, 202, 230, 231, 232, 233, 235, 236, 319, 365, 369, 370, 372, 373, 374, 378, 379, 380, 381, 383, 388, 392, 457, 465
inner experience, xxx, 25, 37, 43, 60, 69, 77, 96, 321, 325, 351
inner observation, 339
inner perception, xl, 55, 64, 158
intention, xxiii, 5, 61, 123, 157, 207, 229, 230, 231, 233, 276, 290, 362, 452
intentional inexistence, xxi, 71, 73, 347
intentional object, 46, 164, 334
intentionality, xv, xvi, xxi, xxvii, xlvii, 67, 344, 347

JEVONS, 51, 52, 61, 147, 148, 156, 336, 340

KANT, lxiii, 33, 34, 35, 39, 40, 41, 42, 43, 194, 306, 327, 328, 329, 332
KERRY, xxxvi, 129, 130, 173
KROMAN, 147, 156, 157
KRONECKER, xliv, 12, 179, 185, 312

LANGE, 35, 36, 37, 38, 40, 41, 42, 43, 44, 45, 47, 157, 194, 329, 330, 331, 332, 333, 334
LEIBNIZ, xxxii, 17, 89, 101, 134, 143, 148, 249, 306, 313, 409
LOCKE, 18, 64, 65, 79, 89, 133, 134, 147, 158, 313
logic, xiii, xv, xvi, xxiii, xxvi, xxviii, xxix, xxxi, xxxvii, xlv, xlvii, xlviii, liii, lviii, lix, lx, lxiv, 6, 18, 34, 41, 62, 70, 82, 90, 124, 128, 145, 156, 157, 181, 201, 206, 248, 272, 307, 308, 359, 375, 390, 396, 412, 449, 465, 467
logical, xiv, xvii, xix, xxiii, xxvi, xxvii, xxxiii, xxxv, xxxvi, xliii, xlv, xlvi, xlvii, liii, lvii, lviii, lix, lxii, 5, 7, 12, 14, 27, 32, 33, 37, 41, 48, 51, 71, 82, 86, 90, 91, 103, 110, 112, 115, 122, 124, 126, 131, 135, 138, 140, 141, 144, 172, 176, 183, 191, 193, 199, 202, 206, 229, 230, 231, 233, 242, 246, 247, 248, 249, 250, 251, 253, 257, 266, 271, 272, 273, 274, 277, 280, 282, 287, 288, 289, 290, 294, 295, 296, 297, 298, 301, 306, 307, 308, 310, 312, 322, 326, 327, 330, 336, 347, 357, 359, 382, 388, 390, 395, 405, 410, 413, 419, 437, 444, 449, 462, 465, 466, 468, 469, 470, 472, 473, 478, 484, 487, 488, 490, 491, 492, 494, 495, 496, 503
logical calculus, 405, 484, 487, 488, 491, 492, 503
logical content, 33, 82, 135, 230, 326
logical difference, 257, 266
logical inclusion, 71, 347
logical point of view, 193, 290, 473
logical requirement, 242, 250, 251

MACH, 223
MEINONG, 205, 346
mere intention, 230
metaphysical, xix, xli, 20, 38, 41, 57, 71, 75, 86, 167, 308, 309, 317, 342, 347, 349, 350, 352, 357
metaphysical combination, 57, 342, 347, 349
metaphysical whole, xli, 75, 167, 349, 352
MILL, J. ST., xlii, 18, 36, 65, 69, 70, 72, 76, 86, 90, 145, 147, 156, 157, 169, 170, 177, 181, 306, 330, 346, 348, 350
MILL, JAMES, 65, 70, 86, 169, 346

moment, xlvi, xlix, l, li, lx, 20, 23, 24, 26, 27, 43, 48, 59, 60, 72, 73, 76, 78, 82, 89, 106, 108, 125, 156, 172, 177, 206, 208, 213, 214, 215, 216, 217, 218, 219, 220, 221, 222, 223, 224, 226, 227, 228, 229, 232, 233, 255, 257, 260, 267, 291, 316, 319, 321, 322, 350, 364, 429

MORELAND, J. P., xliii

multiplicity, xvii, xviii, xx, xxiv, xxvii-xxxv, xlii, xlix, 15-21, 23-26, 49-55, 77, 81-85, 87-89, 100, 104, 106-111, 120, 125, 135-138, 141-143, 146, 164, 207-209, 217, 222, 226, 229, 233, 235, 301, 312-318, 320f., 324-328, 334f., 344, 352-356

nominalism, xxix, xliii, xliv, 134, 136, 179, 182, 185, 187

number, systematic, liv, lv, 247, 252, 271, 275, 276, 277, 279, 281, 282, 283, 285, 286, 288, 289, 293, 294, 295, 407, 495f.
  cardinal, xviii, xxxiv, xxxix, xlv, li, lii, lv, lviii, 7, 11, 30, 31, 37, 81, 85, 86, 87, 89, 119, 120, 121, 127, 131, 180, 184, 301, 312, 330, 369, 407, 411, 413, 414, 415, 416, 417, 458, 465, 469, 494, 503
  imaginary, lvii, lix, 7, 12, 92, 120, 160, 181, 233, 298, 307, 310, 357, 359, 411-416, 419, 427, 428, 431, 433, 434, 437, 439, 440, 443, 444, 451, 468
  ordinal, lviii, 7, 11, 13, 121, 131, 183, 184, 185, 312, 328, 411, 465, 488, 494

one-to-one correlation, xxxiii, xxxiv, 103, 110, 111, 112, 114, 115, 117, 121, 127, 129, 131, 370, 372, 379

origin of geometry, 106; Cp. 258-266 & 284ff.

parts, xvi, xxiii, xxiv, xxvi, xlv, xlvi, lvi, 6, 11, 19, 20, 21, 22, 25, 26, 28, 38, 39, 41, 70, 71, 75, 76, 78, 81, 88, 105, 106, 107, 154, 157, 160, 161, 166, 181, 192, 195, 199, 207, 215, 216, 273, 286, 301, 315, 316, 317, 318, 320, 321, 323, 332, 346, 347, 349, 350, 351, 352, 353, 362, 381, 387, 415, 416, 426

phenomena, xix, xxiii, xxx, xxxix, xl, 16, 18, 22, 23, 24, 32, 36, 38, 57, 65, 67, 70, 71, 72, 73, 75, 77, 78, 93, 95, 125, 136, 144, 146, 223, 259, 267, 314, 316, 318-320, 325, 326, 330, 331, 332, 342, 344, 346-351, 357

phenomenal, xx, xxi, xxii, xxiv, l, 57, 71, 306, 308, 342, 346, 347

physical phenomena, 18, 71, 73, 347, 348
physical relation, xxi, xxii, xxx, 348
psychical phenomena, 71, 72, 73, 316, 320, 330, 342, 347, 348
psychological, xiv, xv, xvii, xxi, xxvi, xxviii, xxix, xxxi, xxxv, xl, lii, liii, 5, 6, 8, 14, 17, 20, 22, 26, 27, 29, 32, 33, 37, 44, 48, 52, 55, 57, 67, 68, 74, 78, 81, 83, 95, 97, 98, 103, 115, 119, 124, 133, 151, 158, 163, 173, 174, 180, 192, 202, 203, 215, 225, 229, 231, 259, 301, 308, 311, 313, 318, 321, 322, 324, 326, 327, 331, 340, 344, 352, 355, 359, 388
psychological analysis, xxxi, xxxv, 33, 95, 119, 124, 202, 203, 327
psychological characterization, 17, 22, 318
psychological constituent, 355
psychological foundation, xxviii, 52, 55, 97, 151, 340, 355
psychological origination, xxxi, lii, 133
psychological precondition, 29, 33, 37, 68, 78, 324, 326, 331, 352
psychologically well-characterized, 76, 350
psychology, xxiv, xxviii, xlvi, 6, 14, 18, 48, 69, 71, 115, 124, 157, 226, 308, 311, 346, 347, 389
purely logical, xxvi, xxxv, lviii, 112, 140, 410, 419, 462, 465, 466, 468, 470, 472

quasi-quality, 214, 216, 218, 222

reciprocal correlation, 104, 110, 111, 115, 130
reflexion, xxiii, xxvii, xli, l, lxi, 18, 21, 23, 25, 39, 43, 46, 51, 54, 59, 60, 64, 70, 72, 76, 77, 81, 84, 87, 89, 90, 150, 164, 208, 265, 278, 317, 319, 321, 334, 336, 339, 348, 349, 350, 351, 352, 354, 355
relation, 69-77, 346-350, 478
RUSSELL, xxix, xlviii

SARTRE, xxxi
SCHRÖDER, lix, 103, 104, 111, 115, 120, 374, 379
SCHUHMANN, lxi, 409, 453
SCHUPPE, 51, 55, 61, 170, 177
separately and specifically noticed, xix, xxxi, 24, 27, 28, 31, 58, 60, 68, 75, 78, 81, 202, 319, 322, 323, 325, 331, 342, 343, 345, 349, 351, 355
SIGWART, 34, 51, 52, 59, 61, 63, 89, 90, 164, 170, 336, 343
"something" 84f., 88f., 99, 353ff.

space, 35, 36, 37, 38, 40, 41,
42, 43, 44, 45, 47, 48, 49, 68,
157, 158, 226, 291, 308, 309,
310, 311, 329, 330, 331, 332,
333, 334, 357, 415, 426, 455,
459, 462, 472

species, xlv, xlvi, lv, 16, 23, 68,
76, 87, 99, 112, 122, 140,
144, 170, 206, 215, 219, 235,
248, 261, 277, 278, 287, 290,
292, 295, 318, 345, 350, 373,
376, 385, 410, 440, 442, 444,
445, 448, 455, 456, 458, 459,
464, 476, 477, 478, 483, 484,
485, 488, 489, 490, 493, 494,
495, 497, 499, 502, 503

STOLZ, xxxiii, xxxvi, 103,
104, 120, 161, 374

STUMPF, xvi, lviii, 20, 44, 64,
74, 82, 206, 218, 219, 224

symbolic, xiv, xv, xviii, xxviii,
xxxii, xliv, xlv, xlvi, xlvii,
xlviii, xlix, li, lii, liv, lvi, lvii,
lviii, lix, lxiv, 6, 7, 17, 97,
109, 111, 113, 135, 137, 159,
160, 179, 181, 186, 196, 197,
200, 203, 205, 206, 207, 208,
209, 210, 211, 212, 215, 223,
224, 226, 230, 231, 233, 235,
236, 237, 238, 240, 242, 245,
246, 248, 251, 253, 254, 255,
264, 267, 268, 272, 273, 275,
276, 277, 278, 279, 280, 281,
282, 285, 287, 290, 292, 293,
294, 296, 299, 301, 376, 395

synthesis, 31, 33, 35, 36, 38, 39,
40, 41, 42, 43, 47, 69, 113,
130, 327, 329, 330, 331, 332,
335, 345, 359, 400, 403

TAYLOR, xli

testimony of experience, l, 111,
215

testimony of inner experience,
37, 77, 325

time, xxvi, xli, xlvii, xlix, lviii,
lxii, 7, 14, 20, 24, 26, 28, 29,
32, 33, 34, 35, 36, 37, 45, 57,
63, 68, 69, 74, 76, 105, 114,
124, 131, 134, 138, 142, 157,
158, 174, 176, 180, 199, 206,
210, 211, 212, 213, 236, 242,
256, 258, 264, 269, 280, 289,
292, 293, 294, 296, 299, 301,
308, 309, 311, 313, 316, 319,
321, 323, 324, 325, 326, 327,
328, 329, 331, 332, 333, 341,
342, 344, 350, 355, 357, 365,
366, 378, 383, 394, 395, 400,
401, 415, 433, 440, 468, 471,
483

totality, xv, xvii, xviii, xix, xx,
xxi, xxiii, xxiv, xxv, xxviii,
li, 15, 17, 19, 21, 23, 24, 25,
30, 31, 32, 33, 41, 74-77,
97-99, 146, 174, 177, 192,
208, 212, 314-320, 324-327,
343, 345, 349, 351, 353-356,
359-366

transition through the
imaginary, 428, 431, 451,
452

truth, xxxiii, xl, 28, 37, 46, 55,
  122, 123, 124, 132, 171, 200,
  202, 216, 276, 282, 311, 323,
  334, 340, 394, 422, 428, 430,
  436, 439, 468, 469, 471, 491,
  492
TYLOR, 88, 262, 267

UEBERWEG, 148, 169, 179
unconscious, 41, 43, 60, 209,
  212, 213, 215, 268, 269, 332
unification, xix, xx, xxvii, xxx,
  17, 21, 23, 38, 46, 47, 53, 68,
  72, 75, 76, 77, 81, 97, 148,
  153, 161, 162, 164, 166, 194,
  222, 255, 271, 317, 318, 334,
  337, 345, 348, 349, 350, 351,
  353, 355, 356, 359, 362, 393
unit, xxiv, xxxviii, xxxix, xl,
  xli, liii, 15, 26, 31, 36, 39,
  50, 63, 79, 86, 88, 90, 91, 92,
  98, 100, 109, 114, 120, 123,
  133, 134, 135, 136, 139, 141,
  142, 143, 146, 147, 149, 150,
  152, 155, 156, 157, 158, 159,
  160, 161, 162, 163, 164, 165,
  172, 173, 174, 185, 192, 193,
  194, 195, 198, 199, 201, 216,
  232, 235, 237, 239, 240, 241,
  242, 245, 246, 249, 253, 258,
  263, 264, 278, 279, 280, 281,
  282, 285, 286, 289, 290, 313,
  321, 329, 332, 355, 359, 361,
  362, 363, 364, 365, 366, 367,
  369, 370, 371, 372, 373, 375,
  376, 377, 379, 380, 381, 382,
  383, 386, 387, 391, 392, 393,
  394, 407, 415, 416, 441, 478,
  485, 488, 495, 503
unity, xviii, xix, xxiii, xxviii,
  xxxvi, xxxviii, xxxix, xli,
  xliv, xlvi, 6, 14, 15, 21, 23,
  24, 31, 33, 39, 40, 41, 42, 50,
  52, 55, 58, 59, 62, 64, 74, 77,
  78, 93, 97, 112, 113, 125,
  133, 135, 141, 162, 164, 165,
  166, 167, 182, 191, 211, 213,
  218, 219, 222, 224, 225, 228,
  230, 232, 233, 255, 256, 301,
  313, 314, 317, 319, 320, 325,
  327, 332, 336, 340, 343, 350,
  360, 362, 400, 460, 489, 490

WEIERSTRASS, 13, 416
wholes, xvii, xviii, 19, 20, 21,
  22, 30, 33, 107, 151, 161,
  214, 315, 316, 317, 318, 324,
  325, 326, 499
WUNDT, 31, 48, 90, 91, 92,
  93, 202

YI, BYEONG-UK, xxxviii

ZIMMERMANN, 194

EDMUND HUSSERL
COLLECTED WORKS
EDITOR: RUDOLF BERNET

1. E. Husserl: *Ideas Pertaining to a Pure Phenomenology and to a Phenomenological Philosophy.* Third Book. Phenomenology and the Foundation of the Sciences. Translated by T. Klein and W. Pohl. 1980　　　　　　　　　　ISBN 90-247-2093-1
2. E. Husserl: *Ideas Pertaining to a Pure Phenomenology and to a Phenomenological Philosophy.* First Book. General Introduction to a Pure Phenomenology. Translated by F. Kersten. 1982　　　　ISBN Hb: 90-247-2503-8; Pb: 90-247-2852-5
3. E. Husserl: *Ideas Pertaining to a Pure Phenomenology and to a Phenomenological Philosophy.* Second Book. Studies in the Phenomenology of Constitution. Translated by R. Rojcewicz and A. Schuwer. 1989
　　　　　　　　　　　　　　　　ISBN Hb: 0-7923-0011-4; Pb: 0-7923-0713-5
4. E. Husserl: *On the Phenomenology of the Consciousness of Internal Time (1893–1917).* Translated by J.B. Brough. 1991
　　　　　　　　　　　　　　　　ISBN Hb: 0-7923-0891-3; Pb: 0-7923-1536-7
5. E. Husserl: *Early Writings in the Philosophy of Logic and Mathematics.* Translated by D. Willard. 1994　　　　　　　　　　　　　　ISBN 0-7923-2262-2
6. E. Husserl: *Psychological and Transcendental Phenomenology and the Confrontation with Heidegger (1927–1931).* The *Encyclopaedia Britannica* Article, the Amsterdam Lectures, 'Phenomenology and Anthropology', and Husserl's Marginal Notes in *Being and Time* and *Kant and the Problem of Metaphysics.* Translated by T. Sheehan and R.E. Palmer. 1997　　　　　　　　　　　　　ISBN 0-7923-4481-2
7. E. Husserl: *Thing and Space: Lectures of 1907.* Translated by R. Rojcewicz. 1997
　　　　　　　　　　　　　　　　　　　　　　　　ISBN 0-7923-4749-8
8. E. Husserl: *The Idea of Phenomenology.* Translated by L. Hardy. 1999
　　　　　　　　　　　　　　　　　　　　　　　　ISBN 0-7923-5691-8
9. E. Husserl: *Analyses Concerning Passive and Active Synthesis: Lectures on Transcendental Logic.* Translated by A. J. Steinbock. 2001
　　　　　　　　　　　　　　　　ISBN Hb: 0-7923-7065-1; Pb: 0-7923-7066-X
10. E. Husserl: *Philosophy of Arithmetic.* Psychological and Logical Investigations - with Supplementary Texts from 1887-1901. Translated by D. Willard. 2003
　　　　　　　　　　　　　　　　ISBN Hb: 1-4020-1546-1; Pb: 1-4020-1603-4

KLUWER ACADEMIC PUBLISHERS – DORDRECHT / BOSTON / LONDON